Bare-Metal Embedded C Programming

Develop high-performance embedded systems with C
for Arm microcontrollers

Israel Gbati

Bare-Metal Embedded C Programming

Copyright © 2024 Packt Publishing

Group Product Manager: Kunal Sawant

Publishing Product Manager: Akash Sharma

Book Project Manager: Prajakta Naik

Senior Editor: Nithya Sadanandan

Technical Editor: Vidhisha Patidar

Copy Editor: Safis Editing

Proofreader: Nithya Sadanandan

Indexer: Tejal Soni

Production Designer: Vijay Kamble

Senior DevRel Marketing Executive: Shrinidhi Manoharan

Business Development Executive: Kriti Sharma

First published: September 2024

Production reference: 2040725

Published by Packt Publishing Ltd.

Grosvenor House

11 St Paul's Square

Birmingham

B3 1RB, UK.

ISBN 978-1-83546-081-8

www.packtpub.com

To my dearest sister, Salome Gbati — a beacon of unwavering strength and grace. Your resilience is a quiet, unshakable force, reminding me that even in the face of life's greatest challenges, we rise, we endure, and we thrive.

Israel Gbati

Foreword

Around the time Israel and I first began working together, we were preparing to showcase our non-contact gesture trackpad at the British Invention Show. With only two weeks to get ready, most of our time was spent soldering components and assembling mechanical systems for the presentation. The firmware, however, was left until the final stretch. With just one day remaining, we braced for an all-nighter, taking turns debugging the code with the persistence of parents tending to a restless child. In the final hours, with our eyes half-closing, Israel made a critical breakthrough, pulling everything together just in time. Though we arrived late to the event, we walked away with the Platinum Award.

I've had the privilege of working alongside Israel for nearly a decade, acting as his tag team partner in this fascinating world of hardware, where electronics and mechanical design harmoniously blend with firmware and software to create technology at the cutting edge of innovation.

Israel truly practices what he preaches. The teachings in this book are actively tested and applied in the field to advance technology aimed at improving the quality of life for individuals facing various health challenges. There is nothing more relevant today than an engineering text written by an active practitioner, especially in the midst of the most fast-paced and dynamic engineering environment in human history. What you hold in your hands is the culmination of years of dedicated effort, driven by both knowledge and real-world applications that extend far beyond theory into the practical challenges of technological entrepreneurship.

I hope this book serves as a valuable resource and inspiration for all who seek to push the boundaries of engineering and technology.

Georgios Papanikolaou

Chief Operating Officer and Head of Hardware Engineering, BiostealthAI

Contributors

About the author

Israel Gbati is a distinguished firmware engineer boasting over a decade of hands-on experience in the field. Throughout his career, he has shared his profound knowledge to more than 100,000 professionals, helping to shape the next generation of experts. In addition to his engineering expertise, Israel is an entrepreneur and an award-winning inventor, recognized for his exceptional inventions. He holds a Bachelor's degree in Mechanical Engineering and Automation, complemented by a double Masters degree in Global Innovation Design from Imperial College London and the Royal College of Arts. Israel is the founder of EmbeddedExpertIO and the cofounder of BiostealthAI, further demonstrating his leadership and commitment to advancing technological innovation.

To my incredible colleagues at BiostealthAI and EmbeddedExpertIO, thank you for embodying the true spirit of professionalism and for your unwavering dedication to our shared mission. Your commitment has not only made our work rewarding but has also created an environment where collaboration and innovation thrive for the greater good.

A special note of gratitude to Mohamed Alezzabi, Olivier Tsiakaka, Ph.D., Georgios Papanikolaou, Desmond Boakye Tanoh, M.D., Husamuldeen Al-Daffaie, and Muhammad Sohan Mollah. Your expertise, support, and camaraderie have been the cornerstone of this journey. Together, we continue to push the boundaries of what's possible.

About the reviewer

Akshay Phadke started his journey as a Software Engineer after graduating with a Master of Science in Electrical and Computer Engineering from the Georgia Institute of Technology in 2016. He has developed data-intensive applications and experiences across different industries, including Networking and Telecommunications, Enterprise Software, and Finance. His expertise encompasses multiple areas of Software Development such as Big Data, Data and Platform Engineering, CI/CD and DevOps, Developer Productivity and Tooling, Infrastructure and Observability, and Full Stack Web Development. Akshay's professional interests include Open Source Software, Distributed Systems, and Building and Scaling Products in a Startup Environment.

Table of Contents

3

Understanding the Build Process and Exploring the GNU Toolchain   63

4

Developing the Linker Script and Startup File   83

8

System Tick (SysTick) Timer   173

9

General-Purpose Timers (TIM)   183

10

The Universal Asynchronous Receiver/Transmitter Protocol   195

11

Analog-to-Digital Converter (ADC)   219

12

Serial Peripheral Interface (SPI)   241

18

19

Preface

In a tech-driven world where embedded systems power nearly every modern device and innovation, the ability to develop efficient and reliable firmware is a prized skill. The journey from writing basic code to mastering low-level firmware development can be daunting, but the rewards are substantial. Whether it's a home appliance, an industrial control system, or a sophisticated IoT device, embedded systems serve as the silent, hardworking engines behind modern technology.

This book, *Bare-Metal Embedded C Programming*, was born out of a desire to help you not only write functional firmware but also to deeply understand the underlying mechanisms that govern how microcontrollers work at their core. My goal is to take you on an in-depth, technical journey into the heart of ARM-based microcontroller firmware development, specifically focusing on the STM32 family. This is not a book for the faint of heart, nor is it one for those looking for quick shortcuts. Instead, it is designed for individuals who are ready to step away from the comforts of pre-built libraries and tools to develop the skills necessary for writing efficient, bare-metal code from scratch.

So, what exactly is *bare-metal* programming? Simply put, it's the art of writing firmware that interacts directly with the hardware—without the abstraction layers provided by third-party libraries. This approach requires precision, a deep understanding of microcontroller architecture, and the ability to read and manipulate registers to achieve the exact behavior you want from your hardware.

Why I Wrote This Book

As someone with years of experience in embedded systems development, I've often noticed a gap in the way firmware development is taught. Many texts and courses focus on high-level development, promoting the use of pre-built libraries that abstract away the complexities of hardware interaction. While this approach is undoubtedly convenient and practical in many cases, it leaves a void for those who truly wish to understand how things work at the lowest level. I believe that understanding the "bare-metal" aspect of embedded systems development is essential for becoming a truly skilled firmware engineer.

This book is my effort to fill that gap. Through step-by-step guidance, I'll show you how to build your own drivers, manipulate registers, and write code that takes full control of the microcontroller. This is not just about learning a new skill—it's about achieving mastery.

Who this book is for

If you're a developer, engineer, or a student eager to dive deep into the world of microcontroller firmware development, this book is for you. You'll find it especially valuable if you're the kind of person who prefers to understand what's happening under the hood, rather than relying on copy-paste solutions from online forums. Whether you're transitioning from other platforms or seeking to build a strong foundation in bare-metal development, this book will give you the hands-on experience you need.

What This Book Covers

Chapter 1, Setting Up the Tools of the Trade

This chapter introduces the essential tools you'll need for development. From navigating datasheets to setting up your Integrated Development Environment (IDE), this chapter lays the groundwork for everything that follows.

Chapter 2, Constructing Peripheral Registers from Memory Addresses

In this chapter, we dive into the core of bare-metal programming. You'll learn how to define and access peripheral registers directly from memory addresses, using the official microcontroller documentation as your guide.

Chapter 3, Understanding the Build Process and Exploring the GNU Toolchain

In this chapter, we take a closer look at the embedded C build process. You'll explore how to compile and link code manually using the GNU Toolchain, gaining complete control over how your firmware is created.

Chapter 4, Developing the Linker Script and Startup File

In this chapter, you will learn how to write a custom linker script to define how your firmware is placed in the microcontroller's memory, including allocating sections like code, data, and stack. Additionally, you'll develop a startup file that configures the microcontroller's initial state, sets up the stack, initializes memory, and jumps to your main code.

Chapter 5, The "Make" Build System

Automating the build process is a critical part of embedded development. This chapter teaches you how to use the Make build system to streamline your workflow by creating custom Makefiles that automate repetitive tasks.

Chapter 6, The Common Microcontroller Software Interface Standard (CMSIS)

CMSIS simplifies development on ARM Cortex microcontrollers. In this chapter, you'll learn how to leverage CMSIS to write efficient code that takes advantage of the microcontroller's features while maintaining simplicity.

Chapter 7, The General-Purpose Input/Output (GPIO) Peripheral

GPIO allows your microcontroller to interact with external devices. This chapter guides you through developing both input and output drivers for GPIO, one of the most frequently used peripherals in embedded systems.

Chapter 8, System Tick (SysTick) Timer

Timing is essential in embedded systems, and the SysTick timer provides an easy way to generate precise time delays and system ticks. This chapter walks you through developing SysTick drivers for use in your embedded applications.

Chapter 9, General-Purpose Timers (TIM)

This chapter introduces you to the general-purpose timers (TIM) in STM32 microcontrollers, teaching you how to develop timer drivers for tasks that require precise timing.

Chapter 10, The Universal Asynchronous Receiver/Transmitter Protocol

Communication is a key aspect of embedded systems. This chapter focuses on the UART protocol, one of the most widely used communication protocols. You'll learn how to develop UART drivers, enabling your microcontroller to send and receive data from external devices.

Chapter 11, Analog-to-Digital Converter (ADC)

Many embedded applications require converting analog signals into digital data that your microcontroller can process. This chapter covers how to configure the ADC peripheral, allowing you to read and convert analog inputs into meaningful digital values.

Chapter 12, Serial Peripheral Interface (SPI)

SPI is a high-speed communication protocol commonly used in embedded systems. This chapter guides you through developing SPI drivers, enabling efficient communication between your microcontroller and other peripherals, such as sensors or memory devices.

Chapter 13, Inter-Integrated Circuit (I2C)

I2C is another popular communication protocol for connecting devices, it is often used for short-distance communication in embedded systems. This chapter covers the development of I2C drivers, allowing your microcontroller to communicate with multiple devices over a shared bus.

Chapter 14, External Interrupts and Events (EXTI)

Responsiveness is critical in embedded systems, and external interrupts allow your system to react to changes in its environment. This chapter covers how to configure and manage external interrupts and events (EXTI) for timely and efficient responses to external stimuli.

Chapter 15, The Real-Time Clock (RTC)

For systems that require accurate timekeeping, the RTC peripheral is indispensable. In this chapter, you'll learn how to set up and use the RTC to track time in low-power systems, even when the microcontroller is in sleep mode.

Chapter 16, Independent Watchdog (IWDG)

Stability is crucial for embedded systems, and the Independent Watchdog Timer (IWDG) ensures that your system can recover from unexpected malfunctions. This chapter teaches you how to configure the IWDG to automatically reset your microcontroller if it stops responding, ensuring reliable operation.

Chapter 17, Direct Memory Access (DMA)

Direct Memory Access (DMA) allows data transfers to occur independently of the CPU, significantly improving system efficiency. This chapter covers how to configure and use DMA for memory-to-memory transfers, as well as for peripherals like ADC and UART, offloading the work from the CPU.

Chapter 18, Power Management and Energy Efficiency in Embedded Systems

Power management is essential for energy-efficient systems, especially in battery-powered devices. In this final chapter, you'll learn techniques for reducing power consumption, including how to use sleep modes, wake-up sources, and optimize firmware to achieve the best balance between performance and energy efficiency.

To get the most out of this book

To fully benefit from this book, it's important to have a general familiarity with the C programming language. While we'll cover the specifics of embedded systems programming in detail, having a basic understanding of how code operates will make the material easier to follow. Familiarity with microcontrollers is certainly helpful but not a strict requirement. Everything you need will be introduced as we progress. Whether you're a beginner or an experienced developer, this book will guide you step-by-step through the fascinating world of bare-metal embedded programming.

Software/hardware covered in the book	Operating system requirements
STM32CubeIDE	Windows
GNU Arm Embedded Toolchain	
NUCLEO-411 Development Board	
10k Potentiometer	

Software/hardware covered in the book	Operating system requirements
OpenOCD	
Notepad++	
RealTerm	

If you are using the digital version of this book, we advise you to type the code yourself or access the code from the book's GitHub repository (a link is available in the next section). Doing so will help you avoid any potential errors related to the copying and pasting of code.

Download the example code files

You can download the example code files for this book from GitHub at `https://github.com/PacktPublishing/Bare-Metal-Embedded-C-Programming`. If there's an update to the code, it will be updated in the GitHub repository.

We also have other code bundles from our rich catalog of books and videos available at `https://github.com/PacktPublishing/`. Check them out!

Conventions used

There are a number of text conventions used throughout this book.

`Code in text`: Indicates code words in text, database table names, folder names, filenames, file extensions, pathnames, dummy URLs, user input, and Twitter handles. Here is an example: "Copy the path to the `openocd bin` folder."

A block of code is set as follows:

```
// 22: Set PA5(LED_PIN) high
GPIOA_OD_R |= LED_PIN;
```

When we wish to draw your attention to a particular part of a code block, the relevant lines or items are set in bold:

```
// 1: Define base address for peripherals
#define PERIPH_BASE         (0x40000000UL)
// 2: Offset for AHB1 peripheral bus
```

Any command-line input or output is written as follows:

```
monitor flash write_image erase 4_makefile_project.elf
```

Bold: Indicates a new term, an important word, or words that you see onscreen. For instance, words in menus or dialog boxes appear in **bold**. Here is an example: "Right-click on **This PC**, and then choose **Properties**."

> **Tips or important notes**
> Appear like this.

Get in touch

Feedback from our readers is always welcome.

General feedback: If you have questions about any aspect of this book, email us at customercare@ packtpub.com and mention the book title in the subject of your message.

Errata: Although we have taken every care to ensure the accuracy of our content, mistakes do happen. If you have found a mistake in this book, we would be grateful if you would report this to us. Please visit www.packtpub.com/support/errata and fill in the form.

Piracy: If you come across any illegal copies of our works in any form on the internet, we would be grateful if you would provide us with the location address or website name. Please contact us at copyright@packt.com with a link to the material.

If you are interested in becoming an author: If there is a topic that you have expertise in and you are interested in either writing or contributing to a book, please visit authors.packtpub.com.

Share Your Thoughts

Once you've read *Bare-Metal Embedded C Programming*, we'd love to hear your thoughts! Scan the QR code below to go straight to the Amazon review page for this book and share your feedback.

https://packt.link/r/183546081X

Your review is important to us and the tech community and will help us make sure we're delivering excellent quality content.

1

Setting Up the Tools of the Trade

In the world of embedded systems, crafting efficient firmware begins with a clear comprehension of the tools available. This chapter will guide you in establishing a robust development environment, ensuring that you are equipped with all the necessary tools for a comprehensive firmware development experience.

Central to our discussion is the concept of datasheets. Consider these as the detailed blueprints for any microcontroller, encompassing its capabilities, specifications, and intricate details. However, the challenge often isn't merely understanding a datasheet but also sourcing the correct datasheets tailored to your specific microcontroller. To address this, I will assist you in pinpointing and understanding both datasheets and user manuals important to our chosen microcontroller.

As we progress, we'll delve into the intricacies of setting up our **Integrated Development Environment (IDE)** and acknowledging its critical function within the development life cycle. Furthermore, you'll gain insights into configuring the GNU Arm Embedded Toolchain and OpenOCD. These tools will later empower us to craft our firmware, bypassing the need for an IDE altogether.

In this chapter, we will explore the following main topics:

- Essential development tools for microcontrollers
- The development board
- Datasheets and manuals – unraveling the details
- Navigating the STM32CubeIDE

Getting the most out of this book – get to know your free benefits

Unlock exclusive **free** benefits that come with your purchase, thoughtfully crafted to supercharge your learning journey and help you learn without limits.

Here's a quick overview of what you get with this book:

Next-gen reader

Our web-based reader, designed to help you learn effectively, comes with the following features:

- **Multi-device progress sync**: Learn from any device with seamless progress sync.
- **Highlighting and notetaking**: Turn your reading into lasting knowledge.
- **Bookmarking**: Revisit your most important learnings anytime.
- **Dark mode**: Focus with minimal eye strain by switching to dark or sepia mode.

Figure 1.1: Illustration of the next-gen Packt Reader's features

Interactive AI assistant (beta)

Our interactive AI assistant has been trained on the content of this book, so it can help you out if you encounter any issues. It comes with the following features:

- **Summarize it**: Summarize key sections or an entire chapter.
- **AI code explainers**: In the next-gen Packt Reader, click the **Explain** button above each code block for AI-powered code explanations.

Note: The AI assistant is part of next-gen Packt Reader and is still in beta.

Figure 1.2: Illustration of Packt's AI assistant

DRM-free PDF or ePub version

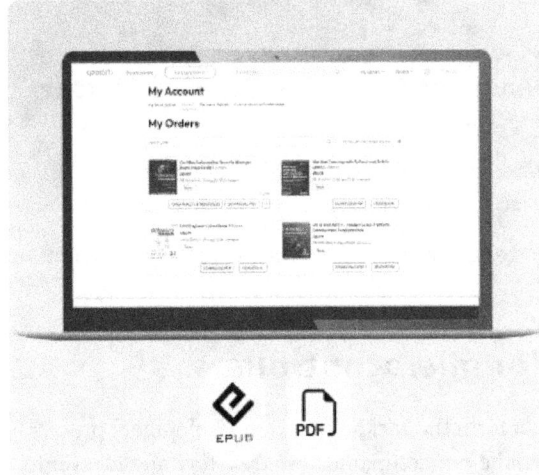

Learn without limits with the following perks included with your purchase:

 Learn from anywhere with a DRM-free PDF copy of this book.

 Use your favorite e-reader to learn using a DRM-free ePub version of this book.

Figure 1.3: Free PDF and ePub

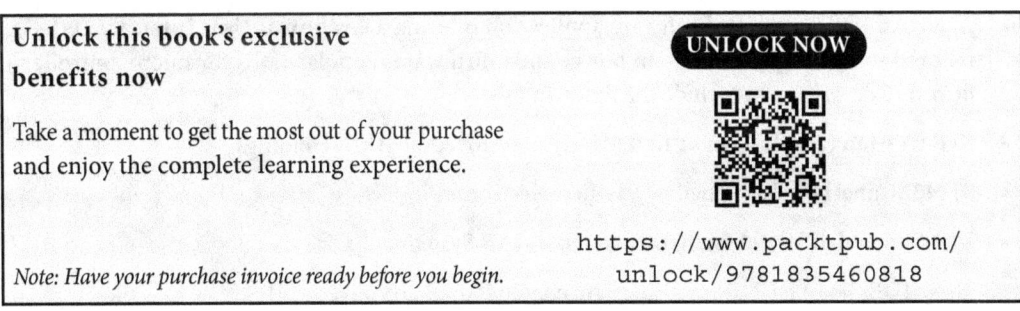

Unlock this book's exclusive benefits now

Take a moment to get the most out of your purchase and enjoy the complete learning experience.

UNLOCK NOW

https://www.packtpub.com/
unlock/9781835460818

Note: Have your purchase invoice ready before you begin.

Technical requirements

The following are the prerequisites for the chapter:

- STM32CubeIDE: `https://www.st.com/en/development-tools/stm32cubeide.html`
- GNU Arm Embedded Toolchain (`gcc-arm-none-eabi-10.3-2021.10-win32.exe`): `https://developer.arm.com/downloads/-/gnu-rm`
- OpenOCD: `https://github.com/xpack-dev-tools/openocd-xpack/releases`
- Notepad++: `https://notepad-plus-plus.org/downloads/v8.5.8/`

- STM32F11 reference manual: `https://www.st.com/resource/en/reference_manual/rm0383-stm32f411xce-advanced-armbased-32bit-mcus-stmicroelectronics.pdf`

- STM32F411 datasheet: `https://www.st.com/resource/en/reference_manual/rm0383-stm32f411xce-advanced-armbased-32bit-mcus-stmicroelectronics.pdf`

- NUCLEO-F411 user manual: `https://www.st.com/resource/en/user_manual/um1724-stm32-nucleo64-boards-mb1136-stmicroelectronics.pdf`

- Cortex-M4 generic user guide: `https://developer.arm.com/documentation/dui0553/latest/`

Essential development tools for microcontrollers

In this section, we will explore the essential tools that form the backbone of our development process. Understanding these tools is important, as they will be our companions in transforming ideas into functioning firmware.

When selecting tools for firmware development, we have two primary options.

- IDEs: An IDE is a unified software application offering a **Graphical User Interface** (**GUI**) tailored to crafting software – in our context, firmware. Popular IDEs for microcontroller firmware development include the following:

- **Keil uVision (also known as Keil MDK)**: Developed by ARM Holdings

- **STM32CubeIDE**: Developed by STMicroelectronics

- **IAR Embedded Workbench**: Developed by IAR Systems

  These IDEs boast a GUI-centric design, enabling users to conveniently create new files, build, compile, and step through code lines interactively. For the demonstrations and exercises in this book, we'll use the STM32CubeIDE. It has all the requisite features and is generously available for free, without any code size constraints.

- **Toolchains**: At its core, a toolchain is a cohesive set of development tools, sequenced in distinct stages, to produce the final firmware build for the target microcontroller. This approach bypasses the comfort of a GUI. Instead, firmware is written using basic text editors such as Notepad or Notepad++, with the command line used to execute the various phases of the build process –commands such as `assemble`, `compile`, and `link` are often used. In this book, we'll use the open source GNU Arm Embedded Toolchain. Based on the renowned open source **GNU Compiler Collection** (**GCC**), this integrates a GCC compiler tailored for ARM, the **GNU Debugger** (**GDB**) debugger, and several other invaluable utilities.

In the following section, we will carefully go through the process of setting up our preferred IDE, the STM32CubeIDE.

Setting up the STM32CubeIDE

Throughout this book, we'll use both the STM32CubeIDE and the GNU Arm Embedded Toolchain to develop our firmware. Leveraging an IDE such as STM32CubeIDE enables us to easily analyze and compare the linker script and startup files, autogenerated by the IDE, against those we'll construct from the ground up.

Let's start by downloading and installing STM32CubeIDE:

1. Launch your web browser and navigate to st.com.

2. Click on **STM32 Developer Zone**, and then select **STM32CubeIDE**.

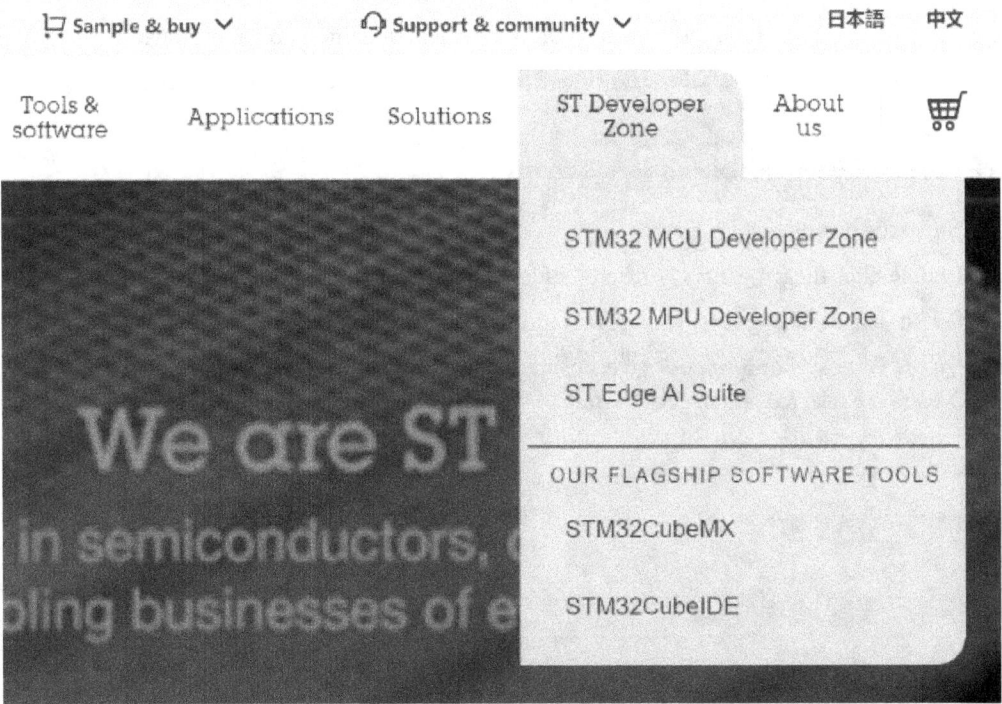

Figure 1.1: The home page of st.com

3. Scroll down to the **All software versions** section of the page and click on **Download Software**. You'll need to log into your ST account before proceeding with the download.

Figure 1.2: The All software versions section of the stm32cubeide page

4. If you don't have an account, click on **Login/Register** to sign up. If you already have one, simply log in.

5. Complete the registration form with your first name, last name, and email address.

6. Click on **Download** to start the download process. A .zip file will be downloaded into your Downloads folder.

Let's install the STM32CubeIDE:

1. Unzip the downloaded package.

2. Double-click the st-stm32cubeide file to initiate the installer.

3. Retain default settings by clicking **Next** throughout the setup process.

4. On the **Choose Components** page, ensure that both **SEGGER J-Link drivers** and **ST-LINK drivers** are selected. Then, click **Install**.

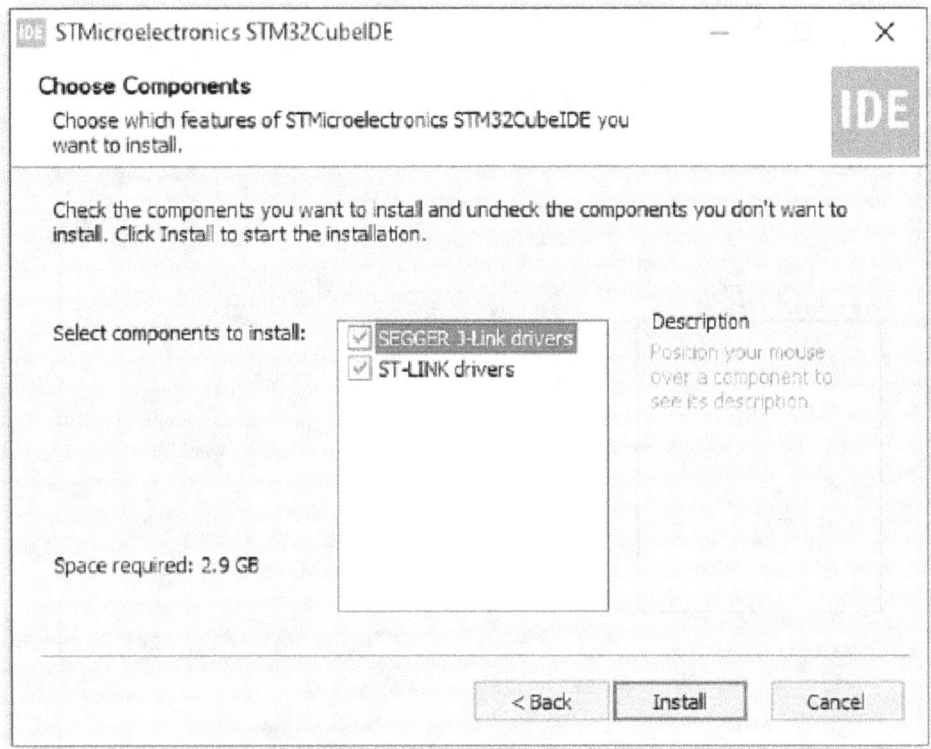

Figure 1.3: The installer showing the Choose Components page

Having successfully installed STM32CubeIDE on our computer, we will now proceed to configure our alternate development tool, the GNU Arm Embedded Toolchain.

Setting up the GNU Arm Embedded Toolchain

In this section, we will go through the process of setting up the GNU Arm Embedded Toolchain – an important tool for developing firmware for ARM-based microcontrollers:

1. Launch your web browser and navigate to `https://developer.arm.com/downloads/-/gnu-rm`.

2. Scroll down the page to find the download link appropriate for your operating system. For those of you using Windows, like myself, opt for the `.exe` version. For Linux or macOS users, choose the corresponding `.tar` file for your operating system.

3. After the download completes, double-click the installer to begin the installation process.

4. Read through the license agreement. Then, choose to install in the default folder location by clicking **Install**.

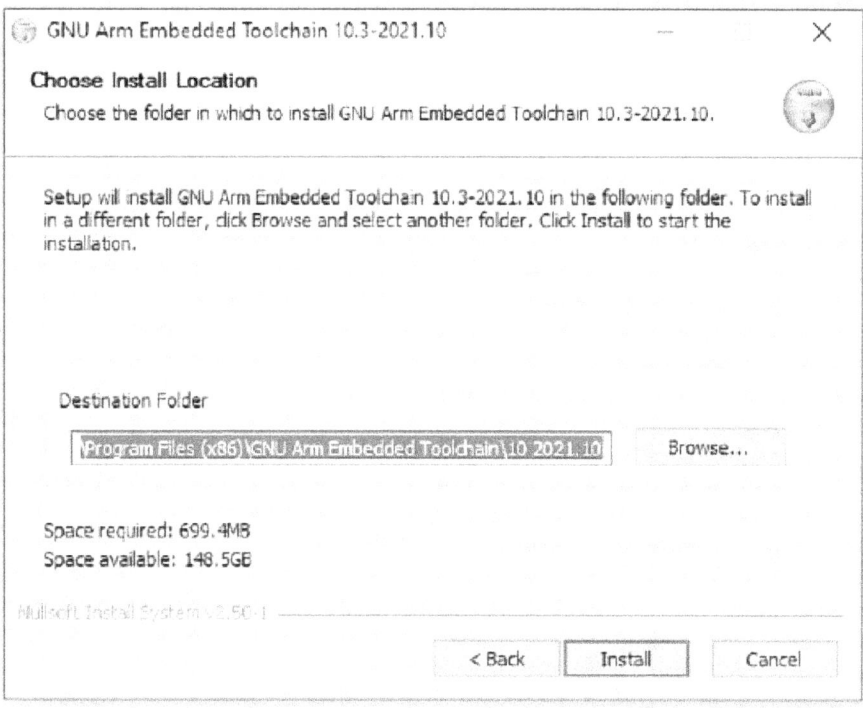

Figure 1.4: The GNU Arm Embedded Toolchain installer

5. When the installation is complete, ensure that you check the **Add path to environment variable** option.

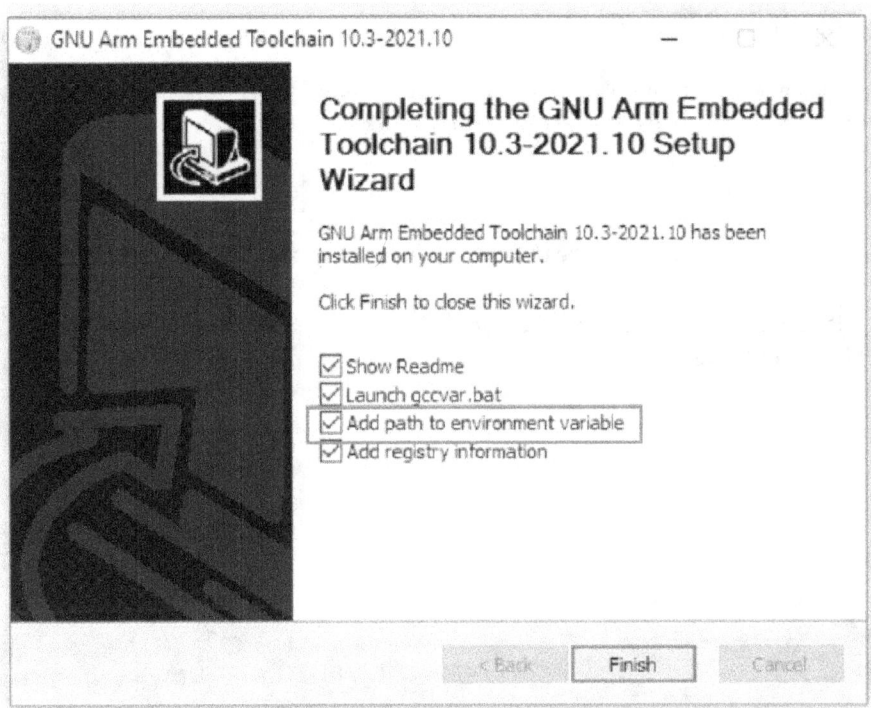

Figure 1.5: The installer showing the Add path to environment variable option

6.   Click **Finish** to finalize the setup.

Setting up OpenOCD

For firmware development with the GNU Arm Toolchain, OpenOCD plays an integral role, facilitating both the downloading of firmware into our microcontroller and the debugging of code in real time.

Let's set up OpenOCD:

1.   Launch your web browser and navigate to `https://openocd.org/pages/getting-openocd.html`.

2.   Scroll to the section titled **Unofficial binary packages**.

3.   Click on the link for multiplatform binaries.

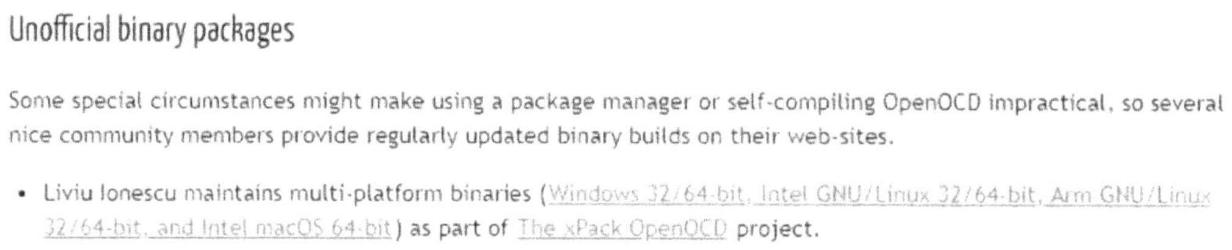
Unofficial binary packages
Some special circumstances might make using a package manager or self-compiling OpenOCD impractical, so several nice community members provide regularly updated binary builds on their web-sites.
• Liviu Ionescu maintains multi-platform binaries (Windows 32/64-bit, Intel GNU/Linux 32/64-bit, Arm GNU/Linux 32/64-bit, and Intel macOS 64-bit) as part of The xPack OpenOCD project.

Figure 1.6: The Unofficial binary packages section

4. Identify the latest version compatible with your operating system. For an exhaustive list, click on **Show all 14 assets**.

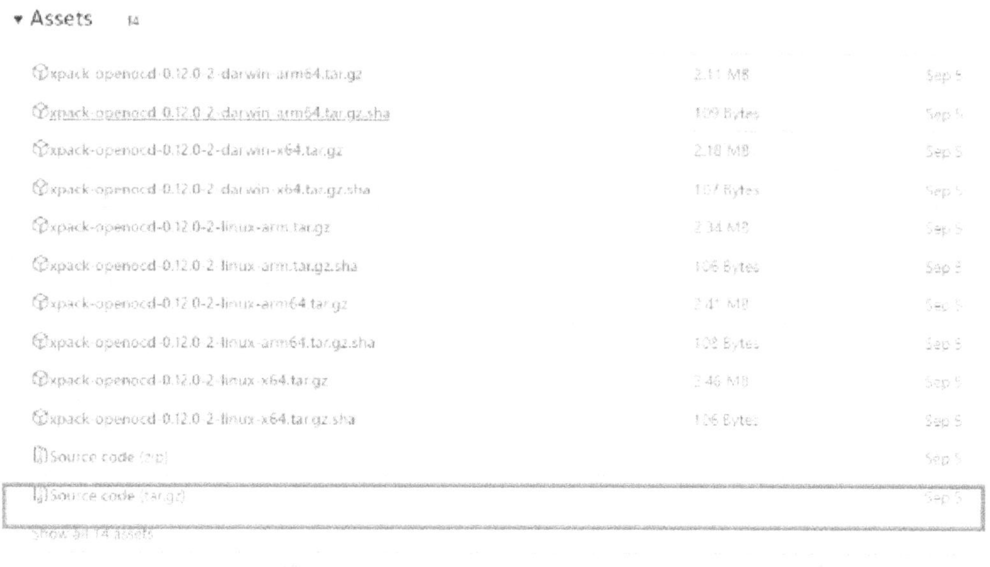
Assets 14
xpack-openocd-0.12.0-2-darwin-arm64.tar.gz
xpack-openocd-0.12.0-2-darwin-arm64.tar.gz.sha
xpack-openocd-0.12.0-2-darwin-x64.tar.gz
xpack-openocd-0.12.0-2-darwin-x64.tar.gz.sha
xpack-openocd-0.12.0-2-linux-arm.tar.gz
xpack-openocd-0.12.0-2-linux-arm.tar.gz.sha
xpack-openocd-0.12.0-2-linux-arm64.tar.gz
xpack-openocd-0.12.0-2-linux-arm64.tar.gz.sha
xpack-openocd-0.12.0-2-linux-x64.tar.gz
xpack-openocd-0.12.0-2-linux-x64.tar.gz.sha
Source code (zip)
Source code (tar.gz)

Figure 1.7: The OpenOCD packages

5. For Windows users, download the `win32-x64.zip` version. For Linux or macOS users, download the corresponding `.tar` file for your operating system.
6. Once downloaded, unzip the package.

7. Navigate to the extracted folder, and then the `bin` subfolder. Here, you'll find the `openocd.exe` application. This is the application we shall call in the command prompt together with the specific script of our chosen microcontroller, in order to debug or download code onto the microcontroller.

Within the `xpack-openocd-0.12.0-2 | openocd | scripts` directory structure, you'll find scripts tailored for various microcontrollers and development boards.

Next, we need to add OpenOCD to our environment variables:

1. Begin by relocating the entire `openocd` folder to your `Program Files` directory.

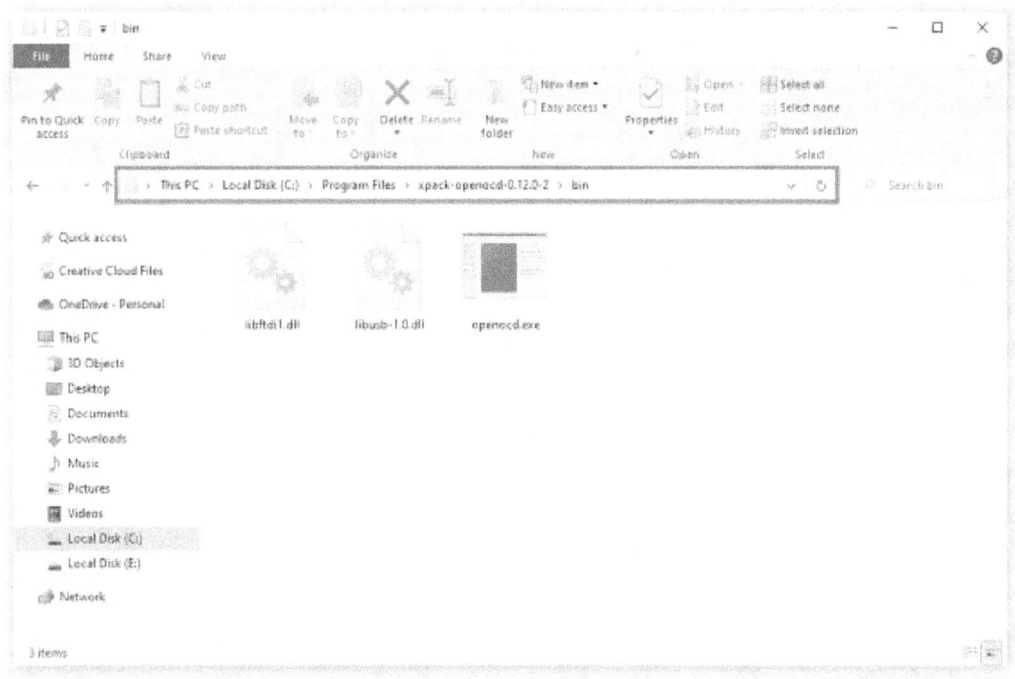

Figure 1.8: OpenOCD moved to Program Files, showing the path to the bin folder

2. Copy the path to the `openocd bin` folder.

3. Right-click on **This PC**, and then choose **Properties**.

4. Search for and select **Edit the system environment variables**.

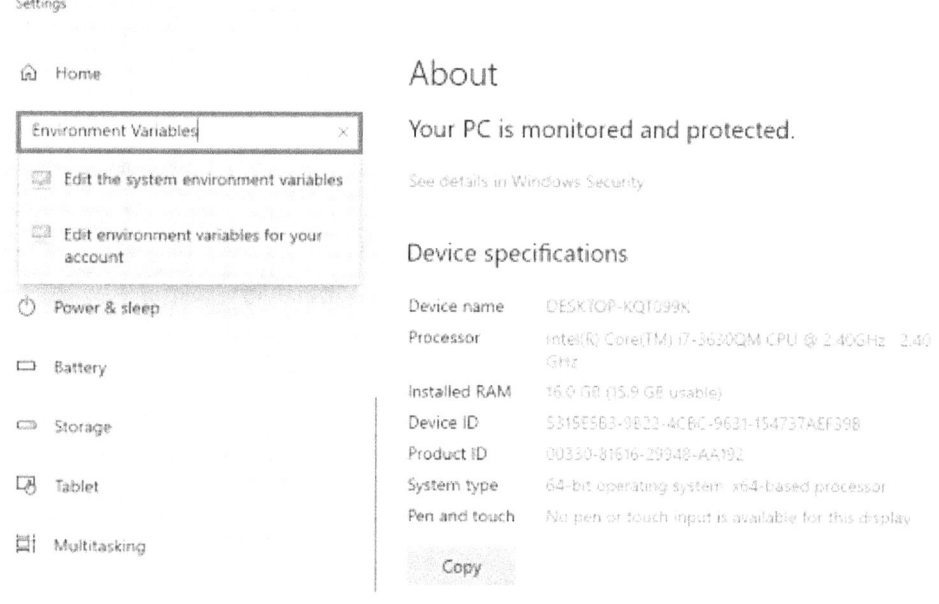

Figure 1.9: The This PC properties page

5. Click the **Environment Variables** button in the **System Properties** pop-up window.

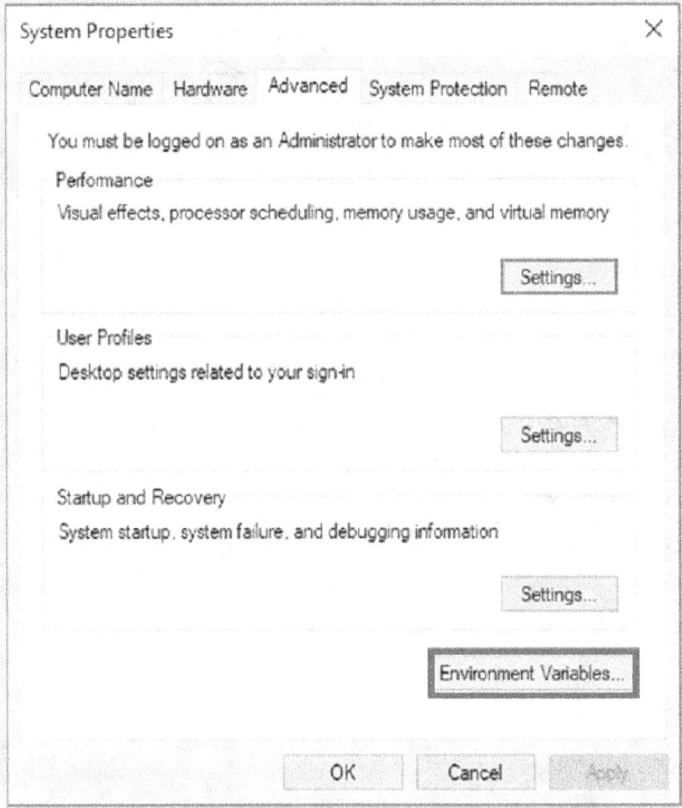

Figure 1.10: The System Properties pop-up window

6. Under the **User variables** section of the **Environment Variables** popup, double-click the **Path** entry.

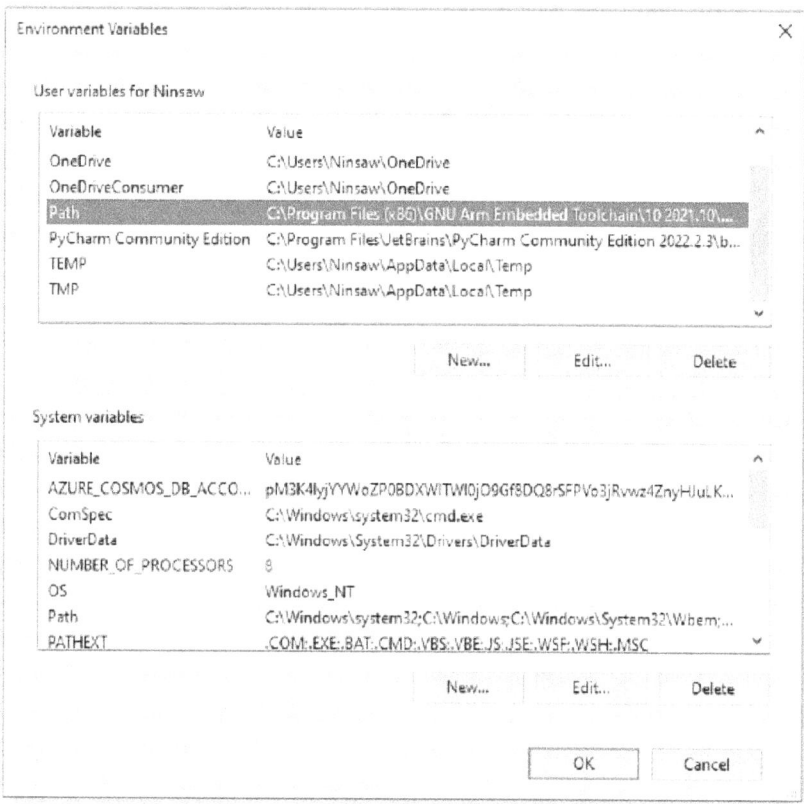

Figure 1.11: The Environment Variables popup

7. In the **Edit environment variable** popup, click on **New** to create a row for a new path entry.

8. Paste the previously copied OpenOCD path into this new row.

9.   Confirm your changes by clicking **OK** on the various pop-up windows.

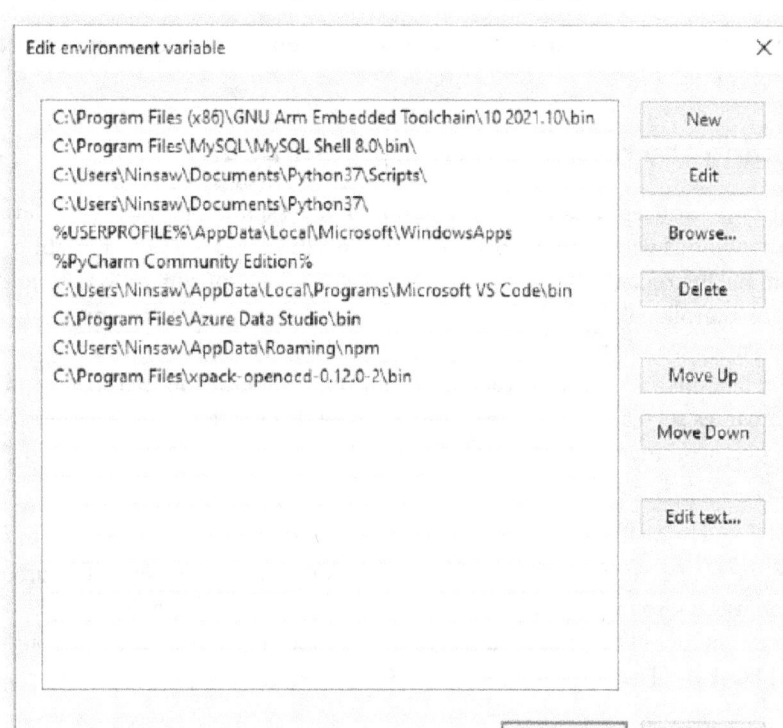

Figure 1.12: The Edit environment variable popup

Finally, we have successfully completed the setup process. We have configured two essential, standalone tools to develop firmware for STM32 microcontrollers – the STM32CubeIDE for an IDE, and the GNU Arm Embedded Toolchain, complemented by OpenOCD, to develop and debug our firmware without an IDE.

Next, we will turn our attention to our development board.

The development board

In this segment of the chapter, we will delve into the specifications and features of the development board selected for this book.

Understanding the role of a development board

Firstly, let's clarify the concept of a development board. It's essential to note that a development board is not synonymous with a microcontroller. While a development board might derive part of its name from the microcontroller mounted on it, it would be a misnomer to refer to the board itself as the microcontroller. A development board allows us to validate our firmware on the exact microcontroller variant we aim to deploy in our final product. Consequently, the firmware tested on our development board is assured to operate identically on the microcontroller in the end product. This is why companies such as STMicroelectronics offer a diverse range of development boards, tailored to each microcontroller in their portfolio.

It's also essential to contrast the role of a development board with prototyping boards, such as Arduino. While prototyping boards (which might not house the microcontrollers intended for the final product) serve as preliminary testing platforms, development boards elevate this process. They enable us to rigorously test concepts and the performance evaluation of our firmware on the designated microcontroller meant for bulk product manufacturing. For the purposes of this book, our focus will be on the NUCLEO-F411 development board.

An overview of the NUCLEO-F411 Development Board

The NUCLEO-F411 development board is equipped with an STM32F411RE microcontroller, capable of a peak operating frequency of 100MHz. It comes with a generous 512 Kbytes of flash memory and 128 Kbytes of SRAM. Furthermore, the board is equipped with several columns of berg pins, allowing us to effortlessly make connections using jumper wires to interface with a variety of modules and components – from sensors and motors to LEDs. For quick and straightforward input/output firmware tests, the board also features a built-in user button and LED, eliminating the immediate need for external components.

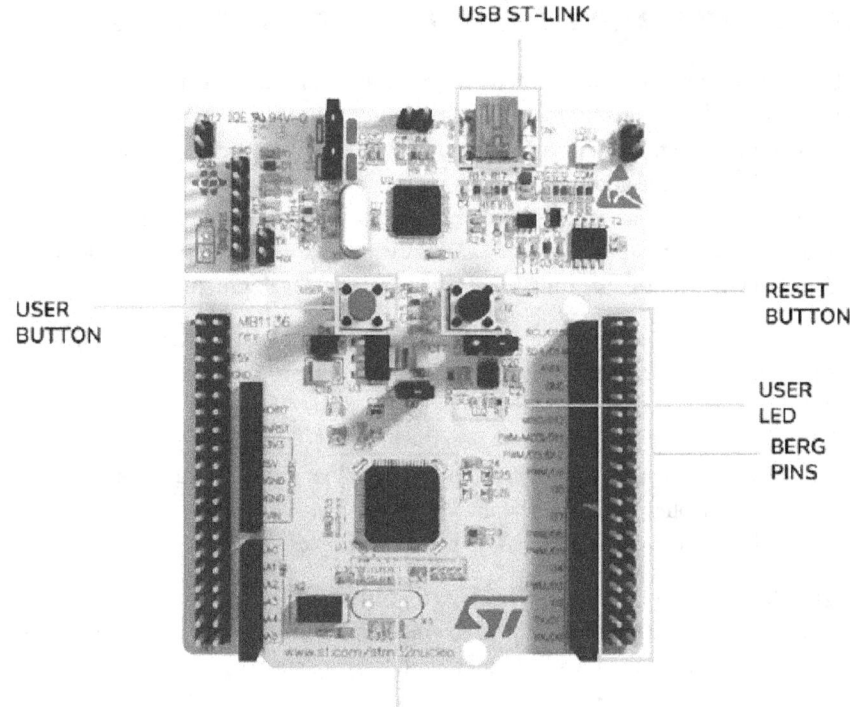

USB ST-LINK

USER
BUTTON

RESET
BUTTON

USER
LED

BERG
PINS

STM32F411RE MICROCONTROLLER

Figure 1.13: The NUCLEO-F411 development board

🔍 **Quick tip**: Need to see a high-resolution version of this image? Open this book in the next-gen Packt Reader or view it in the PDF/ePub copy.

🔒 **The next-gen Packt Reader** and a **free PDF/ePub copy** of this book are included with your purchase. Unlock them by scanning the QR code below or visiting `https://www.packtpub.com/unlock/9781835460818`.

Now that we're familiar with the development board, let's delve into the various types of documentation that are essential for a comprehensive understanding and programming of the development board.

Datasheets and manuals – unraveling the details

Our main objective in this book is to write firmware code that interacts directly with the registers of our microcontroller. This means there's no abstraction or intermediary library between our code and the target microcontroller. To achieve this, it's important to grasp the internal architecture of the microcontroller, understand the addresses of each register we interact with, and know the functions of relevant bits within those registers. This is where datasheets and manuals come in. Manufacturers provide these documents for users to understand their products, which in our case refers to the microcontroller core architecture, the microcontroller, and the development board.

Two distinct companies play roles in the making of our development board. The first is ARM Holdings, which licenses processor and microcontroller core architecture designs to semiconductor manufacturing firms such as STMicroelectronics, Texas Instruments, and Renesas. These manufacturers then produce the physical microcontroller or processor based on the licensed designs from ARM, often with their custom additions. This explains why two different microcontrollers from separate manufacturers might share the same microcontroller core. For instance, both the TM4C123 from Texas Instruments and STM32F4 from STMicroelectronics are based on the ARM Cortex-M4 core.

Since our chosen development board, the NUCLEO-F411 from STMicroelectronics, is based on the ARM Cortex-M4 microcontroller core, in the following sections, we'll delve into the documentation for the board, its integrated microcontroller, and the underlying core.

Understanding STMicroelectronics' documentation

A significant reason for the popularity of STM32 microcontrollers is STMicroelectronics' continued commitment to providing comprehensive support. This includes well-organized documentation and various firmware development resources.

STMicroelectronics has a range of documents, each following a specific naming convention. Let's discuss those relevant to our work:

- **Reference Manual (RM)**: All RMs start with the letters RM, followed by a number. For instance, the RM for our microcontroller is RM0383. This document details every register in our microcontroller, clarifying each bit's role and providing insights on register configurations.

- **Datasheet**: The datasheet is named after the microcontroller, so for the STM32F411 microcontroller, the datasheet is simply called STM32F411. This document provides a functional overview of the microcontroller, a complete memory map, a block diagram showcasing the microcontroller's peripherals and connecting buses, as well as the pinout and electrical characteristics of the microcontroller.

- **UM (User Manual)**: Starting with the letters UM and followed by a number, such as UM1724 for our NUCLEO-F411, this document focuses on the development board. It describes how components on our board, such as LEDs and buttons, are connected to specific ports and pins of the microcontroller.

The generic user guide by ARM

ARM provides documents for every microcontroller and processor core they design. Important to our discussion is the generic user guide for our microcontroller core. As we're using the STM32F411, which is based on the ARM Cortex-M4 core, we'll refer to the Cortex-M4 generic user guide.

This means that if we were using an STM32F7 microcontroller, which is based on the ARM Cortex-M7 core, then we would need to get the Cortex-M7 generic user guide. The naming convention of this document is simply the name of the microcontroller core + the phrase `generic user guide`.

As the name implies, this document provides information generic to the specific microcontroller core. This means that the information provided in the Cortex-M4 generic user guide applies to all microcontrollers based on the Cortex-M4 core, irrespective of the manufacturers of those microcontrollers. In contrast, the information provided in the STMicroelectronics documentation applies to only STMicroelectronics' microcontrollers.

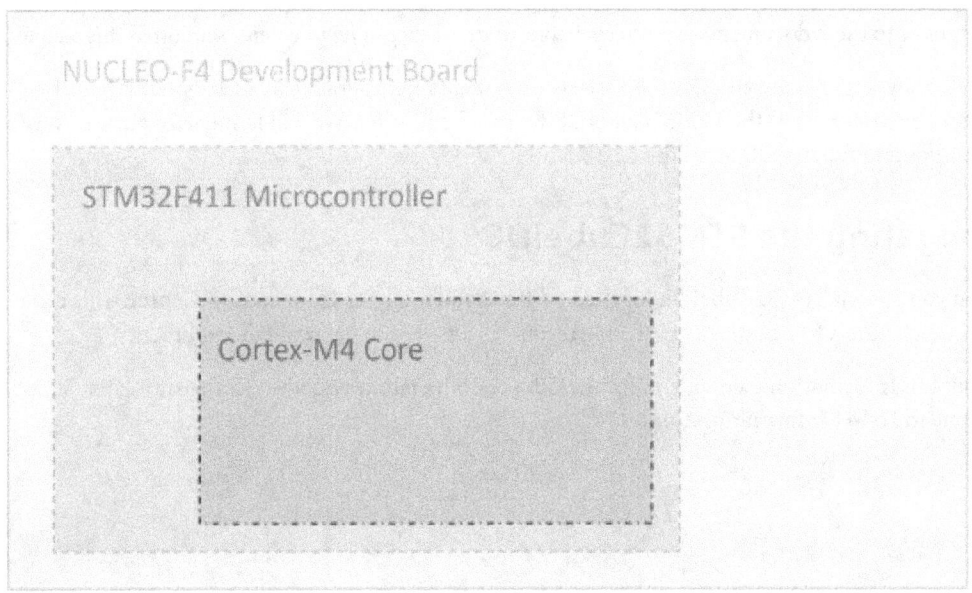

Figure 1.14: The relationship between the development board, microcontroller, and microcontroller core

Why do we need the generic user guide?

The generic user guide provides information on the core peripherals of the processor core. As the term suggests, these core peripherals are consistent across all microcontrollers, based on a specific core. The Cortex-M4 core has five core peripherals – the System Timer, Floating-Point Unit, System Control Block, Memory Protection Unit, and the Nested Vectored Interrupt Controller. When developing bare-metal drivers for these peripherals, the generic user guide is the definitive source for the essential details.

Additionally, the guide provides information on the microcontroller core's Instruction Set, as well as the Programmer's Model, Exception Model, fault handling, and power management.

Getting the documents

To obtain the aforementioned documents, you can use the following search phrases on Google:

- **RM**: Either `STM32F11 Reference Manual` or `RM0383`.
- **Datasheet**: `STM32F411 Datasheet`.
- **UM**: `Nucleo-F11 User Manual` or `UM1724`.
- **Generic user guide**: `Cortex-M4 Generic User Guide`

Direct links to these documents are also available, in the *Technical requirements* section of this chapter.

Before analyzing the key areas of the various documents to program our development board, let's first take a closer look at the STM32CubeIDE we installed earlier. We will familiarize ourselves with its features and functionalities in the next section.

Navigating the STM32CubeIDE

When you launch STM32CubeIDE for the first time, you'll see the **Information Center**. This center offers quick access to a number of valuable resources for STM32 firmware development.

To exit the Information Center, simply click the **X** on its tab. If you wish to revisit it later, simply navigate to **Help | Information Center**.

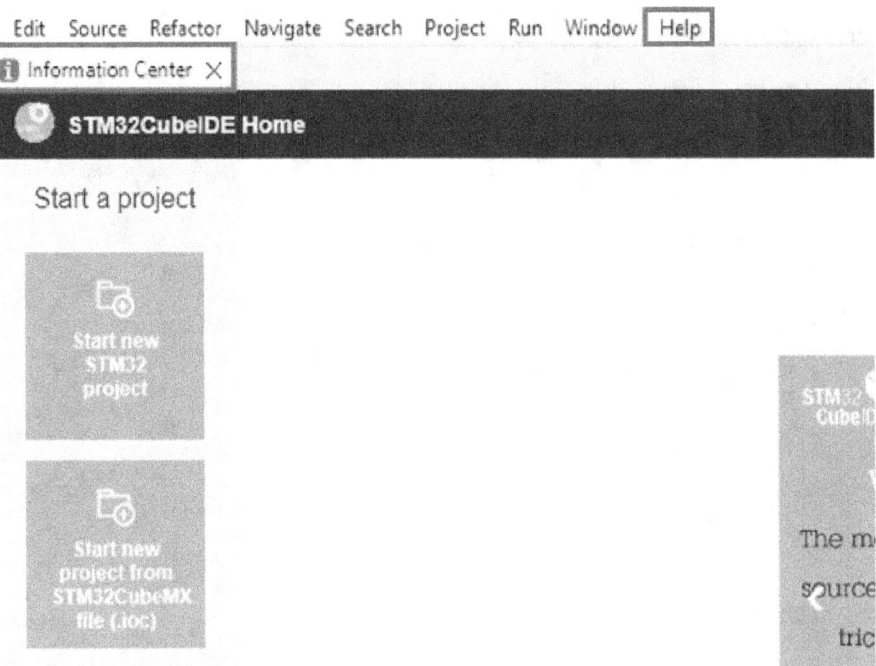

Figure 1.15: Information Center

The STM32CubeIDE is based on the Eclipse framework, and therefore, the layout and elements are similar to those of other Eclipse-based IDEs.

Let's go through the process of creating a new project:

1. Either click **Create a New STM32 project** in the empty **Project Explorer** pane or select **File | New | STM32 Project**.

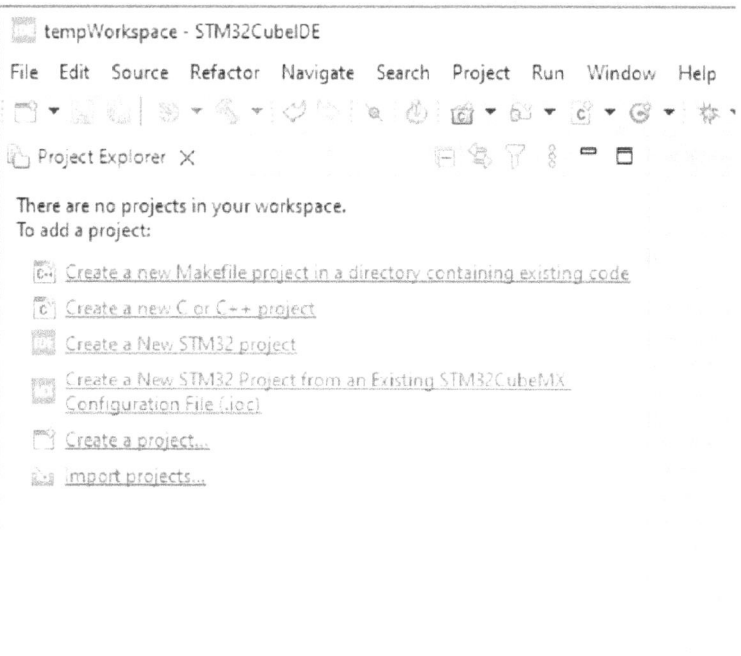

Figure 1.16: A workspace showing an empty Project Explorer pane

2. You will be presented with the **Target Selection** window to select the microcontroller or development board for your project.

3. Click the **Board Selector** tab.

4. Enter NUCLEO-F411 into the **Commercial Part Number** field.

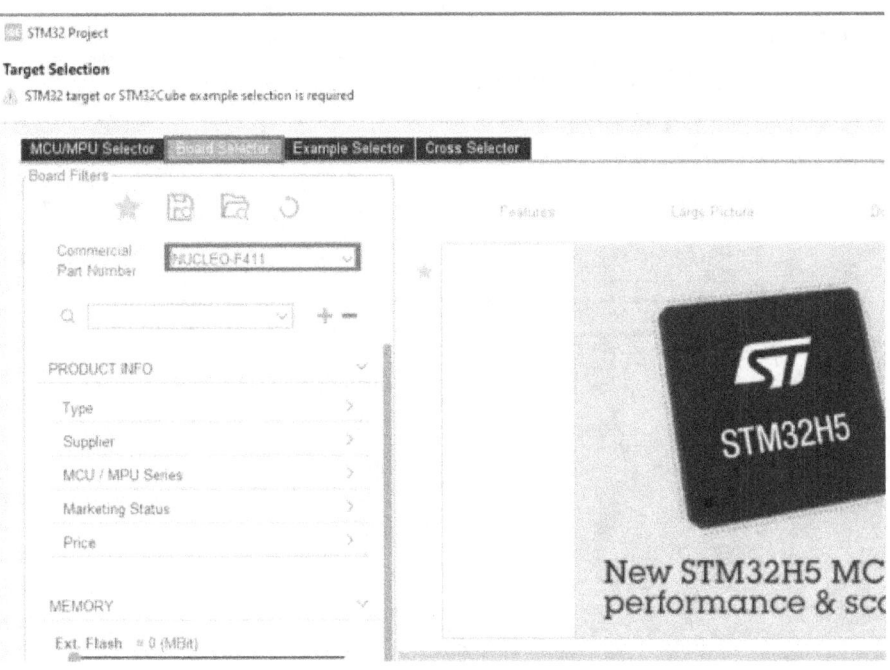

Figure 1.17: The Target Selection window

5. From the displayed board list, select **NUCLEO-F11RE**, and then click **Next**.

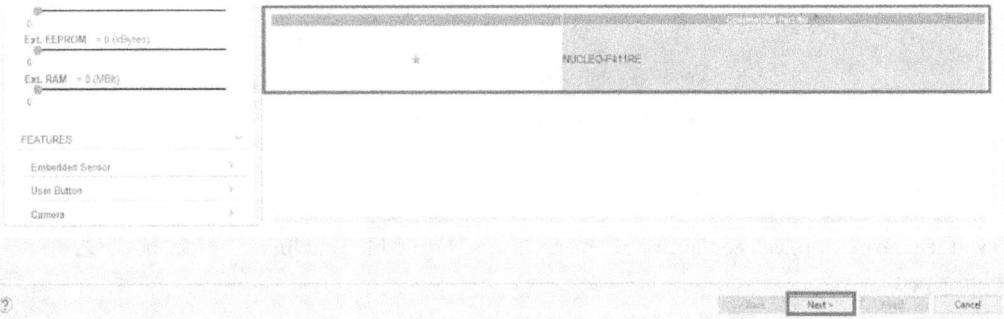

Figure 1.18: The board list with NUCLEO-F411 selected

6.   Give the project a name.

7.   For **Targeted Project Type**, select **Empty**.

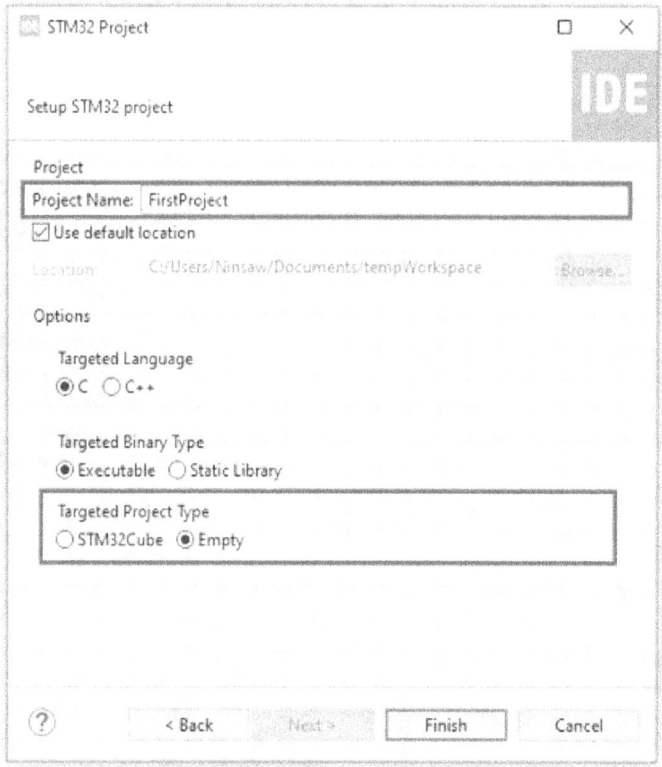

Figure 1.19: The Setup STM32 project window

8.   Click **Finish** to create the project.

You will see the new project, containing all the necessary startup files and linker scripts, in the **Project Explorer** pane.

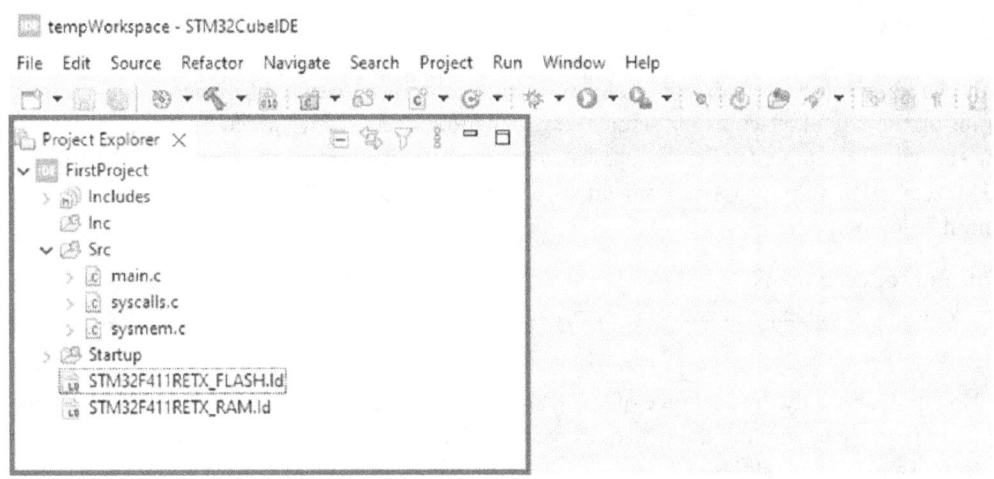

Figure 1.20: The Project Explorer pane showing a new project

Understanding the control icons

The most frequently used control icons are the **New**, **Build**, and **Debug** icons.

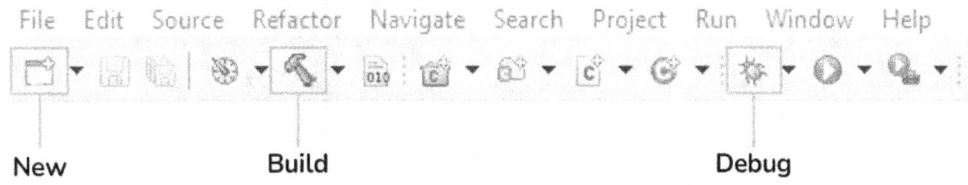

Figure 1.21: The control icons

Let's look closely at the functions of these icons:

- **The New icon**: This icon allows us to create various files, including source code, header files, projects, libraries, and more. This function is also accessible via **File | New**.

- **The Build icon**: Used for building projects. This functionality can also be accessed by right-clicking on a project and selecting **Build Project**.

- **The Debug icon**: This launches a debug configuration to facilitate project debugging. This functionality is also available by right-clicking a project and selecting **Debug Project**.

These control icons provide quick access to essential functions, significantly enhancing productivity and streamlining the development process.

Summary

In this chapter, we set out to create a robust environment for embedded firmware development, focusing on the careful selection and setup of essential tools. Each tool we selected plays a crucial role in the efficient development of firmware for microcontrollers. We explored the installation processes of STM32CubeIDE, the GNU Arm Embedded Toolchain, and OpenOCD, laying a solid groundwork for our development activities.

We then introduced the NUCLEO-F411 development board, equipped with an STM32F411RE microcontroller, as our experimental platform, and we spent time identifying some of the components on the board.

We also emphasized the importance of knowing the different types of datasheets and reference manuals, equipping us with the ability to quickly access detailed information about a microcontroller's architecture and functionalities.

Moving on to the next chapter, we will leap into developing our first bare-metal firmware, using only the documentation we have compiled as our guide.

Unlock this book's exclusive benefits now

This book comes with additional benefits designed to elevate your learning experience.

Note: Have your purchase invoice ready before you begin.

https://www.packtpub.com/
unlock/9781835460818

2

Constructing Peripheral Registers from Memory Addresses

Bare-metal programming is all about working directly with the registers in the microcontroller without going through a library, allowing us to gain a deeper understanding of the microcontroller's capabilities and limitations. This approach enables us to optimize our firmware for speed and efficiency, which are two very important parameters in embedded systems where resources are often limited.

In this chapter, our journey begins with an exploration of various firmware development methodologies, highlighting the different levels of abstraction each offers. We then proceed to learn how to identify the ports and pins associated with key components on our development board. This step is crucial for establishing a proper interface with the microcontroller's peripherals.

Next, we delve into defining the addresses of some peripherals using the microcontroller's official documentation. This will allow us to create the addresses of the various registers in those peripherals.

In the latter sections of the chapter, our focus shifts to practical application. We will use the register addresses that we've created to configure PA5 to activate the user **light-emitting diode** (**LED**) of the development board.

In this chapter, we're going to cover the following main topics:

- The different types of firmware development
- Locating and understanding the development board's components
- Defining and creating registers through documentation insights
- Register manipulation – from configuration to running your first firmware

By the end of this chapter, you will have a solid foundation in both navigating and programming STM32 microcontrollers at the register level. You will be equipped to write your initial bare-metal firmware, relying solely on information gathered from the documentation and the **integrated development environment** (IDE).

Technical requirements

All the code examples for this chapter can be found on GitHub at `https://github.com/ PacktPublishing/Bare-Metal-Embedded-C-Programming`.

The different types of firmware development

There are several ways to develop the firmware for a particular microcontroller depending on the resources provided by the microcontroller's manufacturer. When it comes to developing firmware for STM32 microcontrollers from **STMicroelectronics**, we can use the following:

- **Hardware Abstraction Layer (HAL):** This is a library provided by STMicroelectronics. It simplifies the process by offering high-level APIs for configuring every aspect of the microcontroller. What is great about HAL is its portability. We can write code for one STM32 microcontroller, and easily adapt it for another, thanks to the uniformity of their APIs.

- **Low Layer (LL):** Also from STMicroelectronics, the LL library is a leaner alternative to HAL, offering a faster, more expert-oriented approach that's closer to the hardware.

- **Bare-Metal C programming**: With this approach, we dive right into the hardware, accessing the microcontroller's registers directly using the C language. It's more involved but offers a deeper understanding of the microcontroller's workings.

- **Assembly language:** This is similar to Bare-Metal C, but instead of C, we use assembly language to interact directly with the microcontroller's registers.

Let's compare the four firmware development methods we've discussed: HAL, LL, Bare-Metal C, and assembly language. Each method has its unique style and level of abstraction, impacting how we interact with the microcontroller's hardware. We will use the example of configuring a **general-purpose input/output** (GPIO) pin as an output to illustrate these differences.

HAL

The following code snippet demonstrates how to initialize GPIOA pin 5 as an output using HAL:

```
#include "stm32f4xx_hal.h"
GPIO_InitTypeDef GPIO_InitStruct = {0};
// Enable the GPIOA Clock
__HAL_RCC_GPIOA_CLK_ENABLE();
```

```
// Configure the GPIO pin
GPIO_InitStruct.Pin = GPIO_PIN_5;
GPIO_InitStruct.Mode = GPIO_MODE_OUTPUT_PP;
HAL_GPIO_Init(GPIOA, &GPIO_InitStruct);
```

💡 **Quick tip**: Enhance your coding experience with the **AI Code Explainer** and **Quick Copy** features. Open this book in the next-gen Packt Reader. Click the **Copy** button (**1**) to quickly copy code into your coding environment, or click the **Explain** button (**2**) to get the AI assistant to explain a block of code to you.

```
                                              Copy     Explain
function calculate(a, b) {
    return {sum: a + b};                       1         2
};
```

🔓 **The next-gen Packt Reader** is included for free with the purchase of this book. Unlock it by scanning the QR code below or visiting `https://www.packtpub.com/unlock/9781835460818`.

\Let's analyze the snippet:

- `#include "stm32f4xx_hal.h"`: This line includes the HAL library specific to the STM32F4 series, providing access to the HAL functions and data structures

- `GPIO_InitTypeDef GPIO_InitStruct = {0}`: Here, we declare and initialize an instance of the `GPIO_InitTypeDef` structure, which is used to configure the GPIO pin properties

- `__HAL_RCC_GPIOA_CLK_ENABLE()`: This macro call enables the clock for GPIO port A, ensuring that the GPIO peripheral is powered and can function

- `GPIO_InitStruct.Pin = GPIO_PIN_5`: This line sets the pin to be configured, in this case, pin 5 of port A

- `GPIO_InitStruct.Mode = GPIO_MODE_OUTPUT_PP`: Over here, we configure the pin as an output pin

- `HAL_GPIO_Init(GPIOA, &GPIO_InitStruct)`: Finally, we initialize the GPIO pin (PA5) with the configuration settings specified in `GPIO_InitStruct`

This snippet shows the ease and readability of the HAL approach in performing common hardware interfacing tasks. We can summarize the benefits and drawbacks of the HAL approach as follows:

- **Level of abstraction**: High

- **Ease of use**: Easier for beginners due to its high-level abstraction

- **Code verbosity**: More verbose, with several lines of code required for simple tasks

- **Portability**: Excellent across different STM32 devices

- **Performance**: Slightly slower due to additional abstraction layers

LL

This is how we initialize GPIOA pin 5 as an output pin using the LL library:

```
#include "stm32f4xx_ll_bus.h"
#include "stm32f4xx_ll_gpio.h"
// Enable the GPIOA Clock
LL_AHB1_GRP1_EnableClock(LL_AHB1_GRP1_PERIPH_GPIOA);
// Configure the GPIO pin
LL_GPIO_SetPinMode(GPIOA,LL_GPIO_PIN_5,LL_GPIO_MODE_OUTPUT);
```

Let's break it down, line by line:

- `#include "stm32f4xx_ll_bus.h"` and `#include "stm32f4xx_ll_gpio.h"` include the necessary LL library files for handling the bus system and GPIO functionality, respectively

- `LL_AHB1_GRP1_EnableClock(LL_AHB1_GRP1_PERIPH_GPIOA)`: This function call enables the clock for GPIO port A

- `LL_GPIO_SetPinMode(GPIOA, LL_GPIO_PIN_5, LL_GPIO_MODE_OUTPUT)`: Finally, we set the mode of GPIOA pin 5 to output mode

The LL library provides a more direct and lower-level approach to hardware interaction compared to HAL. This is often preferred in scenarios where finer control over hardware and performance is required.

The benefits and drawbacks of the LL approach are as follows:

- **Level of abstraction**: Medium
- **Ease of use**: Moderate, with a balance between abstraction and direct control
- **Code verbosity**: Less verbose than HAL, offering a more straightforward approach to hardware interaction
- **Portability**: Good, but slightly less than HAL
- **Performance**: Faster than HAL, as it's closer to the hardware

Bare-Metal C

Let's see how to accomplish the same task using the Bare-Metal C approach:

```
#define GPIOA_MODER (*(volatile unsigned long *)(GPIOA_BASE + 0x00))
#define RCC_AHB1ENR (*(volatile unsigned long *)(RCC_BASE + 0x30))

// Enable clock for GPIOA
RCC_AHB1ENR |= (1 << 0);

// Set PA5 to output mode
GPIOA_MODER |= (1 << 10); // Set bit 10 (MODER5[1])
```

Let's break down the bare-metal c snippet:

- `#define GPIOA_MODER (*(volatile unsigned long *) (GPIOA_BASE + 0x00))` and `#define RCC_AHB1ENR (*(volatile unsigned long *) (RCC_BASE + 0x30))` define pointers to specific registers within the microcontroller's memory. GPIOA_MODER points to the GPIO port A mode register, and RCC_AHB1ENR points to the **reset and clock control** (**RCC**) AHB1 peripheral clock **enable register** (**ER**). The use of the volatile keyword ensures that the compiler treats these as memory-mapped registers, preventing optimization-related issues.
- `RCC_AHB1ENR |= (1 << 0)`: This line of code enables the clock for GPIOA. It does this by setting the first bit (bit 0) of the RCC_AHB1ENR register. The bitwise OR assignment (|=) ensures that only the specified bit is changed without altering other bits in the register.
- `GPIOA_MODER |= (1 << 10)`: This line sets PA5 to output mode. In the GPIO port mode register (GPIOA_MODER), each pin is controlled by two bits. For PA5, these are bits 10 and 11 (MODER5[1:0]). The code sets bit 10 to 1 (and leaves bit 11 as 0, assuming it was already 0), configuring PA5 as a general-purpose output mode.

With this approach, we can observe the granularity and direct control provided. By directly manipulating the microcontroller's registers, it offers very high efficiency and performance:

- **Level of abstraction**: Low

- **Ease of use**: Challenging for beginners, as it requires in-depth hardware knowledge

- **Code verbosity**: Less verbose, direct

- **Portability**: Limited, as the code is often specific to a particular hardware setup

- **Performance**: Very high, as it allows for direct and optimized hardware manipulation

Assembly language

Finally, let's analyze the assembly language implementation for configuring PA5 as an output pin:

```
EQU GPIOA_MODER, 0x40020000
EQU RCC_AHB1ENR, 0x40023800

; Enable clock for GPIOA
LDR R0, =RCC_AHB1ENR
LDR R1, [R0]
ORR R1, R1, #(1 << 0)
STR R1, [R0]

; Set PA5 as output
LDR R0, =GPIOA_MODER
LDR R1, [R0]
ORR R1, R1, #(1 << 10)
STR R1, [R0]
```

Let's break it down:

- EQU GPIOA_MODER, 0x40020000 and EQU RCC_AHB1ENR, 0x40023800 define constants for the memory addresses of the GPIOA mode register (GPIOA_MODER) and the RCC AHB1 peripheral clock ER (RCC_AHB1ENR). 'EQU' is used in assembly to equate a label to a value or address.

The rest of the assembly instructions perform two main tasks:

- Enable the clock for GPIOA:

  - LDR R0, =RCC_AHB1ENR: Load the address of the RCC_AHB1ENR register into the R0 register

  - LDR R1, [R0]: Load the value of the RCC_AHB1ENR register into the R1 register

- `ORR R1, R1, #(1 << 0)`: Perform a bitwise `OR` operation to set bit 0 of R1, turning on the clock for GPIOA

- `STR R1, [R0]`: Store the updated value back into the `RCC_AHB1ENR` register

- Set PA5 as output:

  - `LDR R0, =GPIOA_MODER`: Load the address of the `GPIOA_MODER` register into R0

  - `LDR R1, [R0]`: Load the current value of the `GPIOA_MODER` register into R1

  - `ORR R1, R1, #(1 << 10)`: Use a bitwise `OR` operation to set bit 10 of R1, configuring PA5 as an output

  - `STR R1, [R0]`: Store the updated value back into the `GPIOA_MODER` register

The assembly language approach allows for extremely detailed and direct control over the microcontroller. We often use this in projects where high performance is crucial, and every aspect of the hardware needs to be precisely managed.

The assembly language approach offers us the following:

- **Level of abstraction**: Lowest

- **Ease of use**: Most challenging, requiring a thorough understanding of the microcontroller's architecture

- **Code verbosity**: Can be verbose for complex tasks, due to low-level nature

- **Portability**: Very limited, as it is highly specific to the microcontroller's architecture

- **Performance**: Highest, as it allows for the most optimized and direct control possible

The following diagram shows each method and its closeness to the microcontroller's architecture:

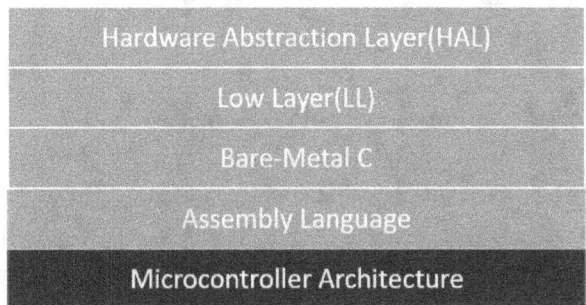

Figure 2.1: Firmware development methods, arranged in order of proximity
to the microcontroller's architecture

Now that we have explored the diverse approaches to firmware development for STM32 microcontrollers, we are ready to delve into the realm of bare-metal C programming.

We will begin our exploration by understanding how the main components of our development board are connected to specific pins of the onboard microcontroller. This initial step is crucial for gaining insight into the hardware layout and preparing us for detailed programming tasks.

Locating and understanding the development board's components

In this section, our focus is to pinpoint the specific ports and pins on the microcontroller to which the user LED, user push button, berg pins, and Arduino-compatible headers are connected on the development board. Understanding these connections is crucial for our programming tasks. To accurately identify these connections, we will consult the *NUCLEO-F411 User Manual*.

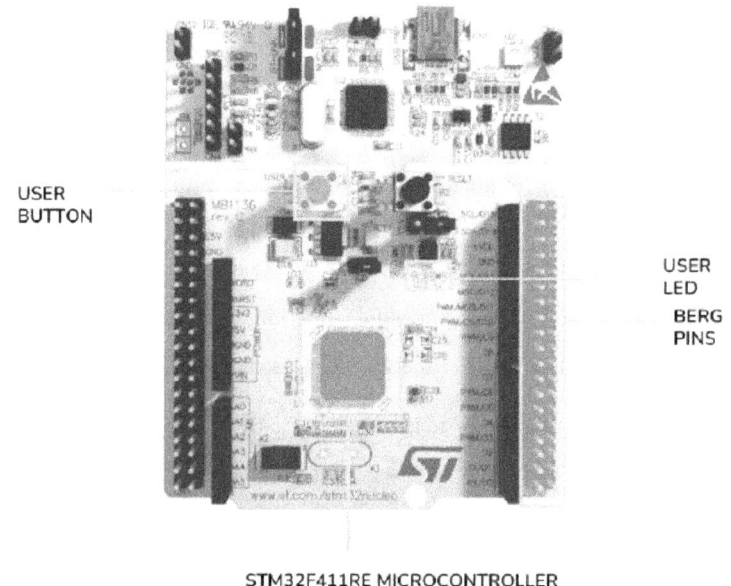

Figure 2.2: Development board showing components of interest

Now, let's locate the microcontroller pin connected to the User LED on the development board.

Locating the LED connection

Our first step is to navigate through the table of contents to find the section dedicated to LEDs. This can be done quickly by locating *Figure 2.3* in the manual, which shows the page number for the *LEDs* section and allows us to jump directly to it.

Click on the page number to jump to the LEDs section.

Figure 2.3: This is the part of the NUCLEO-F411 User Manual's table of
contents showing the page number of the LEDs section

In the *LEDs* section, we find that the User LED, labeled **User LD2**, is linked to the ARDUINO® signal **D13**. This corresponds to either pin PA5 or PB13, depending on the specific STM32 target of our board.

It's important to note the dual naming convention used in the User Manual due to the board's compatibility with the Arduino IDE. In the Arduino naming scheme, pins are categorized as either analog (preceded by "A") or digital (preceded by "D"). For example, digital pin 3 is denoted as D3. Conversely, the standard STM32 convention starts with a "P," followed by a letter indicating the port and then the pin number within that port, such as PA5 for the 5th pin of port A.

To determine whether pin D13 of our development board is PA5 or PB13 of the onboard microcontroller, we refer to *Tables 11 to 23* in the manual. These tables map the ARDUINO® connector pins to standard STM32 pins for each development board covered in the document. Specifically, we look at *Figure 2.5* showing *Table 16*, which pertains to our development board model.

Navigate to *Table 11* by clicking on it as shown in *Figure 2.4*. This action will take you to the initial table in the sequence of tables. Then, scroll down until you get to *Table 16*.

> **User LD2**: the green LED is a user LED connected
> to STM32 I/O PA5 (pin 21) or PB13 (pin 34) depen
> *Table 11* to *Table 23* when:
> - the I/O is HIGH value, the LED is on
> - the I/O is LOW, the LED is off

Figure 2.4: This is the part of the NUCLEO-F411 User Manual that
shows the two possible connections of the User LED

Upon reviewing *Table 16*, we find that D13 indeed corresponds to PA5. This indicates that the User LED on our NUCLEO-F411RE development board is connected to pin PA5 of the onboard microcontroller.

Table 16. ARDUINO® connectors on NUCLEO-F401RE and NUCLEO-F411RE (continued)

Connector	Pin	Pin name	STM32 pin	Function
	6	D13	PA5	SPI1_SCK

Figure 2.5: Table 16 shows that D13 corresponds to PA5

Another useful component found on the development board is the User Push button. Let's find the pin connection of this component.

Locating the User Push button

The User Push button on the development board is an important component for input handling in many embedded experiments, and understanding its connection to the microcontroller is crucial for effective programming and interaction.

To locate the connection details of the User Push button on our board, we will navigate through the table of contents to find the section dedicated to *Push-buttons*.

This can be done quickly by locating *Figure 2.6* in the manual, which shows the page number for the *Push-buttons* section and allows us to jump directly to it by clicking on the page number.

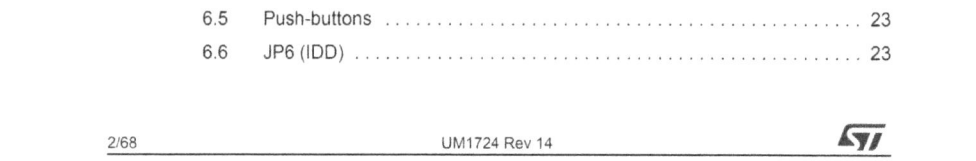

Figure 2.6: This part of the NUCLEO-F411 User Manual provides
the page number for the Push-buttons section

Upon navigating to the *Push-buttons* section of the manual, we locate the relevant information about our User Push-button, identified as **B1 User**. The manual tells us that this button is connected to pin **PC13** of the onboard microcontroller.

Locating the berg pins and Arduino-compatible headers

In the *LEDs* sections, we learned that our NUCLEO-F411 development board features two primary naming systems for its pins: the Arduino naming system and the standard STM32 naming system. While bare-metal programming primarily utilizes the standard STM32 naming, the pins on the board itself are labeled according to the Arduino system. It is important to know the actual port names and pin numbers of these exposed pins so that we can properly connect and program external components such as sensors and actuators.

Figure 2.7 shows our NUCLEO-F411 development board with the columns of berg pins highlighted.

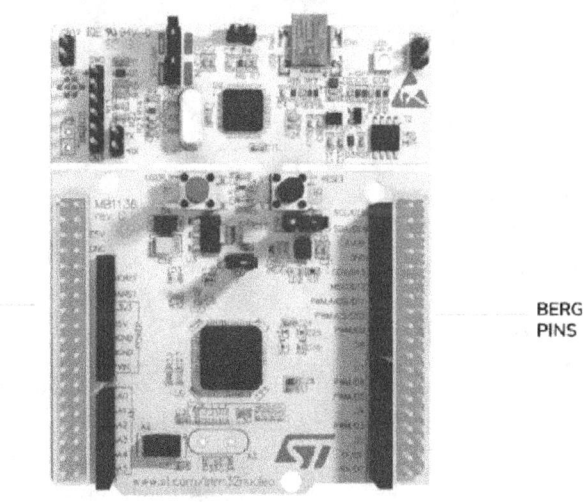

Figure 2.7: This is the NUCLEO-F411 development board with berg pins highlighted

The Arduino header pins of the development board are located at the sides of the berg pins columns. This is highlighted in *Figure 2.8*.

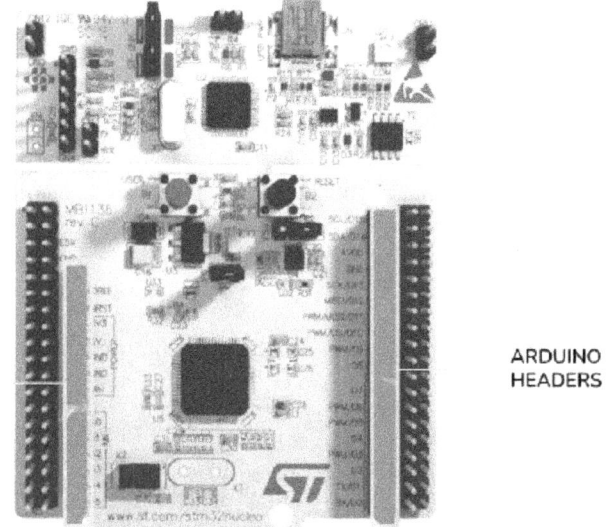

ARDUINO
HEADERS

ARDUINO
HEADERS

Figure 2.8: This NUCLEO-F411 development board with Arduino headers highlighted

Let's start by finding the microcontroller pin connections of the Arduino header pins.

Arduino-compatible headers

To identify the standard STM32 names for the pins on the Arduino header, we navigate to the section titled *ARDUINO® connectors* using the table of contents. This directs us to *Table 11*, which provides the mappings of Arduino pins to STM32 pins. For our specific NUCLEO-F411 development board, we focus on *Table 16*, which offers the relevant mapping for our model.

Next, let's locate the connections of the berg pins.

The berg pins

Similar to our approach with other components, to find details about the berg pins, we again consult the table of contents in the manual and locate the section called *Extension connectors*. This section includes figures illustrating the pinouts for various NUCLEO boards. We then scroll to find the pinout corresponding to our specific NUCLEO model. Here, the pinout of our development board is presented in the standard STM32 naming system.

Over here, we also discover that the manual refers to the columns of male header berg pins as the ST morpho connector. This means that whenever the *morpho connector* term is used, it is referring to these male header berg pins.

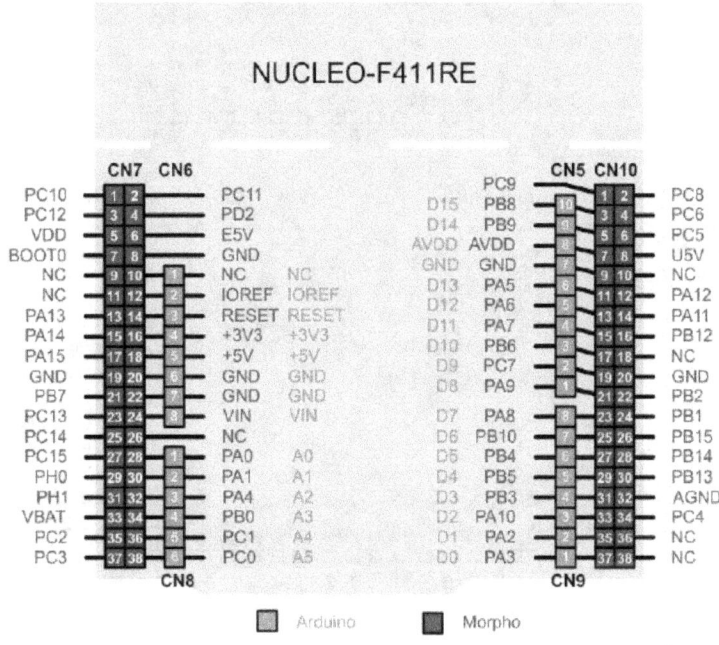

Figure 2.9: Pinout of the NUCLEO-F411 development board

In this section, we learned that NUCLEO development boards use two naming systems. Firstly, the Arduino naming system, which is visibly marked on the board, and secondly, the standard STM32 naming system, detailed in the documentation. We discovered that the standard STM32 naming system is particularly relevant for our purposes, as it directly correlates to the pin names of the onboard microcontroller.

The next section will guide us on how to access and manipulate the relevant memory locations of the onboard microcontroller. Our focus will be on configuring pin PA5 as an output pin. This will allow us to control the LED connected to PA5.

Defining and creating registers through documentation insights

In the previous section, we established that the User LED is connected to pin PA5. This means that it is linked to pin number 5 on GPIO PORTA. In other words, to get to the LED, we have to go through PORTA and then locate pin number 5 of that port.

As illustrated in *Figure 2.10*, the microcontroller has exposed pins on all four sides. These pins are organized into distinct groups known as **ports**. For instance, pins in PORTA are denoted with the *PA* prefix, while those in PORTB start with *PB*, and so forth. This systematic arrangement allows us to easily identify and access specific pins for programming and hardware interfacing tasks.

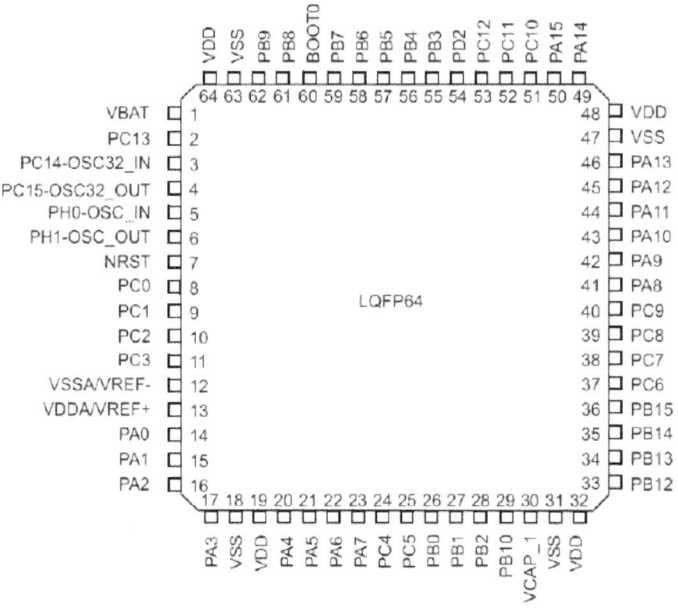

Figure 2.10: STM32F411 pinout

In the next section, we will go through the steps to locate the precise address of GPIO PORTA.

Locating GPIO PORTA

To effectively interact with any part of our microcontroller, it's essential to know the memory address of that specific part. Our next step is to explore the memory map of the microcontroller. By doing so, we can locate the address of GPIO PORTA in a step-by-step manner.

Since our focus is now on the onboard microcontroller rather than the development board, we need to refer to the microcontroller's datasheet, specifically the `stm32f411re.pdf` document.

Let's start by navigating to the table of contents of the document. There, we'll find a section entitled *Memory Mapping*. Click on the corresponding page number to jump to that section.

Over here, we find a comprehensive diagram that illustrates the entire memory map of the microcontroller. A relevant excerpt of this diagram is presented in *Figure 2.11*, which shows the overall memory layout.

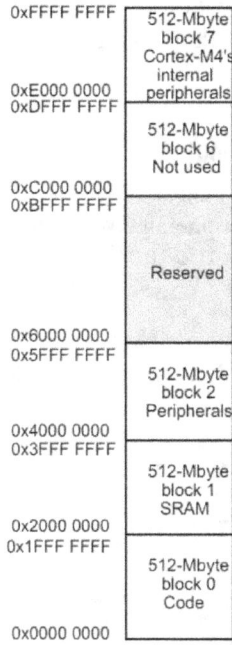

Figure 2.11: Memory map

The memory map shows that everything inside the microcontroller is addressed from 0x0000 0000 to 0xFFFF FFFF. We're interested in the part about peripherals because that's where we find GPIOA.

In the context of microcontrollers, a peripheral refers to a hardware component that is not part of the **central processing unit** (**CPU**) but is connected to the microcontroller to extend its capabilities. Peripherals perform specific functions and can include a wide range of components, such as the following:

- **GPIO ports**: These are used for interfacing with external devices such as LEDs, switches, and sensors. They can be programmed to either receive input signals or send output signals.

- **Communication interfaces**: These include serial communication interfaces such as a **universal asynchronous receiver-transmitter** (**UART**), **Serial Peripheral Interface** (**SPI**), and **Inter-Integrated Circuit** (**I2C**), which enable the microcontroller to communicate with other devices, sensors, or even other microcontrollers.

- **Timers and counters**: Timers are used for measuring time intervals or generating time-based events, while counters can be used to count events or pulses.

- **Analog-to-digital converters** (**ADCs**): ADCs convert analog signals (such as those from a temperature sensor) into digital values that the microcontroller can process.

The peripherals mentioned here represent a selection of the common types found in microcontrollers. As we progress through subsequent chapters, we will delve deeper into these and other peripherals.

In *Figure 12.11*, the memory map shows that the address range for all the microcontroller's peripherals spans from 0x40000000 to 0x5FFFFFFF. This means that GPIO PORTA's address lies within this specified range.

> **Peripherals base address = 0x40000000**
>
> Let's note down the start of the peripheral address, which we will refer to as PERIPH_BASE, indicating the base address for the peripherals. We will need this for calculating the address of GPIO PORTA. PERIPH_BASE = 0x40000000.

Figure 12.12 shows a zoomed-in view of the peripherals section in the memory map. Here, we observe that the peripheral memory is segmented into five distinct blocks: APB1, APB2, AHB1, AHB2, and the Cortex-M internal peripherals block, which is located at the top.

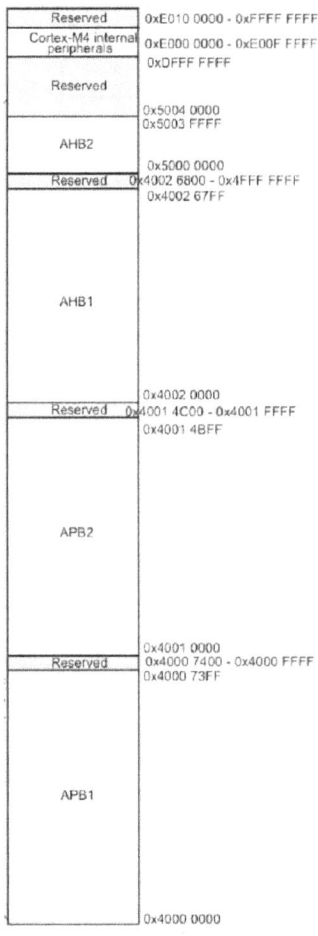

Figure 2.12: Peripherals memory map

These blocks, except for the Cortex-M internal peripherals, are named after the bus systems they interface with – namely, the **Advanced Peripheral Bus (APB)** and the **Advanced High-Performance Bus (AHB)**:

- **APB1 and APB2**: These buses cater to lower bandwidth peripherals, providing a more efficient means of communication for devices that do not require high-speed data transfer.

- **AHB1 and AHB2**: These are designed for high-speed data transfer and are used to connect high-bandwidth peripherals. They enable faster and more efficient data, control, and address communication.

On *pages 54 to 56* of the datasheet, we find a table delineating the boundary addresses for each bus and the associated peripherals. A segment of this table is shown in *Figure 2.13*, where we find that GPIOA is allocated a boundary address from 0x40020000 to 0x4002 03FF and is connected to the AHB1 bus. Therefore, this indicates that the addresses for all registers related to GPIO PORTA are encompassed within this address range.

Bus	Boundary address	Peripheral
AHB2	0x5000 0000 - 0x5003 FFFF	USB OTG FS
AHB1	0x4002 6800 - 0x4FFF FFFF	Reserved
	0x4002 6400 - 0x4002 67FF	DMA2
	0x4002 6000 - 0x4002 63FF	DMA1
	0x4002 5000 - 0x4002 4FFF	Reserved
	0x4002 3C00 - 0x4002 3FFF	Flash interface register
	0x4002 3800 - 0x4002 3BFF	RCC
	0x4002 3400 - 0x4002 37FF	Reserved
	0x4002 3000 - 0x4002 33FF	CRC
	0x4002 2000 - 0x4002 2FFF	Reserved
	0x4002 1C00 - 0x4002 1FFF	GPIOH
	0x4002 1400 - 0x4002 1BFF	Reserved
	0x4002 1000 - 0x4002 13FF	GPIOE
	0x4002 0C00 - 0x4002 0FFF	GPIOD
	0x4002 0800 - 0x4002 0BFF	GPIOC
	0x4002 0400 - 0x4002 07FF	GPIOB
	0x4002 0000 - 0x4002 03FF	GPIOA

Figure 2.13: Boundary address and Bus of GPIOA

GPIOA base address = PERIPH_BASE + 0x20000 = 0x40020000

From the table, we find that the starting address for the GPIOA boundary is 0x40020000. This reveals that adding an offset of 0x20000 to the PERIPH_BASE address (which is 0x40000000) results in the base address of GPIOA, calculated as 0x40000000 + 0x20000 = 0x40020000. The term "offset value" refers to the value added to derive a specific address from a base address. In this case, the offset value for GPIOA from the PERIPH_BASE address is 0x20000.

Understanding the concept of offset values is crucial for accurately calculating desired addresses in microcontroller programming. This understanding enables precise navigation and manipulation within the system's memory map.

Clock gating

Having identified the exact address of GPIOA, our next step is to enable clock access to it before configuring its registers. This step is necessary because, by default, the clock to all unused peripherals is disabled to conserve power.

Modern microcontrollers use a power-saving technique known as **clock gating**. In simple terms, clock gating involves selectively turning off the clock signal to certain parts of the microcontroller when they're not in use. The clock signal is an essential part of microcontroller operations, as it drives the sequential logic by providing a regular pulse that synchronizes the activities of the microcontroller's circuits. However, when a particular part of the microcontroller, such as a peripheral, is not actively being used, the clock signal to that part is disabled. This disabling prevents unnecessary power consumption by idle circuits. Therefore, before using any peripheral, it's required to first enable clock access to it.

As shown in *Figure 2.14*, there's a peripheral listed as RCC, which has a boundary address range from 0x40023800 to 0x40023BFF. The functions of this peripheral include enabling and disabling clock access to other peripherals.

RCC base Address = PERIPH_BASE + 0x23800 = 0x40023800

From the boundary address information, we can see that the RCC _Base address is obtained by adding an offset of 0x23800 to PERIPH_BASE.

Bus	Boundary address	Peripheral
AHB2	0x5000 0000 - 0x5003 FFFF	USB OTG FS
AHB1	0x4002 6800 - 0x4FFF FFFF	Reserved
	0x4002 6400 - 0x4002 67FF	DMA2
	0x4002 6000 - 0x4002 63FF	DMA1
	0x4002 5000 - 0x4002 4FFF	Reserved
	0x4002 3C00 - 0x4002 3FFF	Flash interface register
	0x4002 3800 - 0x4002 3BFF	RCC
	0x4002 3400 - 0x4002 37FF	Reserved
	0x4002 3000 - 0x4002 33FF	CRC
	0x4002 2000 - 0x4002 2FFF	Reserved
	0x4002 1C00 - 0x4002 1FFF	GPIOH
	0x4002 1400 - 0x4002 1BFF	Reserved
	0x4002 1000 - 0x4002 13FF	GPIOE
	0x4002 0C00 - 0x4002 0FFF	GPIOD
	0x4002 0800 - 0x4002 0BFF	GPIOC
	0x4002 0400 - 0x4002 07FF	GPIOB
	0x4002 0000 - 0x4002 03FF	GPIOA

Figure 2.14: Boundary address and Bus of RCC

Having successfully determined the base addresses for the two essential peripherals needed to configure GPIOA pin 5 (which controls the connected LED), our next step involves using these base addresses to derive the specific register addresses necessary for setting the pin as an output and ultimately activating the LED.

To locate the detailed information about these registers, we will refer to the reference manual (*RM0383*). This document provides comprehensive insights into all registers and their configurations.

The AHB1 ER

Our reference manual, RM0383, is a comprehensive document spanning over 800 pages, and some STM32 reference manuals even exceed 1,500 pages. The objective is not to read the entire manual cover to cover, but rather to develop the skill to efficiently locate specific information as needed. Previously, we established that GPIOA is connected to the AHB1 bus. We also learned that activating this peripheral requires enabling clock access through the RCC peripheral.

The RCC peripheral in our microcontroller includes a specific register dedicated to enabling the clock for each bus. In STM32 microcontrollers, the naming of registers follows a straightforward pattern: *peripheral acronym + underscore + register acronym*. For example, the register responsible for controlling clock access to the AHB1 bus is named RCC_AHB1ENR.

Let's explain this a bit more:

- **RCC** stands for **Reset and Clock Control**
- **AHB1** stands for **Advanced High-Performance Bus 1**
- **ENR** stands for **Enable Register**

This systematic naming convention simplifies the process of identifying and accessing the appropriate registers.

To find the information about the RCC_AHB1ENR register, we begin by opening the reference manual. Next, we navigate to the table of contents and search for the section titled *RCC AHB1 Peripheral Clock Enable Register (RCC_AHB1ENR)*. Once located, we click on the page number provided alongside this section title to directly jump to the relevant part of the document.

Let's begin by examining the details presented at the top of the page, as illustrated in *Figure 2.15*. This section provides key information about the register, including the following:

- **Register name**: The full name of the register is provided along with its abbreviation, namely **RCC AHB1 Peripheral Clock Enable Register (RCC_AHB1ENR)**.
- **Address offset**: The offset from the base address is indicated as 0x30.
- **Reset value**: This is specified as 0x00000000, indicating the value the register holds upon reset. In other words, the default value of the register.
- **Access type**: The register supports various access types – it can be accessed without wait states and allows word, half-word, and byte access.
- **Register diagram**: A detailed diagram of the register is included, showing all 32 bits along with labels for each bit.

6.3.9   RCC AHB1 peripheral clock enable register (RCC_AHB1ENR)

Address offset: 0x30

Reset value: 0x0000 0000

Access: no wait state, word, half-word and byte access.

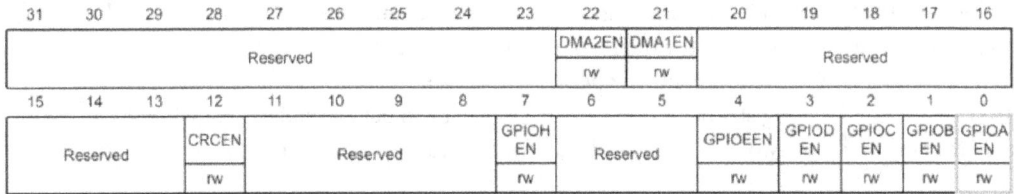

31	30	29	28	27	26	25	24	23	22	21	20	19	18	17	16
Reserved									DMA2EN	DMA1EN	Reserved				
									rw	rw					

15	14	13	12	11	10	9	8	7	6	5	4	3	2	1	0
Reserved			CRCEN	Reserved				GPIOH EN	Reserved		GPIOEEN	GPIOD EN	GPIOC EN	GPIOB EN	GPIOA EN
			rw					rw			rw	rw	rw	rw	rw

Figure 2.15: RCC AHB1 ER

RCC_AHB1ENR Address = RCC_BASE + 0x30 = 0x40023830

From this information, we can accurately calculate the address of the RCC_AHB1ENR register. We do this by adding the RCC_AHB1ENR offset to the RCC_BASE address. The formula is as follows: RCC_BASE + RCC_AHB1ENR Offset = 0x40023800 + 0x30 = 0x40023830.

The same section of the reference manual also includes a detailed description of each bit within the register. We are particularly interested in the bit named 'GPIOAEN', which stands for **GPIOA Enable**. This is identified as bit number 0. Further down, at the start of the next page in the document, we find a precise description of bit 0, as depicted in *Figure 2.16*. This description explains that setting bit 0 to 0 disables the GPIOA clock while setting it to 1 enables the GPIOA clock.

Bit 0 **GPIOAEN:** IO port A clock enable
Set and cleared by software.
0: IO port A clock disabled
1: IO port A clock enabled

Figure 2.16: GPIOAEN bit

Having understood the steps required to enable clock access to GPIOA, our next section will focus on learning how to set and clear bits within a register.

Setting and clearing bits in registers

In bare-metal programming, manipulating individual bits within registers is a fundamental operation. This manipulation is important for configuring hardware settings, controlling peripheral devices, and optimizing the performance of embedded systems. Let's start by understanding bits and registers.

A **bit** is the basic unit of information in computing and digital communications, which can have a value of either 0 or 1. **Registers**, on the other hand, are small-sized storage locations within microcontrollers, used to store data temporarily for various operations. Registers are typically a collection of bits (such as 8-bit, 16-bit, or 32-bit registers), and each bit in a register can be manipulated individually. In bare-metal programming, the two most frequently used bit operations are setting a bit and clearing a bit.

Setting a bit

Setting a bit means changing its value to 1. We will often use this to activate or enable a specific function within a microcontroller.

How it's done: The bitwise OR operation (|) is used for setting a bit:

```
register |= 1 << bit_position;
```

In this operation, 1 is shifted left (<<) to bit_position and then ORed with the current value of the register. The left shift operation creates a binary value where only the target bit is 1, and all others are 0. The OR operation then sets the target bit in the register to 1, leaving the rest unchanged.

Let's assume register initially holds the 0011 (binary) value and we want to set the third bit (bit position 2, 0-indexed). The *bit-shifted value* would be 0100 (binary for 1 << 2). The OR operation is then as follows:

```
0011: original register value
(0001 <<2) = 0100: bit-shifted value
0011
OR
0100
------
0111 (resulting value)
```

In this example, the *first*, *second*, and *fourth* bits of the original register value retain their value while the value of the third bit is changed to 1.

Let's see the opposite of setting a bit.

Clearing a bit

Conversely, clearing a bit means changing its value to 0, typically to deactivate a function.

How it's done: The combination of bitwise AND (&) and NOT (~) operations is used for clearing a bit:

```
register &= ~(1 << bit_position);
```

Here, 1 is left-shifted to `bit_position`, and then a bitwise NOT operation is applied to create a binary number where all bits are 1, except the target bit. The AND operation with the register clears the target bit, leaving others as they are.

Let's assume `register` initially holds the 0111 (binary) value and we want to clear the third bit (bit position 2, 0-indexed). *The bit-shifted value for the mask* would be 0100 (binary for 1 << 2). To clear the bit, we use the bitwise AND operation with the bitwise NOT of the bit-shifted value. The operation is as follows:

```
0111: original register value
(0001 << 2) = 0100: bit-shifted value
~0100 = 1011: bitwise NOT of bit-shifted value
0111
AND
1011
----
0011 (resulting value)
```

In this operation, the third bit of the register is cleared (changed to 0), while the other bits retain their original values.

Let's summarize the key points from the last two sections. Firstly, we understood that enabling clock access to GPIOA requires setting `bit0` in the RCC_AHB1ENR register. Secondly, we explored how to set and clear bits in registers using bitwise operations. Moving forward, in the next section, we will focus on configuring GPIOA pin 5 (PA5) as an output pin. This step is crucial in our progress toward activating the LED connected to PA5. This will take us a step closer to activating the LED connected to PA5.

The GPIO port mode register (GPIOx_MODER)

The **GPIO port mode register** in STM32 microcontrollers is a specialized register used for setting the mode of each GPIO pin. To locate information about this register, we navigate to the table of contents of the reference manual and look for the section titled *GPIO Port Mode Register (GPIOx_MODER)*. By clicking on the associated page number, we are directly taken to the section.

Figure 2.17 shows the top section of the page. Here, we can observe the following details:

- **Port applicability (x = A…E and H)**: This notation indicates that the register information is relevant to multiple ports. Specifically, it applies to GPIOA_MODER through GPIOE_MODER, as well as GPIOH_MODER. This means the same register structure and configuration are consistent across these GPIO ports.

- **Address offset – 0x00**: This tells us that the register's address is the same as the base address of the corresponding GPIO port. In other words, for each GPIO port, the MODER register is located at the very beginning of the port's memory space. Therefore, the GPIOA MODE register address is the same as the GPIOA base address.

- **Reset values**:

  - **Port A**: The default value for the GPIOA_MODER register is 0xA800 0000. This value represents the initial configuration state of the GPIOA pins upon reset.

  - **Port B**: The GPIOB_MODER register has a default reset value of 0x0000 0280.

  - **Other ports**: The reset value for the MODER registers of all other specified GPIO ports is 0x0000 0000:

```
GPIOA MODE Register Address = GPIOA_BASE = 0x40020000
```

8.4.1   GPIO port mode register (GPIOx_MODER) (x = A..E and H)

Address offset: 0x00

Reset values:
- 0xA800 0000 for port A
- 0x0000 0280 for port B
- 0x0000 0000 for other ports

31	30	29	28	27	26	25	24	23	22	21	20	19	18	17	16
MODER15[1:0]		MODER14[1:0]		MODER13[1:0]		MODER12[1:0]		MODER11[1:0]		MODER10[1:0]		MODER9[1:0]		MODER8[1:0]	
rw	rw	rw	rw	rw	rw	rw	rw	rw	rw	rw	rw	rw	rw	rw	rw

15	14	13	12	11	10	9	8	7	6	5	4	3	2	1	0
MODER7[1:0]		MODER6[1:0]		MODER5[1:0]		MODER4[1:0]		MODER3[1:0]		MODER2[1:0]		MODER1[1:0]		MODER0[1:0]	
rw	rw	rw	rw	rw	rw	rw	rw	rw	rw	rw	rw	rw	rw	rw	rw

Figure 2.17: GPIO port mode register

We also observe that this is a 32-bit register, with its bits organized in pairs. For example, bit0 and bit1 together form a pair known as MODER0, bit2 and bit3 form MODER1, bit4 and bit5 form MODER2, and so on. Each of these pairs corresponds to a single pin of the GPIO port. Specifically, MODER0 controls the configuration of PIN0 of the corresponding port, MODER1 controls PIN1, and this pattern continues similarly for the other pins.

Given our objective of configuring the mode of PIN5, we need to focus on MODER5. In the register, MODER5 comprises bit10 and bit11. These two bits are the required bits for setting the operational mode of PIN5.

The reference manual provides a truth table, illustrated in *Figure 2.18*, which explains the combinations of the two MODER bits necessary to configure a pin. This table is an invaluable resource for understanding how to set the bits for the desired pin configuration, whether it's as an input, output, or an alternate function mode.

Bits 2y:2y+1 **MODERy[1:0]:** Port x configuration bits (y = 0..15)

These bits are written by software to configure the I/O direction mode.
00: Input (reset state)
01: General purpose output mode
10: Alternate function mode
11: Analog mode

Figure 2.18: The MODER bits configuration

As shown in *Figure 2.18*, the MODER register within the GPIO port is composed of pairs of bits, designated as 2y:2y+1 MODERy[1:0], where *y* ranges from 0 to 15, representing each of the 16 pins in the port (PIN0 to PIN15).

In this equation, *y* represents the pin number. For Pin 5, *y = 5*. Plugging this value into the equation gives us the bit positions in the MODER register that correspond to Pin 5:

The bit positions for MODER5 are calculated as *2*y* and *(2*y) +1*.

Substituting *y = 5*, we get the following:

2*5 = 10 and *2*5 + 1 = 11*, which are 10 and 11, respectively.

So, bits 10 and 11 in the MODER register (MODER5[1:0]) are the bits that control the mode of Pin 5. By setting these bits to specific values (00, 01, 10, or 11), we can configure Pin 5 as an input, general-purpose output, alternate function, or analog mode, respectively.

The GPIO MODER register supports four distinct bit combinations, each defining a different operational mode for the corresponding pin:

- 00: When both bits are 0, the corresponding pin is set as an input pin. This is the standard mode for pins to receive data from external sources.

- 01: Setting the bits to 01 sets the pin function to general-purpose output. In this mode, the pin can send data out, for instance, to light up an LED.

- 10: The 10 state configures the pin for alternate functions. Each pin can serve specific additional purposes (such as PWM output and I2C communication lines), and this mode enables those functions.

- **11**: When the bits are set to `11`, the pin operates in analog mode. This mode is typically used for ADC, useful in reading values from analog sensors.

From this, we understand that to configure PA5 as an output, we must set bit 10 of the `GPIOA_MODER` register to `0` and bit 11 to `1`.

Let's summarize our progress toward activating the LED connected to PA5:

- **Enabling clock access to GPIOA**: We've learned how to enable the clock for GPIOA, by setting `bit0` of the `RCC_AHB1ENR` register to `1`. This step is essential to power the GPIOA for operation.

- **Configuring PA5 for output**: We have just learned how to set PA5 as a general-purpose output pin.

These two steps effectively configure PA5 as an output pin. The final task involves controlling the output state of the pin – setting it to either `1` or `0`, which corresponds to on or off, respectively. This translates to sending either 3.3v or 0v to PA5, thus turning the connected LED on or off. To manage the output state of a pin, we need to interact with the **Output Data Register** (**ODR**). Locating and configuring this register will be the focus of the next section.

So far, we have the following information for our quest to activate the LED connected to PA5:

- We know how to enable clock access to GPIOA through the `RCC_AHB1ENR` register

- We know how to configure the PIN5 of GPIOA as a general-purpose output pin.

These two steps make PA5 act as an output pin. The final step is to be able to set the output state of the pin. The state can be either `1` or `0` corresponding to on or off, corresponding to sending 3.3v or 0v to PA5, and finally corresponding to turning on or turning off the LED connected to PA5. To set the output of a pin, we need to access the ODR; this shall be the focus of the next section.

GPIO Port Output Data Register (GPIOx_ODR)

The GPIO Port **ODR** in STM32 microcontrollers is used for controlling the output state of each GPIO pin. To find information about this register, we refer to the table of contents in the microcontroller's reference manual and locate the section titled *GPIO Port Output Data Register (GPIOx_ODR)*. Clicking on the page number corresponding to this section takes us straight to the required information.

Figure 2.19 displays the top part of the page, where we can observe the following details:

- **Port applicability (x = A…E and H)**: This indicates that the register information is relevant to multiple ports. Specifically, it applies to `GPIOA_ODR` through `GPIOE_ODR`, as well as `GPIOH_ODR`. This means the same register structure and configuration are consistent across these GPIO ports.

- **Address offset – 0x14**: This tells us that the address of the ODR register is offset by `0x14` from the base address of its respective GPIO port. This means that, for each GPIO port, the ODR register can be found at this offset from the port's base memory address.

- **Reset values – 0x00000000**: This indicates that, by default, all bits in the ODR are set to 0 upon a reset. This default state ensures that all GPIO pins are initially in a low output state.

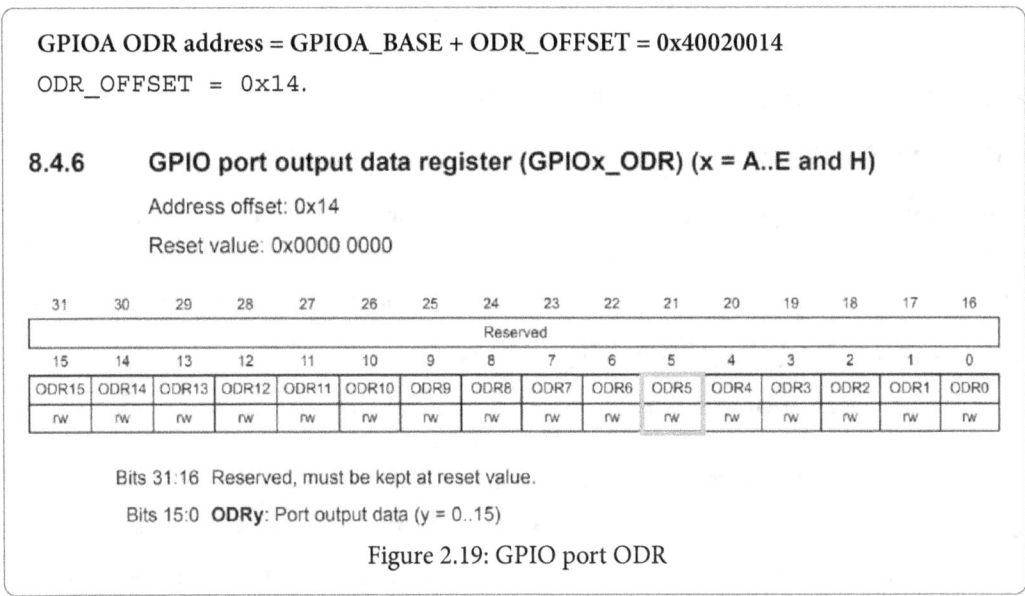

GPIOA ODR address = GPIOA_BASE + ODR_OFFSET = 0x40020014

ODR_OFFSET = 0x14.

8.4.6   GPIO port output data register (GPIOx_ODR) (x = A..E and H)

Address offset: 0x14

Reset value: 0x0000 0000

31	30	29	28	27	26	25	24	23	22	21	20	19	18	17	16
							Reserved								

15	14	13	12	11	10	9	8	7	6	5	4	3	2	1	0
ODR15	ODR14	ODR13	ODR12	ODR11	ODR10	ODR9	ODR8	ODR7	ODR6	ODR5	ODR4	ODR3	ODR2	ODR1	ODR0
rw	rw	rw	rw	rw	rw	rw	rw	rw	rw	rw	rw	rw	rw	rw	rw

Bits 31:16 Reserved, must be kept at reset value.

Bits 15:0 **ODRy**: Port output data (y = 0..15)

Figure 2.19: GPIO port ODR

Let's delve into the structure of the bits:

- **Bits 31:16 (reserved)**: These bits are reserved and should not be used for any operation.

- **Bits 15:0 (ODRy: port output data, y = 0..15)**: These bits, ranging from 0 to 15, correspond to each of the 16 pins in the GPIO port. They are directly programmable and can be both read and written by software. Changing the value of these bits alters the output state of the corresponding GPIO pin.

Let's consider our final task of activating the LED connected to PA5:

- **GPIOA_ODR bit for PA5**: Given that PA5 corresponds to the 5th pin in the GPIOA port, we target the 5th bit (bit 5) in the GPIOA_ODR register.

- **Setting PA5 high**: To set PA5 high, we need to write 1 to bit 5 of the GPIOA_ODR register. This can be achieved using a bitwise OR operation, as follows:

```
GPIOA_ODR |= 1 << 5;
```

In this operation, we shift 1 left by 5 positions (resulting in a binary value where only the 6th bit is 1) and then OR it with the current value of the GPIOA_ODR register. This action sets PA5 high without altering the state of other pins in the port.

Setting PA5 high will supply voltage to the connected LED, effectively turning it on.

Now that we understand how to set the state of GPIO output pins, in the next section, we will combine all the pieces of information we've acquired and develop our first firmware.

Register manipulation – from configuration to running your first firmware

In this section, we will apply the knowledge acquired throughout this chapter to develop our first bare-metal firmware.

We begin by creating a new project, a process we covered in *Chapter 1*. Here is a summary of the steps:

1. **Start a new project**: Go to **File | New | STM32 Project** in your IDE.

2. **Select the target microcontroller**: A **Target Selection** window will appear, prompting you to choose the microcontroller or development board for your project.

3. **Use the board selector**: Click on the **Board Selector** tab.

4. **Search for our board**: Input NUCLEO-F411 in the **Commercial Part Number** field.

5. **Select our board**: From the list of boards that appear, choose **NUCLEO-F411RE** and then click **Next**.

6. **Name our project**: Assign a name to your project, for instance, RegisterManipulation.

7. **Project configuration**: In the project options, select an Empty project setup.

8. **Final step**: Click **Finish** to create your project.

Once the project is created, open the main.c file in your project workspace. Clear all pre-existing text in this file to start with a clean slate for our code.

For a clearer understanding, we will structure our code into two distinct sections. The first section will be titled Register Definitions, and the second will be named Main Function.

Register Definitions

This section of the code defines constants and macros for memory addresses and bit masks. Here are all the memory addresses and bit masks required for controlling the LED connected to PA5:

```
// 1: Define base address for peripherals
#define PERIPH_BASE        (0x40000000UL)
// 2: Offset for AHB1 peripheral bus
#define AHB1PERIPH_OFFSET  (0x00020000UL)
// 3: Base address for AHB1 peripherals
#define AHB1PERIPH_BASE    (PERIPH_BASE + AHB1PERIPH_OFFSET)
```

```
// 4: Offset for GPIOA
#define GPIOA_OFFSET         (0x0000UL)
// 5: Base address for GPIOA
#define GPIOA_BASE           (AHB1PERIPH_BASE + GPIOA_OFFSET)
// 6: Offset for RCC
#define RCC_OFFSET           (0x3800UL)
// 7: Base address for RCC
#define RCC_BASE             (AHB1PERIPH_BASE + RCC_OFFSET)
// 8: Offset for AHB1EN register
#define AHB1EN_R_OFFSET      (0x30UL)
// 9: Address of AHB1EN register
#define RCC_AHB1EN_R  (*(volatile unsigned int *)(RCC_BASE
+ AHB1EN_R_OFFSET))
// 10: Offset for mode register
#define MODE_R_OFFSET        (0x00UL)
// 11: Address of GPIOA mode register
#define GPIOA_MODE_R  (*(volatile unsigned int *)(GPIOA_BASE + MODE_R_
OFFSET))
// 12: Offset for output data register
#define OD_R_OFFSET    (0x14UL)
// 13: Address of GPIOA output data register
#define GPIOA_OD_R    (*(volatile unsigned int *)(GPIOA_BASE +  OD_R_
OFFSET))
// 14: Bit mask for enabling GPIOA (bit 0)
#define GPIOAEN         (1U<<0)
// 15: Bit mask for GPIOA pin 5
#define PIN5            (1U<<5)
// 16: Alias for PIN5 representing LED pin
#define LED_PIN         PIN5
```

In the previous sections, we gathered base addresses, offsets, and bit masks from the documentation. These remain unchanged in our code. However, a notable aspect of the code snippet is the use of the UL suffix at the end of each hexadecimal value as well as the use of keywords such as volatile. Let's delve into the significance of these in the context of C and C++ programming.

The UL suffix

When we see a number in the code ending with UL, it's more than just a part of the number – it's a clear instruction to the compiler about the type and size of the number:

- **Unsigned**: The U in UL indicates that the number is unsigned. In other words, it's a positive number with no sign to indicate it could be negative. This designation allows the number to represent a wider range of positive values compared to a signed integer of the same size.

- **Long integer**: The L signifies that the number is a long integer. This is important because the size of a long integer can vary based on the system and the compiler. Typically, a long integer is larger than a regular int – often 32 bits, but sometimes 64 bits on certain systems.

The UL suffix collectively ensures that these values are treated as unsigned long integers. This is important for firmware development and other forms of low-level programming, where the exact size and "signedness" of an integer can significantly impact program behavior and memory management. Using UL leads to more predictable, platform-independent code, ensuring that the values behave consistently across different compilers and systems.

The use of "(*(volatile unsigned int *)"

In the context of bare-metal programming, particularly in our current code snippet, we notice that the address of each register is prefixed with "(*(volatile unsigned int *)".

Let's break down what each part of this notation means and why it's used:

- **Type casting to a pointer**: The (unsigned int *) expression is a **type cast**. It tells the compiler to treat the subsequent address as a pointer to an unsigned integer. In C and C++, pointers are variables that store the memory address of another variable. In the context of our firmware, this casting means that we are directing the compiler to treat a certain address as the location of an unsigned integer. However, this unsigned integer is not just any number; it corresponds directly to the state of a 32-bit hardware register. Each bit in this 32-bit integer mirrors a specific bit in the register, thereby allowing direct control and monitoring of the hardware's state through standard programming constructs.

- **Dereferencing the pointer**: The leading asterisk (*) in *(unsigned int *) is used to dereference the pointer. Dereferencing a pointer means accessing the value stored at the memory address the pointer is pointing to. Essentially, it's not just about knowing where the data is (the address); it's about actually accessing and using that data.

- volatile: The volatile keyword tells the compiler that the value at the pointer can change at any time, without any action being taken by the code. This is often the case with hardware registers, where the value can change due to hardware events, external inputs, or other aspects of the system outside the program's control.

Without volatile, the compiler might optimize out certain reads and writes to these addresses under the assumption that the values don't change unexpectedly. This is because when a compiler processes code, it looks for ways to make the program run more efficiently. This includes removing redundant operations or simplifying code paths. This process occurs during the compilation of the program before it's run. If the compiler determines that a variable (including a memory-mapped hardware register) doesn't change its value, it might optimize the code by eliminating repeated reads from or writes to that variable. For example, if a value is read from a register and then read again later, the compiler might assume the value hasn't changed and use the previously read value instead of accessing the register a second time. The use of `volatile` is a directive to tell the compiler not to apply certain optimizations to accesses of the marked variable, maintaining the integrity of operations.

In our code snippet, we encounter two additional terms that are very common in bare-metal programming: **bit mask** and **alias**. Let's take a closer look at each of these concepts to gain a clearer understanding of the roles they play:

- **Bit masks**: A bit mask is a binary number used as a template to manipulate specific bits within another binary number, usually a register value, while leaving other bits unaffected. We use bit masks to set, clear, or toggle individual bits in a register. They work by applying bitwise operations (such as AND, OR, XOR) along with the mask to achieve the desired result. For example, a bit mask might be used to turn on a specific LED connected to a microcontroller pin without altering the state of other pins. A bit mask is created by setting the bits we want to manipulate to 1 while keeping others at 0. This mask is then combined with the register value using the appropriate bitwise operation.

  For example, a `0b00000100` mask used with OR will set the third bit of the target register.

- **Alias**: An alias is simply a name given to a bit or a group of bits to make the code more readable and easier to manage. We use aliases to refer to specific hardware features or configurations represented by bits in a register. By using a descriptive name, we can make the code more understandable and maintainable. For example, naming a bit mask that controls an LED as `LED_PIN` makes the code self-explanatory. We define aliases using preprocessor directives such as `#define` in C or C++. For example, `#define LED_PIN (1U<<5)` creates an `LED_PIN` alias for the bit mask that represents the sixth pin (zero indexing) in a GPIO port ODR.

Now, let's analyze the second section of the code.

Main Function

This section of our code outlines the primary operations for controlling the LED connected to PA5 of the microcontroller:

```
// Line 17: Start of main function
int main(void)
{
    //  18: Enable clock access to GPIOA
    RCC_AHB1EN_R |= GPIOAEN;

    GPIOA_MODE_R |= (1U<<10);  //  19: Set bit 10 to 1
    GPIOA_MODE_R &= ~(1U<<11); //  20: Set bit 11 to 0

    //  21: Start of infinite loop
    while(1)
    {
```

```
    // Line 22: Set PA5(LED_PIN) high
    GPIOA_OD_R |= LED_PIN;
}  //  23: End of infinite loop

}  //  24: End of main function
```

Let's break down each part of the code for a clearer understanding:

- *Line 17*: This marks the beginning of the main function. This is the entry point of our program where the execution starts:

  ```
  // Line 17: Start of main function
  int main(void)
  {
  ```

- *Line 18*: This enables the clock for GPIOA. As we learned earlier, the RCC_AHB1EN_R register controls the clock to the AHB1 bus peripherals. The |= GPIOAEN operation sets the bit associated with GPIOA (GPIOAEN) in the RCC_AHB1EN_R register, ensuring that the GPIOA peripheral has the necessary clock enabled for its operations:

  ```
  // Line 18: Enable clock access to GPIOA
  RCC_AHB1EN_R |= GPIOAEN;
  ```

- *Line 19–20*: These lines configure PA5 as an output pin. Setting bit 10 to 1 and bit 11 to 0 in GPIOA_MODE_R configures PA5 in general-purpose output mode:

  ```
  GPIOA_MODE_R |= (1U<<10);  // Line 19: Set bit 10 to 1
  GPIOA_MODE_R &= ~(1U<<11); // Line 20: Set bit 11 to 0
  ```

- *Line 21–23*: These lines initiate an infinite loop, which is a common practice in embedded systems for continuous operation:

  ```
  // Line 21: Start of infinite loop
  while(1)
  {
      // Line 22: Set PA5(LED_PIN) high
      GPIOA_OD_R |= LED_PIN;
  }  // Line 23: End of infinite loop
  ```

This is what is inside this loop:

- *Line 22*: This line sets PA5 high. This is achieved by setting the respective bit in GPIOA_OD_R. The |= LED_PIN operation ensures that PA5 outputs a high signal (essentially turning the LED on).

- *Line 24*: This marks the end of the main function:

  ```
  }  // Line 24: End of main function
  ```

Now that we have a clear understanding of each line of the code, let's proceed to enter the entire code into the main.c file. Additionally, for your convenience, this complete source code is available in the GitHub repository for the book. You can find it in the folder titled Chapter2.

This is the entire code:

```
//  1: Define base address for peripherals
#define PERIPH_BASE         (0x40000000UL)
//  2: Offset for AHB1 peripheral bus
#define AHB1PERIPH_OFFSET   (0x00020000UL)
//  3: Base address for AHB1 peripherals
#define AHB1PERIPH_BASE     (PERIPH_BASE + AHB1PERIPH_OFFSET)
//  4: Offset for GPIOA
#define GPIOA_OFFSET        (0x0000UL)
//  5: Base address for GPIOA
#define GPIOA_BASE          (AHB1PERIPH_BASE + GPIOA_OFFSET)
//  6: Offset for RCC
#define RCC_OFFSET          (0x3800UL)
//  7: Base address for RCC
#define RCC_BASE            (AHB1PERIPH_BASE + RCC_OFFSET)
//  8: Offset for AHB1EN register
#define AHB1EN_R_OFFSET     (0x30UL)
//  9: Address of AHB1EN register
#define RCC_AHB1EN_R   (*(volatile unsigned int *)(RCC_BASE
+  AHB1EN_R_OFFSET))
//  10: Offset for mode register
#define MODE_R_OFFSET       (0x00UL)
//  11: Address of GPIOA mode register
#define GPIOA_MODE_R  (*(volatile unsigned int *)(GPIOA_BASE + MODE_R_
OFFSET))
//  12: Offset for output data register
#define OD_R_OFFSET    (0x14UL)
//  13: Address of GPIOA output data register
#define GPIOA_OD_R     (*(volatile unsigned int *)(GPIOA_BASE +  OD_R_
OFFSET))
//  14: Bit mask for enabling GPIOA (bit 0)
#define GPIOAEN        (1U<<0)
//  15: Bit mask for GPIOA pin 5
#define PIN5           (1U<<5)
//  16: Alias for PIN5 representing LED pin
#define LED_PIN        PIN5

//  17: Start of main function
int main(void)
{
```

```
//  18: Enable clock access to GPIOA
RCC_AHB1EN_R |= GPIOAEN;

GPIOA_MODE_R |= (1U<<10);  //  19: Set bit 10 to 1
GPIOA_MODE_R &= ~(1U<<11); //  20: Set bit 11 to 0

//  21: Start of infinite loop
while(1)
{
    //  22: Set PA5(LED_PIN) high
    GPIOA_OD_R |= LED_PIN;

}  //  23: End of infinite loop

}  //  24: End of main function
```

To build the project, first select it by clicking on the project name once, and then initiate the build process by clicking on the *build* icon, represented by a hammer symbol, in the IDE.

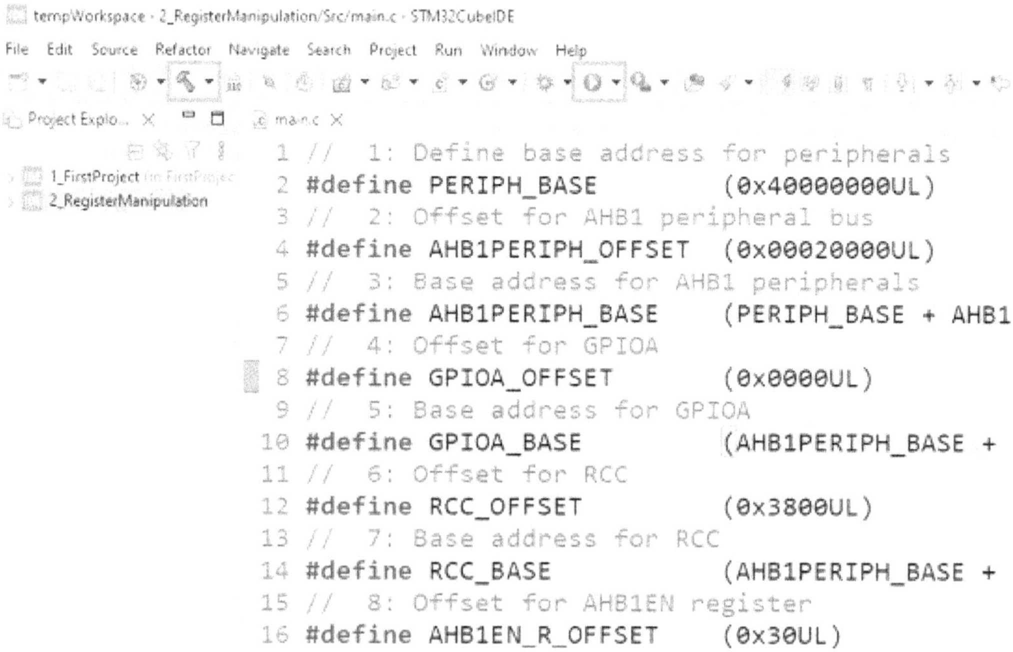

Figure 2.20: The build and run icons

Once the build process is complete, the next step involves uploading the program to the microcontroller. To do this, we make sure the project is selected in the IDE, and then initiate the upload by clicking on the *run* icon, represented by a play symbol.

The first time you run the program, the IDE may prompt you to edit or confirm the launch configurations. A dialog box titled **Edit Configurations** will appear. It's sufficient to accept the default settings by clicking **OK**. *Figure 2.21* shows the **Edit Configuration** dialog box.

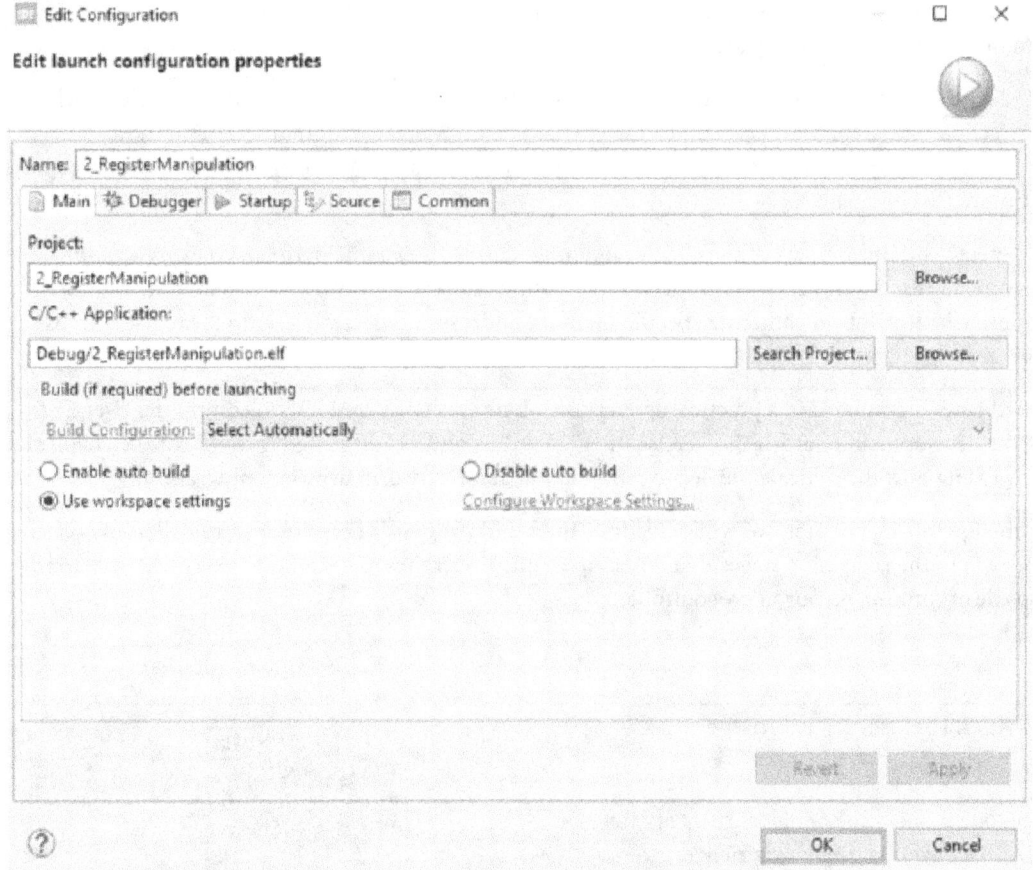

Figure 2.21: The Edit Configuration dialog box

Following this confirmation, the IDE will commence the process of uploading your program to the microcontroller. When the upload is complete, you should see the green LED on the development board light up. This indicates that our bare-metal code is functioning as intended and successfully controlling the hardware.

We have now mastered configuring PA5 to control the LED, starting from the very basics of consulting the documentation for accurate addresses, effectively typecasting these addresses for register access, and creating aliases for specific bits within the registers.

Summary

In this chapter, we delved into the core of bare-metal programming, emphasizing the direct interaction with microcontroller registers. This gave us insight into some of the key registers of our microcontroller and the structure of those registers.

We began by exploring various firmware development approaches, each offering a distinct level of abstraction. These approaches included the HAL, LL, Bare-Metal C programming, and the assembly language. This exploration helped us understand the trade-offs and applications of each approach in firmware development.

We spent a significant part of the chapter defining addresses of some peripherals using the official documentation. This step was important in creating addresses for various registers within those peripherals. It involved gathering specific memory addresses, typecasting them for register access, and creating aliases for bits in the registers.

The chapter culminated in a practical application where we configured PA5 to control the User LED on the development board. This exercise integrated the concepts discussed throughout the chapter, showcasing how theoretical knowledge can be applied in real-world firmware programming.

In the upcoming chapter, we will delve into the build process. This exploration will allow us to understand the sequential stages that our source code undergoes to become an executable program capable of running on our microcontroller.

3

Understanding the Build Process and Exploring the GNU Toolchain

Bare-metal programming is a journey of deep understanding and precision, and in this chapter, we will navigate the complex realm of the embedded firmware build process. Our focus is the GNU Arm Toolchain, an important element in firmware development. Through a blend of theory and hands-on programming exercises, you will gain insights into how **integrated development environments (IDEs)** streamline the build process and how these processes can be manually replicated using the GNU Arm Toolchain.

As the chapter progresses, we delve into the nuances of the compiler and its various options, tailored for Arm Cortex microcontrollers. The programming exercise in this chapter is designed to help you understand and effectively utilize the GNU tools, from compiling and linking to analyzing the depths of the output object files.

In this chapter, we shall cover the following main topics:

- The foundations – understanding the embedded build process
- A tour of GNU binary tools for embedded systems
- From IDE to command-line – watching the build process unfold

By the end of the chapter, you will have an understanding of the embedded firmware build process and have developed the skills to fluidly switch between IDEs and command-line interfaces, enhancing your versatility as a firmware developer.

Technical requirements

All the code examples for this chapter can be found on GitHub at `https://github.com/PacktPublishing/Bare-Metal-Embedded-C-Programming`.

The foundations – understanding the embedded build process

The journey from high-level source code in embedded firmware development to an executable binary image is intricate and multilayered. This process is commonly referred to as the firmware build process and involves several critical stages – pre-processing, compilation, assembly, linking, and locating. Each of these stages plays an important role in transforming human-readable code into machine-executable instructions. *Figure 3.1* shows the entire build process and the tools involved at each stage of the process.

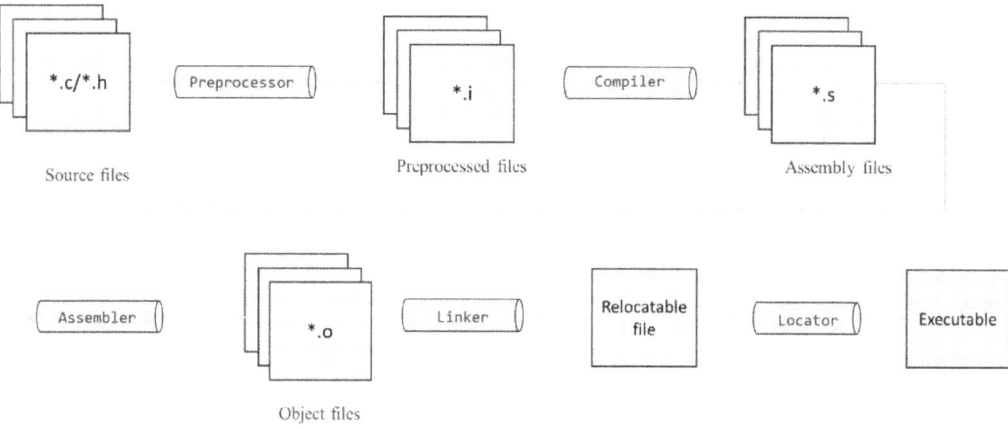

Figure 3.1: The build process, detailing the input and output files for each stage and the specific tools used

Let's examine the stages.

The pre-processing stage

Pre-processing is the initial stage in the firmware build process. In this stage, the source code undergoes a series of transformations to prepare it for compilation. Typically, source files in embedded systems are written in C (.c files) and accompanied by header files (.h files). The preprocessor is the specialized tool in the build process whose task is to handle these input files.

During pre-processing, the preprocessor executes several key operations:

- **Stripping comments**: Comments, which are crucial for human readability but irrelevant for machines, are removed from code.

- **Evaluating preprocessor directives**: Lines starting with the # symbol, known as preprocessor directives, are processed. These directives often include macro definitions (#define), conditional compilation instructions (#ifdef, #ifndef, #endif), and file inclusion commands (#include). The preprocessor replaces these directives with their defined values or corresponding code segments.

- **Output generation**: The output of this stage is a set of intermediate files, typically with the .i extension. These files represent the transformed source code, devoid of comments and with all directives evaluated.

The next stage is the compilation stage.

The compilation stage

Compilation starts immediately after pre-processing. The compiler's role is to take .i files and convert them into architecture-specific assembly code. This phase is where the high-level constructs of C are translated into the lower-level, more granular assembly instructions understood by the target processor architecture.

This stage involves the following:

- **Input**: The pre-processed .i files

- **Process**: The compiler analyzes the code structure, optimizes it for performance and space, and translates it into assembly language

- **Output**: The result of the compilation stage is a set of assembly code files, typically with the .s extension

The next stage makes use of the .s files.

The assembly stage

Assembly is the stage where the .s files containing assembly code are converted into machine code, in the form of **object files**. This stage translates the human-readable assembly instructions into a binary format.

This stage involves the following:

- **Input**: Assembly code files (.s).

- **Process**: The assembler interprets each assembly instruction and converts it into corresponding machine code.

- **Output**: Object files, usually with the .o extension. These files contain binary code and are ready for the next stage of linking.

The linking stage

Linking is the stage where all individual object files are combined to form a cohesive program. This stage also integrates any necessary standard library files and resolves references between different code modules.

This stage involves the following:

- **Input**: Object files (.o) and C standard library files.

- **Process**: The linker stitches together all object files, resolving symbolic references and addresses. It handles tasks such as memory allocation for variables and functions.

- **Output**: The linker generates a **relocatable file**, which is comprehensive but not yet final executable code.

A relocatable file is intermediate output in the firmware build process, which is created during the linking stage. It's a comprehensive file that combines all individual object files (.o) and the necessary library. However, it is not yet a final executable. The concept of **relocation** plays an important role here. Relocation involves adjusting the symbolic addresses in the relocatable file to actual, specific memory locations. This process ensures that when the firmware runs on a target device, each part of the code and data is correctly placed in memory. The relocatable file contains all the necessary components of the firmware, but with addresses that are still "relative" – they need further adjustment during the locating stage to fit the unique memory map of the target microcontroller, leading to the creation of the final executable.

The next stage is where we finally get our executable code.

The locating stage

This is the last stage; it involves converting the relocatable file into the final executable binary. This stage is guided by a linker script, which provides essential information about the memory layout of the target device.

This stage involves the following:

- **Input**: A relocatable file and a linker script.

- **Process**: The locator uses the linker script to place code and data sections into their designated memory locations. It adjusts addresses and offsets to fit the target's memory map.

- **Output**: An executable binary file, typically in formats such as **Executable and Linkable Format (ELF)** or a plain binary format.

With this knowledge in hand, we are well-equipped to delve into the GNU Toolchain for Arm. Our goal is to effectively utilize specific tools within the toolchain to execute the various stages of the build process. This will be the focus of the next section.

A tour of GNU binary tools for embedded systems

In this section, we'll delve into the GNU Bin tools, a suite of tools that come with the installation of the GNU Toolchain for Arm. These tools (commands) are essential for the various stages of the firmware build process, as well as additional tasks such as debugging.

The first command we'll explore is `arm-none-eabi-gcc`.

arm-none-eabi-gcc

Let's break down the components of the command:

- `arm`: This specifies the target architecture.

- `none`: This component indicates the operating system for which the code is being compiled. Here, `none` signifies that the code is meant for a bare-metal environment, meaning it will run directly on the hardware without an underlying operating system.

- `eabi`: This stands for **Embedded Application Binary Interface**. EABI defines a standard for the binary layout of system and user programs, libraries, and so on. It ensures that the compiled code will work correctly on any Arm processor that adheres to the EABI standard.

- `gcc`: This is short for **GNU Compiler Collection**.

This single command compiles, assembles, and links our input code in one go. To use it, type `arm-none-eabi-gcc` in the command prompt or terminal, followed by the source file, then `-o`, and the desired output filename.

See *Figure 3.2* for an example:

Figure 3.2: Usage of the arm-none-eabi-gcc command

In this example, we specify the source file as `main.c` and the output file as `main.o`.

Some common compiler flags

Since we have just introduced the `-o` compiler flag, let's take this opportunity to introduce some of the other commonly used compiler flags. These compiler flags are used to modify command behavior as well as add options to commands:

- `-c`: This flag is used to compile and assemble but not link. When added to the command, it processes the code up to the assembly stage but stops before linking.

- `-o file`: As mentioned earlier, this specifies the name of the output file.

- `-g`: Generates debugging information in the executable.

- `-Wall`: Enables all warning messages, helping us identify potential issues in the code.

- `-Werror`: Treats all warnings as errors, ensuring code quality and stability.

- `-I [DIR]`: Includes a specified directory to search for header files; it's useful for organizing large projects.

- `-ansi` and `-std=STANDARD`: These flags specify which standard version of the c language should be used.

- `-v`: Provides verbose output from GCC, giving us detailed information about the compilation process.

Table 3.1 provides a summary of the flags and example usage for each flag.

Flag	Purpose	Example usage
`-c`	Compile and assemble but don't link	`arm-none-eabi-gcc -c source_file`
`-o file`	Link to the output file, named file	`arm-none-eabi-gcc source_file -o output_file`
`-g`	Generate debugging info in the executable	`arm-none-eabi-gcc -g source_file`
`-Wall`	Enable all warning messages	`arm-none-eabi-gcc -Wall source_file`
`-Werror`	Treat warnings as errors	`arm-none-eabi-gcc -Werror source_file`
`-I [DIR]`	Include a directory for header files	`arm-none-eabi-gcc -I directory_path source_file`
`-ansi`	Use the **American National Standards Institute (ANSI)** standard	`arm-none-eabi-gcc -ansi source_file`

Flag	Purpose	Example usage
`-std`	Specify a standard version (e.g., C11)	`arm-none-eabi-gcc -std=c11 source_file`
`-v`	Verbose output from GCC	`arm-none-eabi-gcc -v source_file`

Table 3.1: Some compiler flags and their example usage

Some architecture-specific flags

In addition to the general compiler flags, there are several architecture-specific flags that enable precise configuration for various processor architectures. These flags are integral for tailoring the build process to specific ARM processors and their respective architectures. Let's delve into some of the most frequently used architecture-specific flags:

- `-mcpu=[NAME]`: Specifies the target ARM processor. Using this option configures the compiler to optimize the code for a specific processor.

- `-march=[NAME]`: Specifies the target ARM architecture. It configures the compiler for a particular ARM architecture version.

- `-mtune=[NAME]`: Similar to `-mcpu`, this specifies the target ARM processor for optimization purposes.

- `-thumb`: Configures the compiler to generate code for the Thumb instruction set, which is a compressed version of the standard ARM instruction set, providing more code density and efficiency.

- `-marm`: Instructs the compiler to generate code for the ARM instruction set.

- `-mlittle-endian/-mbig-endian`: These options specify the endianness for the generated code. Little-endian is the most common format in ARM processors.

Let's see an example involving some of these flags:

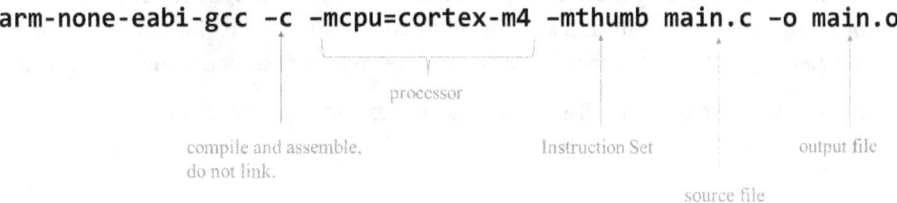

Figure 3.3: Usage of the arm-none-eabi-gcc command with some architecture-specific flags

In this example, we say the following:

- `-c`: Compile and assemble but do not link

- `-mcpu=cortex-m4`: Build for the Cortex-M4 processor

- `-mthumb`: Use the Thumb instruction set

- `-o main.o`: Output the compiled file as `main.o`

Table 3.2 provides a summary of the architecture-specific flags and example usage for each flag.

Flag	Purpose	Example usage
`-mcpu=[NAME]`	Specify the target ARM processor	`-mcpu=cortex-m4`
`-march=[NAME]`	Specify the target ARM architecture	`-march=armv7-m`
`-mtune=[NAME]`	Optimize for a specific ARM processor	`-mtune=cortex-m4`
`-thumb`	Generate code for the Thumb instruction set	`-mthumb`
`-marm`	Generate code for the ARM instruction set	`-marm`
`-mlittle-endian`	Generate code for little-endian mode	`-mlittle-endian`
`-mbig-endian`	Generate code for big-endian mode	`-mbig-endian`

Table 3.2: Some architecture-specific compiler flags and their example usage

Other commands in the GNU Toolchain for Arm

Apart from the `arm-none-eabi-gcc` command, there are other important commands that we will frequently use when building with the GNU toolchain for Arm. Let's examine some of these commands:

- `arm-none-eabi-nm`: The `arm-none-eabi-nm` command is a handy tool for listing the symbols from an object file. Symbols in this context refer to various identifiers in a program, such as function names, variable names, and constants. This tool is invaluable for examining the contents of compiled files, offering us insights into the structure and components of our program. This can be particularly useful for debugging purposes.

- `arm-none-eabi-size`: In embedded firmware development, where memory resources are often limited, understanding the memory footprint of different sections of our code is crucial. This tool provides valuable insights into how much memory the various parts of our code consume, allowing us to make informed decisions about optimization and memory management.

- `arm-none-eabi-objdump`: This tool is used to extract and display detailed information from object files. It offers an in-depth view of the machine instructions, making it an invaluable resource for thorough analysis of object files. This includes capabilities such as disassembling code, presenting section headers, and revealing symbol tables. Its utility becomes crucial when we need to delve into the intricate details of compiled code, providing clarity on a file's structure, content, and operational mechanics. This helps us to both debug and optimize our code.

Disassembling code refers to the process of converting machine code, which is a set of binary instructions that a computer's processor can execute directly, back into assembly language. Assembly language is a more human-readable form of instructions, although it's still quite low-level compared to the C language. *Figure 3.4* presents a comparison of C-language code, its corresponding assembly language translation, and the resulting machine code.

```
//   18: Enable clock access to GPIOA
 RCC_AHB1EN_R |= GPIOAEN;
 4:    4b0c          ldr     r3, [pc, #48]    ; (38 <main+0x38>)
 6:    681b          ldr     r3, [r3, #0]
 8:    4a0b          ldr     r2, [pc, #44]    ; (38 <main+0x38>)
 a:    f043 0301     orr.w   r3, r3, #1
 e:    6013          str     r3, [r2, #0]
```

Figure 3.4: C-language code, its corresponding assembly language code, and the resulting machine code

- `arm-none-eabi-readelf`: This tool provides detailed information about the output ELF file, including section headers, program headers, and symbol tables. It is useful when we work with ELF files, as it offers insights into how an executable is structured and prepared to run on a system.

- : `arm-none-eabi-objcopy`: We use this tool to convert object files from one format to another or to make a copy of an object file.

Table 3.3 provides a summary of these additional tools and example usage for each tool.

Tool	Function	Example Usage
`arm-none-eabi-nm`	Lists symbols from object files	`arm-none-eabi-nm [object file]`
`arm-none-eabi-size`	Lists section sizes of object/executable files	`arm-none-eabi-size [file]`
`arm-none-eabi-objdump`	Dumps information about object files	`arm-none-eabi-objdump [options] [object file]`
`arm-none-eabi-readelf`	Displays information about ELF files	`arm-none-eabi-readelf [options] [ELF file]`
`arm-none-eabi-objcopy`	Converts/copies object files between formats	`arm-none-eabi-objcopy [options] [input file] [output file]`

Table 3.3: Some common commands in the GNU toolchain for Arm and their example usage

In this section, we explored the GNU Binary Tools that are essential for embedded firmware development. These tools, including commands such as `arm-none-eabi-gcc`, play important roles in various stages of the firmware build process and are invaluable for tasks such as compiling, linking, and debugging. The next section will further expand our understanding of these tools; we'll dive into their practical applications, demonstrating their utility in the embedded firmware build process.

From IDE to the command line – watching the build process unfold

In this section, our aim is to understand how the compiler within the IDE handles our code when we initiate a build. Additionally, we will delve into the practical applications of some of the GNU Binary Tools we discussed in the previous section.

Observing the build process from the IDE's perspective

Let's start by revisiting the bare-metal GPIO driver we developed in the previous chapter.

Begin by launching your STM32CubeIDE. To conduct a proper analysis of the build commands that the IDE executes, it's necessary to first clean the project and then build again. The reason for this is straightforward – we've already built the project, in a previous chapter. Without any modifications to the source code since then, a new build attempt would skip the detailed command execution we aim to scrutinize, since the source code hasn't changed.

Let's clean the project:

1. Locate the project in the **Projects** pane.
2. Right-click on the name of the project.
3. A menu will appear. From this menu, select the **Clean Project** option.
4. The IDE will now clear any already compiled data from the project. This action resets the build state of the project to its initial condition, erasing any previous build results.

Now, let's build it again:

1. Right-click once more on the same project's name in the **Projects** pane.
2. Again, a menu will appear. This time, select the **Build Project** option.
3. By selecting this option, the IDE will start the process of building the project from scratch.

To observe the build commands in the STM32CubeIDE, we must find the **Console** pane, which is usually positioned at the bottom area of the IDE's interface. The **Console** pane is an important component that displays real-time outputs and logs for various actions, including the build process. This makes it an invaluable tool to monitor the commands and actions undertaken during the compilation of our projects.

If the **Console** pane is not readily visible in your current IDE layout, you can access it through the menu bar located at the top of the IDE. Simply click on the **Window** menu to reveal a drop-down list of options. From there, navigate to **Show View**, which will expand to show more choices. Among these, select **Console** to bring the pane into view within your workspace.

Once you have the **Console** pane visible, you can monitor the execution of the build commands. Whenever we initiate the build process, each command used during the build – encompassing compilation, linking, and other stages – will be displayed in this pane. *Figure 3.5* shows some of the content of **Console** pane after building our 2_RegisterManipulation bare-metal GPIO driver project. *Figure 3.5* shows some of the content of the **Console** pane after building our 2_RegisterManipulation bare-metal GPIO driver project:

```
  Problems  Tasks  Console ×  Properties
CDT Build Console [2_RegisterManipulation]
 21:38:13 **** Build of configuration Debug for project 2_RegisterManipulation ****
1make -j8 all
2arm-none-eabi-gcc -mcpu=cortex-m4 -g3 -DDEBUG -c -x assembler-with-cpp -MMD -MP -MF"Start
3arm-none-eabi-gcc "../Src/main.c" -mcpu=cortex-m4 -std=gnu11 -g3 -DDEBUG -DNUCLEO_F411RE
4arm-none-eabi-gcc "../Src/syscalls.c" -mcpu=cortex-m4 -std=gnu11 -g3 -DDEBUG -DNUCLEO_F41
5arm-none-eabi-gcc "../Src/sysmem.c" -mcpu=cortex-m4 -std=gnu11 -g3 -DDEBUG -DNUCLEO_F411R
6arm-none-eabi-gcc -o "2_RegisterManipulation.elf" @"objects.list"    -mcpu=cortex-m4 -T"C:
7Finished building target: 2_RegisterManipulation.elf

8arm-none-eabi-size   2_RegisterManipulation.elf
9arm-none-eabi-objdump -h -S  2_RegisterManipulation.elf  > "2_RegisterManipulation.list"
   text    data    bss     dec     hex filename
    720       8    1568    2296     8f8 2_RegisterManipulation.elf
 Finished building: default.size.stdout

 Finished building: 2_RegisterManipulation.list
```

Figure 3.5: The Console pane after building the 2_RegisterManipulation bare-metal GPIO driver project.

Figure 3.5 provides a snapshot of the build process steps, although it only displays a segment of each step. For a comprehensive view of all the steps, including the full lines of commands and responses, you should refer to the **Console** pane in your STM32CubeIDE.

Let's analyze the console pane in *Figure 3.5*, according to the line numbering..

Compilation of assembly and C files

Line (1)

make -j8 all: This is a command to the make build automation tool, requesting it to execute the build. We shall learn about make in the upcoming chapters.

Lines (2)(3)(4)(5)

The following lines are specific `arm-none-eabi-gcc` commands to compile individual source files such as `main.c`, `syscalls.c`, and `sysmem.c`. These commands specify the target CPU (`-mcpu=cortex-m4`) and other compiler flags. Each source file is compiled into an object file (`.o`).

Linking process

Line (6)

The `arm-none-eabi-gcc` command with a list of object files (`@"objects.list"`) and a linker script (`STM32F411RETX_FLASH.ld`) links these object files into an executable file (`2_RegisterManipulation.elf`). The linker script guides how different sections of the code and data are placed in the final executable. We shall discuss linker scripts in the next chapter.

Size calculation:

Line (8)

`arm-none-eabi-size 2_RegisterManipulation.elf`: This command calculates the size of the compiled program, breaking it down into text (code), data (initialized data), and bss (uninitialized data) sections. The output shows the size of these sections in bytes and their total in both decimal (**dec**) and hexadecimal (**hex**) formats.

Creation of a list file:

Line (9)

`arm-none-eabi-objdump -h -S 2_RegisterManipulation.elf > "2_RegisterManipulation.list"`: This command disassembles the executable and outputs a detailed list file. The `-h` flag shows the header information, and `-S` intersperses source code with disassembly. A list file is a detailed textual representation of compiled code, containing both the assembly language instructions and their corresponding machine code, often with annotations of the original high-level source code.

As we can observe from these logs, it is clear that our STM32CubeIDE employs the same GNU Binary Tools previously discussed. In our forthcoming section, we will manually execute these commands via the command line. This approach will teach us how to build our firmware without using an IDE, simply using the source text files, the command-line interface, and our suite of GNU Bin Tools.

Working with the GNU bin tools

In this section, our focus is to execute some of the GNU Bin Tools directly, using our command line. To start I want to show you why I use the words *commands* and *tools* interchangeably to describe the GNU Bin Tools.

Locate the installation folder for the GNU Arm Embedded Toolchain on your computer. On my computer, this is `C:/Program Files(x86)/GNU Arm Embedded Toolchain`.

Once you've located the GNU Arm Embedded Toolchain folder, the next step involves accessing the `bin` folder within it.

Upon opening the `bin` folder, you'll be greeted with a plethora of tools, each represented by an executable file (`.exe`). Looking closely, you will find `arm-none-eabi-gcc.exe`, our compiler, along with other tools we previously discussed. When we input a command corresponding to these tools in the command line, the associated `.exe` file is executed. For instance, entering `arm-none-eabi-gcc` in the command line will run the `arm-none-eabi-gcc.exe` executable.

Now that we have clarified that, it's time to shift our focus toward practical testing. However, before diving into this testing phase, a few essential preparatory steps are required. Let's create a backup of our current project, `2_RegisterManipulation`:

1. **Locate your project**: Navigate to your workspace or the folder where the `2_RegisterManipulation` project is stored.

2. **Copy the project**: Right-click on the `2_RegisterManipulation` project folder, select **Copy**, and then **Paste** within the same directory.

3. **Rename the Backup**: Rename the newly pasted folder to `2_RegisterManipulation-old`.

With the backup in place, our next move is to modify the `main.c` file in the `2_Register Manipulation` project, changing the LED's behavior from a constant *on* state to a blinking one:

1. **Access the source File**: Go to the `2_RegisterManipulation/Src` directory.

2. **Edit the main.c file**: Right-click on the `main.c` file and choose to open it in a simple text editor, such as Notepad.

3. **Update the code**: Find the section of code that sets PA5 (`LED_PIN`) high. It should look like this:

   ```
   //  22: Set PA5(LED_PIN) high
   GPIOA_OD_R |= LED_PIN;
   ```

   Replace this code with the following to toggle the state of PA5 and create a blinking effect:

   ```
   //  22: Toggle PA5(LED_PIN)
   GPIOA_OD_R ^= LED_PIN;
   for(int i = 0; i < 100000; i++){} // Delay loop for visible blinking
   ```

4. **Save the changes**: After updating the code, save the `main.c` file.

Let's go back to our project folder and access the 2_RegisterManipulation/Debug directory through the command prompt. This specific folder is important because it's where STM32CubeIDE automatically places the project's makefile. Understanding the role and structure of **makefiles** is crucial in embedded firmware development, and we will delve into this topic in more detail in upcoming chapters.

We can access the folder through the command prompt in multiple ways:

Windows users can choose between these methods:

- **Method 1 – using the context menu**: Navigate to the 2_RegisterManipulation/Debug folder in Windows Explorer. Once there, hold down the *Shift* key, right-click in an empty space within the folder, and select **Open Command Window Here**. This action will open a Command Prompt window directly in the Debug folder.

- **Method 2 – copying the path**: Alternatively, click on the address bar in Windows Explorer while in the 2_RegisterManipulation/Debug folder and then copy the folder path. Then, open Command Prompt from the **Start** menu or by typing cmd in the **Run** dialog (*Win + R*). In the Command Prompt, type cd (note the space after 'cd'), paste the copied path, and press *Enter*. This will change the directory to the Debug folder.

Users of other operating systems can choose between these methods:

- Navigate to the folder, and open the terminal specific to your operating system.

- Use the cd (change directory) command, followed by the absolute path to the 2_RegisterManipulation/Debug folder to navigate to it. The exact path may vary, based on where the project is located on your system.

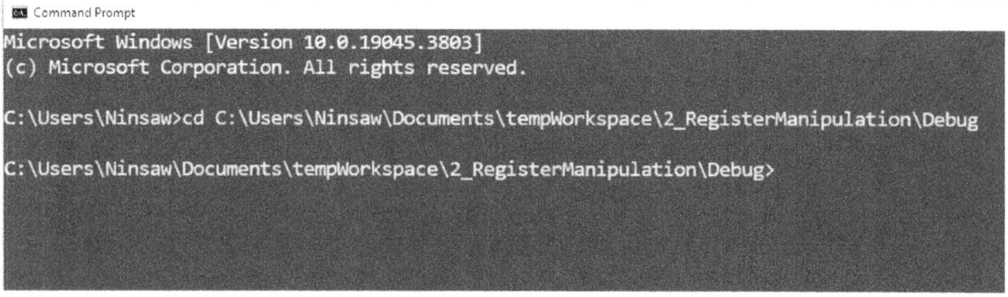

```
Command Prompt
Microsoft Windows [Version 10.0.19045.3803]
(c) Microsoft Corporation. All rights reserved.

C:\Users\Ninsaw>cd C:\Users\Ninsaw\Documents\tempWorkspace\2_RegisterManipulation\Debug

C:\Users\Ninsaw\Documents\tempWorkspace\2_RegisterManipulation\Debug>
```

Figure 3.6: Accessing the Debug folder through the Windows Command Prompt

In this practical exercise, we'll replicate the commands used by STM32CubeIDE, extracting them directly from its console pane. We'll execute them one by one, as depicted in *Figure 3.5*, starting with line number 2 (since our current focus isn't on makefiles).

To do this, follow these steps:

1. Copy line number 2 from the STM32CubeIDE console pane and paste it into the command prompt. This line compiles the startup file using `arm-none-eabi-gcc`, referencing paths specific to my system setup:

   ```
   C:\Users\Ninsaw\Documents\tempWorkspace\2_RegisterManipulation\
   Debug>arm-none-eabi-gcc -mcpu=cortex-m4 -g3 -DDEBUG -c -x
   assembler-with-cpp -MMD -MP -MF"Startup/startup_stm32f411retx.d"
   -MT"Startup/startup_stm32f411retx.o" --specs=nano.specs
   -mfpu=fpv4-sp-d16 -mfloat-abi=hard -mthumb -o "Startup/startup_
   stm32f411retx.o" "../Startup/startup_stm32f411retx.s"
   ```

   Successful execution will not result in any error messages.

2. Proceed with line number 3, which compiles the `main.c` file. However, before pasting this command into the command prompt, remove the `-fcyclomatic-complexity` flag, as it is not supported by some versions of the GNU Toolchain for Arm. Paste the command into a text editor, delete the flag, and then copy the modified command into the command prompt. The command for line number 3 should look like this:

   ```
   C:\Users\Ninsaw\Documents\tempWorkspace\2_RegisterManipulation\
   Debug>arm-none-eabi-gcc "../Src/main.c" -mcpu=cortex-m4
   -std=gnu11 -g3 -DDEBUG -DNUCLEO_F411RE -DSTM32 -DSTM32F4
   -DSTM32F411RETx -c -I../Inc -O0 -ffunction-sections -fdata-
   sections -Wall -fstack-usage -MMD -MP -MF"Src/main.d" -MT"Src/
   main.o" --specs=nano.specs -mfpu=fpv4-sp-d16 -mfloat-abi=hard
   -mthumb -o "Src/main.o"
   ```

3. Follow the same procedure for lines 4 and 5 to compile the `syscalls.c` and `system.c` files, respectively.

4. We will proceed to link these files. To do this, copy the linking command, which is on line number 6 in the STM32CubeIDE console pane, and paste it into the command prompt, like this:

   ```
   C:\Users\Ninsaw\Documents\tempWorkspace\2_RegisterManipulation\
   Debug>arm-none-eabi-gcc -o "2_RegisterManipulation.elf"
   @"objects.list"    -mcpu=cortex-m4 -T"C:\Users\Ninsaw\Documents\
   tempWorkspace\2_RegisterManipulation\STM32F411RETX_FLASH.
   ld" --specs=nosys.specs -Wl,-Map="2_RegisterManipulation.map"
   -Wl,--gc-sections -static --specs=nano.specs -mfpu=fpv4-sp-d16
   -mfloat-abi=hard -mthumb -Wl,--start-group -lc -lm -Wl,--end-
   group
   ```

   A successful linking process will not result in any error messages.

5. We can use the `arm-none-eabi-size` command to display the size of our output `.elf` file. This will give us insights into the sizes of various sections, such as **text**, **data**, and **bss**. We will discuss these sections in detail later in the book, particularly when we delve into writing linker scripts.

To execute this, copy line number 8 from the STM32CubeIDE console pane and paste it into the command prompt. Running the following command:

```
C:\Users\Ninsaw\Documents\tempWorkspace\2_RegisterManipulation\
Debug>arm-none-eabi-size   2_RegisterManipulation.elf
```

It will display the size details of the .elf file:

```
   text    data     bss     dec     hex filename
    744       8    1568    2320     910 2_RegisterManipulation.elf
```

Figure 3.7: Output produced by executing the arm-none-eabi-size command

Observing the results, we can confirm that the output matches exactly what is displayed in the STM32CubeIDE console pane.

At this stage, we can choose to convert our .elf file into the .bin format using the arm-none-eabi-objcopy tool, with the appropriate flags.

Type the following in the command prompt and press *Enter*:

```
arm-none-eabi-objcopy -O binary 2_RegisterManipulation.elf 2_
RegisterManipulation.bin
```

Let's break down this snippet:

- -O binary specifies the output format, which in this case is a binary file

- 2_RegisterManipulation.elf is the source ELF file you are converting

- 2_RegisterManipulation.bin is the name of the output binary file that will be created

This final step marks the completion of our first build process. We have successfully compiled and linked all necessary files, resulting in the creation of our final executable in two formats. The next process involves uploading the firmware to our microcontroller using *OpenOCD*. This will be covered in the next section.

Uploading firmware to the microcontroller using OpenOCD

Open On-Chip Debugger (**OpenOCD**) plays an important role in embedded firmware programming. It not only facilitates the transfer of executable firmware code to a microcontroller but also provides robust debugging capabilities. In this section, we shall go through a step-by-step process to upload the 2_RegisterManipulation executable file into our microcontroller.

We will start by locating the correct OpenOCD script for our development board. OpenOCD comes with a variety of scripts, each tailored to different microcontrollers and development boards. In our case, the focus is on the **st_nucleo_f4 series**. To find the right script, follow these steps:

1. Navigate to the OpenOCD installation directory, typically found in the `Program Files` folder for Windows users. The OpenOCD folder is usually named `xpack-openocd`. Once there, enter the `openocd` subfolder, then the `scripts` subfolder, and finally, the `board` subfolder. You will find a file named `st_nucleo_f4.cfg`; this is the OpenOCD file we have to execute for our NUCLEO-F4 development board.

2. To launch OpenOCD, connect your development board, open the command prompt window, and enter the following command:

   ```
   openocd -f board/st_nucleo_f4.cfg
   ```

   This command starts OpenOCD with the configuration file specific to the STM32 NUCLEO F4 development board. The configuration file, `st_nucleo_f4.cfg`, contains all the necessary settings for OpenOCD to communicate with the development board.

This is a snippet of the output from the command prompt after executing the command:

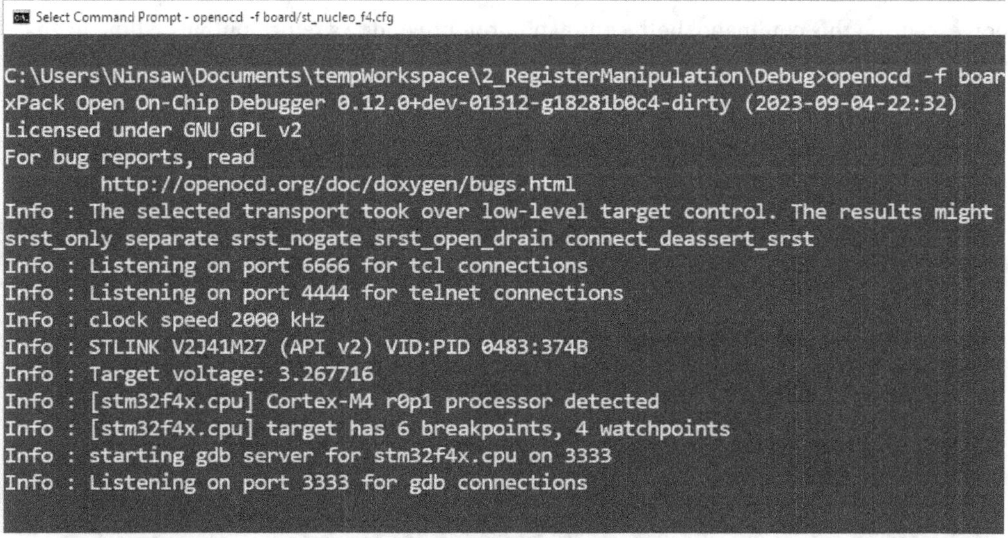

```
Select Command Prompt - openocd  -f board/st_nucleo_f4.cfg

C:\Users\Ninsaw\Documents\tempWorkspace\2_RegisterManipulation\Debug>openocd -f boar
xPack Open On-Chip Debugger 0.12.0+dev-01312-g18281b0c4-dirty (2023-09-04-22:32)
Licensed under GNU GPL v2
For bug reports, read
        http://openocd.org/doc/doxygen/bugs.html
Info : The selected transport took over low-level target control. The results might
srst_only separate srst_nogate srst_open_drain connect_deassert_srst
Info : Listening on port 6666 for tcl connections
Info : Listening on port 4444 for telnet connections
Info : clock speed 2000 kHz
Info : STLINK V2J41M27 (API v2) VID:PID 0483:374B
Info : Target voltage: 3.267716
Info : [stm32f4x.cpu] Cortex-M4 r0p1 processor detected
Info : [stm32f4x.cpu] target has 6 breakpoints, 4 watchpoints
Info : starting gdb server for stm32f4x.cpu on 3333
Info : Listening on port 3333 for gdb connections
```

Figure 3.8: OpenOCD's first output

The information presented here includes the following:

- **Communication ports**: OpenOCD reports that it is listening on port `6666` for Tcl connections and port `4444` for Telnet connections. These ports are used to send commands to OpenOCD and interact with it during debugging sessions.

- **Processor and debug capabilities**: The debugger has identified the Cortex-M4 r0p1 processor in the STM32F4 series microcontroller. Additionally, it notes that the target has six breakpoints and four watchpoints, which are crucial for setting breakpoints and watchpoints during debugging.

- **GDB server initiation**: The last two lines in the output indicate that a GDB server has been started for the STM32F4 series CPU on port 3333. This server allows a **GDB (GNU Debugger)** client to connect for debugging purposes.

With OpenOCD running, the next step involves using the GDB to upload the firmware to the microcontroller. Let's access another command prompt window, still from the Debug folder (as OpenOCD should keep running in the first one), and enter the following command to start the GDB:

```
arm-none-eabi-gdb
```

Once the GDB is open, we establish a connection to the microcontroller by running the following:

```
target remote localhost:3333
```

This command connects GDB to the OpenOCD server running on the local machine (localhost) on port 3333, which is the default port for OpenOCD.

Upon executing this command, both command prompt windows return outputs telling us that debugging has started:

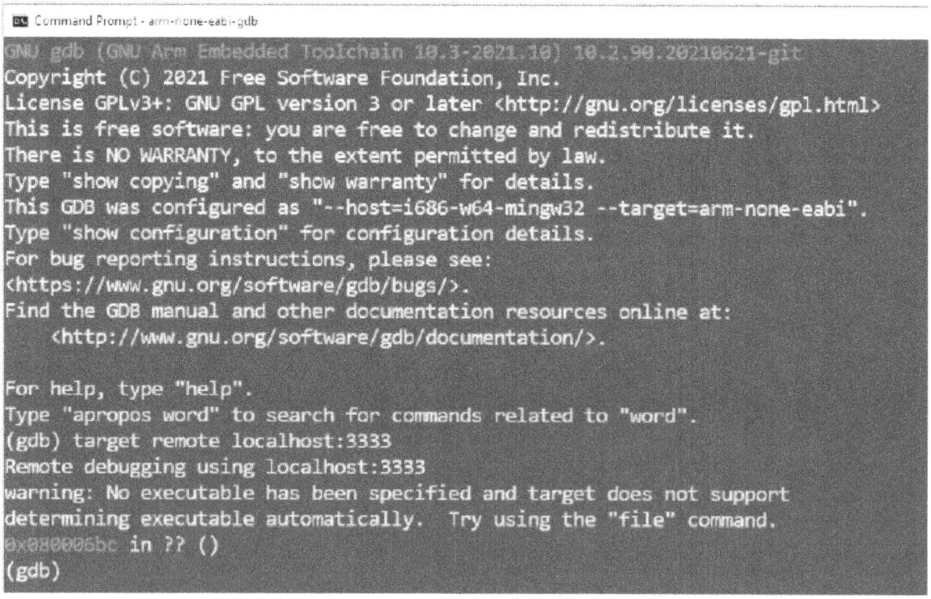

Figure 3.9: Output from the command prompt window running the GDB

This is the output from the command prompt window running `st_nucleo_f4.cfg`:

```
 Command Prompt - openocd -f board/st_nucleo_f4.cfg
            http://openocd.org/doc/doxygen/bugs.html
Info : The selected transport took over low-level target control. The result
srst_only separate srst_nogate srst_open_drain connect_deassert_srst
Info : Listening on port 6666 for tcl connections
Info : Listening on port 4444 for telnet connections
Info : clock speed 2000 kHz
Info : STLINK V2J41M27 (API v2) VID:PID 0483:374B
Info : Target voltage: 3.264567
Info : [stm32f4x.cpu] Cortex-M4 r0p1 processor detected
Info : [stm32f4x.cpu] target has 6 breakpoints, 4 watchpoints
Info : starting gdb server for stm32f4x.cpu on 3333
Info : Listening on port 3333 for gdb connections
Info : accepting 'gdb' connection on tcp/3333
[stm32f4x.cpu] halted due to debug-request, current mode: Thread
xPSR: 0x01000000 pc: 0x080006bc msp: 0x2001ffc0
Info : device id = 0x10006431
Info : flash size = 512 KiB
Info : flash size = 512 bytes
Warn : Prefer GDB command "target extended-remote :3333" instead of "target
```

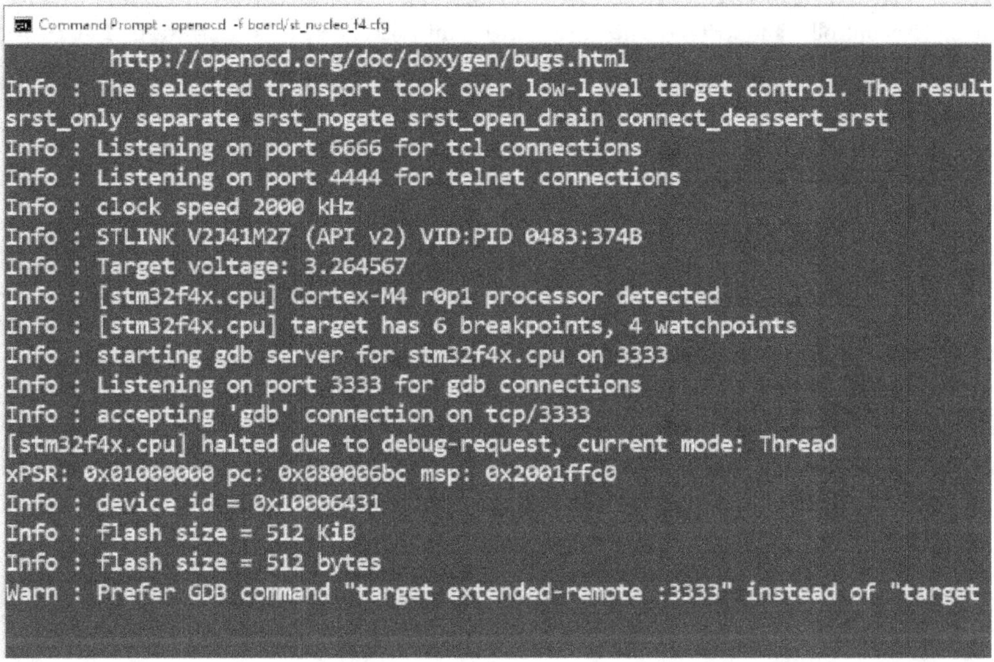

Figure 3.10: Output from the command prompt window running st_nucleo_f4.cfg

Before loading the firmware, we have to reset and initialize the board using the following command:

```
monitor reset init
```

Let's break down the command into its components:

- `monitor`: This prefix is used in the GDB to indicate that the following command is not a GDB command but is meant for the debugging server (in this case, OpenOCD)
- `reset`: This part of the command instructs OpenOCD to reset the target device
- `init`: This tells OpenOCD to execute its initialization sequence for the target device

Then, we load the firmware onto the microcontroller using the following command:

```
monitor flash write_image erase 2_RegisterManipulation.elf
```

This command erases the existing firmware on the microcontroller and writes the new firmware (in this case, `2_RegisterManipulation.elf`) onto it. This command erases the existing firmware on the microcontroller and writes the new firmware (in this case, `2_RegisterManipulation.elf`) onto it:

- `flash write_image`: This is an OpenOCD command that tells it to write an image to the flash memory of the target microcontroller – in effect, programming the microcontroller with a new firmware image.

- `erase`: This option tells OpenOCD to erase the flash memory before writing the new image. Erasing the flash is a common requirement in microcontroller programming, as it clears any previous program and ensures that the new firmware is written to a clean memory space.

After successfully loading the firmware, we reset the board again with the same `reset` command.

```
monitor reset init
```

Then, we resume the execution of the code on the microcontroller with the following:

```
monitor resume
```

Voila! The firmware should now be running on the microcontroller. You should see the LED on the board blinking, indicating the successful upload and execution of the new firmware.

Summary

In this chapter, we explored the intricacies of the embedded firmware build process, with a specific focus on the GNU Toolchain.

We began by getting to know the embedded build process, exploring its multiple stages – pre-processing, compilation, assembly, linking, and locating. Each stage was analyzed, clarifying its significance in transforming human-readable source code into executable machine instructions. We delved into the roles of pre-processing in preparing code, the nuances of compilation and assembly in translating and converting code, and the intricate tasks of linking and locating in forming a cohesive, executable binary.

Transitioning to practical application, the chapter introduced the GNU Binary Tools for Embedded Systems. By revisiting our previously developed bare-metal GPIO driver, we observed the build commands executed by the STM32CubeIDE, replicating these steps manually using our command-line interface. This approach gave us a deeper appreciation of the underlying processes and commands that IDEs automate. In the latter part of the chapter, we went through the step-by-step process of uploading our firmware to the microcontroller using OpenOCD, from locating the correct OpenOCD script for the development board to executing commands to reset, initialize, and run the firmware on the microcontroller. This practical exercise demonstrated the successful application of the theoretical knowledge we gained earlier, marking a significant milestone in our journey.

In the next chapter, we shall learn how to write our own linker scripts and startup files. This important step will represent another significant milestone in our journey toward mastering the art of developing entirely bare-metal firmware from the ground up.

4

Developing the Linker Script and Startup File

In this chapter, we undertake an in-depth exploration of the core components of embedded **bare-metal programming**, focusing on three critical areas: the microcontroller memory model, the writing of the linker script, and the startup file.

First, we'll explore the microcontroller memory model to understand how memory is organized and utilized. This knowledge is important for accurately allocating program code and data sections within the microcontroller memory. Next, we'll go through the intricacies of writing linker scripts. These scripts are essential for correctly mapping our program to the appropriate sections of the microcontroller's memory, ensuring that the executable runs as intended.

Finally, we will learn about the startup file and then proceed to write our own, focusing on initializing the vector table and configuring `Reset_Handler`.

In this chapter, we're going to cover the following main topics:

- Understanding the memory model
- The linker scripts
- Writing the linker script and startup file

Technical requirements

All the code examples for this chapter can be found on GitHub at `https://github.com/PacktPublishing/Bare-Metal-Embedded-C-Programming`.

Understanding the STM32 memory model

While the STM32 memory map consists of various memory areas, our primary focus in developing the linker script and startup file revolves around two critical areas: **flash memory** and **static random access memory (SRAM)**. These areas are of utmost importance because they are directly involved in program storage. In the initial parts of this section, we will learn about the characteristics of these memory areas and the distinct roles they play.

Figure 4.1 shows a section of the stm32f411 memory map, highlighting the flash memory and SRAM.

Reserved	0x2002 0001 - 0x3FFF FFFF
SRAM (128 KB aliased by bit-banding)	0x2000 0000 - 0x2002 0000
Reserved	0x1FFF C008 - 0x1FFF FFFF
Option bytes	0x1FFF C000 - 0x1FFF C007
Reserved	0x1FFF 7A10 - 0x1FFF BFFF
System memory	0x1FFF 0000 - 0x1FFF 7A0F
Reserved	0x0808 0000 - 0x1FFE FFFF
Flash memory	0x0800 0000 - 0x0807 FFFF
Reserved	0x0008 0000 - 0x07FF FFFF
Aliased to Flash, system, memory or SRAM depending, on the BOOT pins	0x0000 0000 - 0x0007 FFFF

Figure 4.1: A section of the STM32F11 memory map, highlighting the flash memory and SRAM areas

Let's start with flash memory.

Flash memory

One of the primary advantages of flash memory is its non-volatile nature. This means that data stored in flash memory remains intact even when the power supply is disconnected. In STM32 microcontrollers (as well as other microcontrollers), flash memory is typically where the executable code is stored and is read-only during normal operation. Flash memory starts at the 0x08000000 address. However, its size varies depending on the specific STM32 microcontroller model.

STM32 microcontrollers come in various series and models, offering a range of flash memory densities to accommodate different application requirements.

> **What is memory density?**
>
> **Memory density** refers to the concentration of memory storage within a given physical space or component. Memory density is often expressed in terms of bits or bytes stored per unit of physical area, such as bits per square millimeter or bytes per square centimeter. It measures how densely or compactly data can be stored within that space. Higher memory density means that we can store more data than we can in a smaller physical space.
>
> Memory size, on the other hand, refers to the total amount of memory (storage capacity) available in a given storage device. It is typically measured in units such as bytes, **kilobytes (KB)**, **megabytes (MB)**, **gigabytes (GB)**, etc.

Now, let's talk about some operational nuances of flash memory.

Flash memory, including STM32 flash memory, has a limited number of program and erase cycles. Each time we write (program) or erase, it consumes one of these cycles. It is important to consider these limitations when designing applications that frequently write to or erase data from flash memory.

STM32 microcontrollers are known for their low power consumption, and this extends to their flash memory operations. Efficient power management ensures that the microcontroller can operate on minimal power while reading from or writing to flash memory, making STM32 devices suitable for battery-powered applications.

To ensure data integrity, STM32 flash memory often includes built-in error correction mechanisms. These mechanisms help identify and correct errors that may occur during data storage and retrieval, enhancing the reliability of the stored firmware.

The following are some key attributes of the STM32 flash memory:

- **Read-only nature**: Primarily used for storing program code
- **Memory start address**: `0x08000000`
- **Variable size**: Dependent on the specific STM32 microcontroller model
- **Vector table location**: `0x08000004`

Let's take a look at the other primary memory areas relevant to writing our linker script and startup file.

SRAM

SRAM is a type of volatile memory, meaning it loses its contents when the power supply is disconnected. It is used in STM32 microcontrollers for temporary data storage during program execution. Unlike flash memory, which is used for long-term storage of program code, SRAM is designed for high-speed access and low latency, making it ideal for storing variables, intermediate data, and managing the stack during runtime.

Like flash memory, the STM32 microcontrollers feature varying sizes of SRAM, tailored to the needs of different applications. The size of the SRAM determines the amount of runtime data that can be handled and affects the overall performance of the microcontroller in handling complex tasks or multitasking. The SRAM in STM32 microcontrollers starts at the `0x20000000` address. Like flash memory, its size varies depending on the specific STM32 microcontroller model.

The following are the key attributes of the STM32 SRAM:

- **Read and write**: Variables and the stack are stored here

- **Memory start address**: `0x20000000`

- **Variable size**: Dependent on the specific STM32 microcontroller model

Before moving on to introduce the linker script, let's touch on one other memory area that is not relevant to the linker script, but is still relevant to our understanding of the memory layout of our microcontroller.

Peripheral memory

Peripheral memory is dedicated to managing and interfacing with the microcontroller's onboard peripherals. These peripherals include components such as timers, communication interfaces (UART, SPI, I2C), and **analog-to-digital converters** (**ADCs**). Peripheral memory is made up of registers that are used to configure and manage these peripherals.

An important aspect of microcontroller architecture is the use of **memory-mapped input/output** (**I/O**). Memory-mapped I/O is a technique where peripheral registers are assigned specific addresses in the system's memory space. This approach allows firmware to interact with hardware peripherals by reading from or writing to these memory addresses, just as it would with regular memory. The peripheral memory area of the STM32 memory map is the memory-mapped area for the peripheral registers.

Now that we are familiar with the major memory areas, we are ready to learn about the linker script.

The linker script

Linker scripts play an important role in the build process, especially in defining the memory layout and allocating various memory sections used by the firmware. They specify where different sections of the firmware, such as code, data, and uninitialized data, are to be placed in the microcontroller's memory.

While linker scripts set up the structure and boundaries for these sections, it is important to note that they do not populate these sections with data. The actual process of initializing data with specific values is handled by the startup code, which runs when the microcontroller boots up. We provide these linker scripts to the linker to effectively guide the organization of memory during the linking phase.

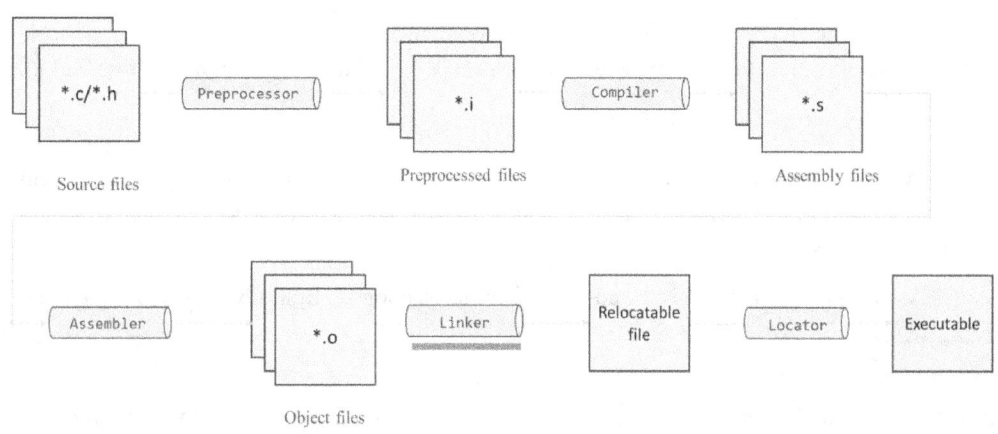

Figure 4.2: The build process with the linker highlighted

Understanding the linking process

In the build process, the linking of object files is an important step that transforms individual pieces of code into functional firmware. The assembler generates object files from source code, each containing code and data sections necessary for the firmware. However, these object files often have unresolved internal references to variables and functions, making them incomplete on their own. For instance, an object file may contain a reference to an `adc_value` variable that is defined elsewhere. It is the linker's job to amalgamate these object files, systematically resolving all such unresolved symbols to create a cohesive output file. To fully appreciate the meticulous work of the linker, we have to understand the attributes assigned to each section by the linker.

Section attributes and their implications

Each section within an object file is identified by a unique name and size, with specific attributes that dictate how they should be treated:

- **Loadable sections**: These sections contain content that must be loaded into memory at runtime. They are essential for the execution of the program and include executable code and initialized data.

- **Allocatable sections**: These sections do not carry content by themselves. Instead, they signal that a certain area of memory should be reserved, typically for uninitialized data that will be defined at runtime.

- **Non-loadable, non-allocatable sections**: Often, a section that is neither loadable nor allocatable contains debugging information or metadata that helps in the development process but is not required for the program's execution.

A crucial aspect of the linking process is the determination of two types of addresses for each allocatable and loadable output section: the **virtual memory address** (**VMA**) and the **load memory address** (**LMA**).

These are the roles of these two addresses:

- **VMA**: This address represents where the section will reside in memory during the execution of the output file. It is the runtime address used by the system to access the section's data or instructions.

- **LMA**: Conversely, the LMA is the address where the section is physically loaded into memory.

> **In most scenarios, the VMA and LMA are identical**
>
> A notable exception occurs when a data section is initially loaded into flash memory but then copied to SRAM upon startup.

To provide a clearer and more comprehensive understanding of the latter stages of the build process, it's essential to delve into another fundamental aspect of our discussion: the specific responsibilities and contributions of the locator within the build process.

Address relocation and the locator

The output file produced by the linker is not immediately suitable for use on a target microcontroller. This is because the addresses assigned to different sections during the linking process do not necessarily correspond to the actual memory layout of the target device. Therefore, these addresses must be relocated to match the target's memory space accurately. This is the job of the locator.

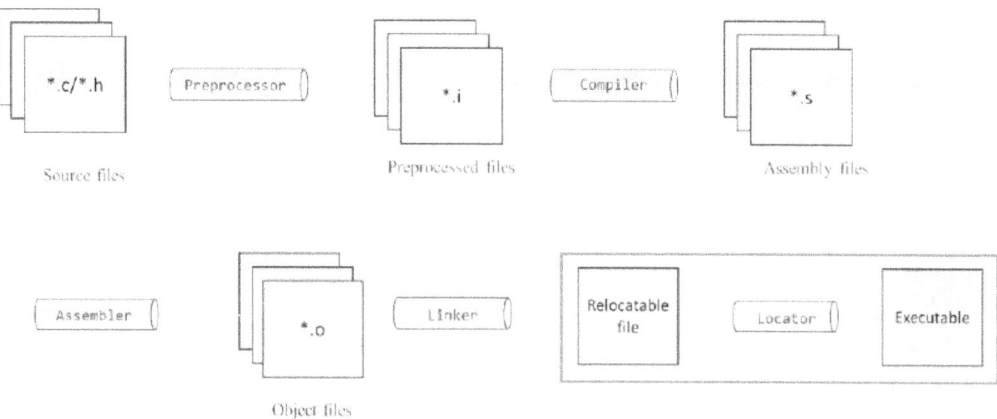

Figure 4.3: The build process, highlighting the relationship between the relocatable file, the locator, and the final executable output

In the GNU toolchain, the locator functionality is integrated into the linker, streamlining the process of address relocation. This capability ensures that the final executable is correctly mapped to the microcontroller's memory, making it ready for execution.

In this section, we examined the build process. From this, we observed that the process of linking object files in embedded systems development involves meticulous organization of code and data sections, symbol resolution, and address relocation.

In the next section, we shall explore the key components of the linker script in detail. This exploration will offer additional insights and deepen our understanding of the core elements discussed in this section.

Key components of the linker script

The key components of a linker script include the **memory layout**, **section definitions**, **options**, and **symbols**, each playing a unique role in ensuring that the firmware is correctly placed and executed within the microcontroller's memory.

Memory layout

This part of the linker script specifies the various memory types available in the microcontroller, such as flash memory and SRAM. It includes their start addresses and sizes, for instance, flash starting at `0x08000000` or SRAM at `0x20000000`.

Section definitions

A critical aspect of the linker script is defining how and where different sections of the program are placed. The `.text` section, containing the program code, is usually positioned at the beginning of flash memory. Following this, the **Block Started by Symbol** (`.bss`) and `.data` sections are allocated in SRAM. The linker script also ensures proper alignment of these sections for efficient memory access and program execution:

- `.text`:

  - **Purpose**: The `.text` section holds the executable instructions of our program. It's where the actual code that the processor executes resides.

  - **Characteristics**: This section is read-only and typically resides in the microcontroller's flash memory, ensuring that the program code is preserved even when the device is powered off.

  - **Size and location**: The size of the `.text` section varies based on the amount of code in your program. In STM32 microcontrollers, it generally starts at a predefined memory address, often in the lower region of the flash memory. For example, `0x00000000` and then relocated to `0x08000000`.

- `.bss`:

  - **Purpose**: The `.bss` section is used for uninitialized global and static variables. Variables in this section don't have initial values when the program starts.

  - **Characteristics**: This section is in the SRAM and is also read-write. It's typically 0-initialized at startup, meaning all variables start with a value of 0.

- `.data`:

  - **Purpose**: The `.data` section contains initialized global and static variables. Unlike variables in `.bss`, these variables have initial values specified in our code.

  - **Characteristics**: It's a read-write section. At runtime, the values in the `.data` section are typically copied from flash memory to SRAM to allow faster access and modification.

  - **Management**: The process of copying these values from flash memory to SRAM is handled by the startup code, executed before the main function of our program.

- **Other sections**:

  - `.rodata`: This section is used for constant data, such as string literals and constant arrays. It's read-only and usually stored in flash memory.

  - `.heap` and `.stack` are sections used for dynamic memory allocation (`malloc`, `free`) and function call stacks, respectively. They are part of SRAM and are crucial for runtime memory management.

The following table summarizes the key sections and their placement in memory.

Section	Purpose	Placed in
`.text`	Holds executable program instructions.	FLASH
`.bss`	Holds uninitialized global/static variables.	SRAM
`.data`	Holds initialized global/static variables with initial values.	FLASH (SRAM at runtime)
`.rodata`	Holds constant data (string literals, constant arrays).	FLASH

Table 4.1: Linker script sections and their placement in memory

Understanding the characteristics of these sections is important for having a properly functioning executable.

Options and symbols

Options in linker scripts are commands or directives that influence the behavior of the linker. A typical linker script includes directives for setting the entry point of the program and directives for defining the memory layout.

Symbols in linker scripts are identifiers that act as placeholders or references to specific memory locations, values, or addresses within the microcontroller's memory space. Symbols can be used to represent the start or end addresses of memory sections or specific variables in the program. For example, a symbol might be defined to represent the beginning of the flash memory or the start of the SRAM region. We can also use symbols to define important constants or values that are used throughout the firmware (such as source code files). These might include hardware addresses, configuration values, or size limits. By using symbols, the code becomes more readable and maintainable, as these values can be changed in one place (the linker script), rather than in multiple locations throughout the code.

Now that we are familiar with the key components of linker scripts, we will proceed to learn about some of the essential directives within these scripts. Each directive in a linker script instructs the linker on how to process and organize the input object files into the final executable.

Linker script directives

In this section, we learn about the essential directives of linker scripts. These directives dictate the memory layout and how various sections—code, data, and others—are allocated within the target microcontroller's memory. We will explore the key directives, their functionality, and how they influence the overall structure and efficiency of the compiled firmware. Let's start with the MEMORY directive.

Memory directive (MEMORY)

The MEMORY directive delineates the microcontroller's memory regions. Each defined block within the MEMORY section represents a distinct area of memory, characterized by its name, start address, and size. This directive allows us to define the memory layout of the target device, specifying different memory regions and their attributes. It plays an important role in guiding the linker on how to allocate sections of the program (code, data, etc.) across the microcontroller's physical memory.

Usage template

The general syntax for the MEMORY directive is as follows:

```
MEMORY
{
  name (attributes) : ORIGIN = origin, LENGTH = length
}
```

- `name`: An identifier we give to the memory region

- `attributes`: This specifies the access permissions for the region, such as read, write, and execute permissions

- `ORIGIN`: This defines the start address of the memory region

- `LENGTH`: This specifies the size of the memory region

Usage example

Consider a microcontroller with flash memory for storing executable code and SRAM for data storage. A linker script might define these memory regions as follows:

```
MEMORY
{
  FLASH (rx) : ORIGIN = 0x08000000, LENGTH = 256K
  SRAM (rwx) : ORIGIN = 0x20000000, LENGTH = 64K
}
```

In this example, two memory regions are defined: `FLASH` and `SRAM`:

- `FLASH` is marked with read (`r`) and execute (`x`) permissions (`rx`), indicating that this region can store executable code but is not writable during program execution. It starts at the `0x08000000` address and extends for `256K` bytes.

- `SRAM` is given read (`r`), write (`w`), and execute (`x`) permissions (`rwx`), allowing it to store data and executable code that can be modified during runtime. It begins at the `0x20000000` address and extends for `64K` bytes.

The `MEMORY` directive, with its comprehensive definition of memory regions and attributes, lays the foundation for efficient and effective memory management in firmware development. Before moving on to the next directive, let's examine all the attributes that can be specified to detail the characteristics and permissions of the memory sections:

- `r`: This attribute allows memory to be read. It is important for sections of memory containing executable code or constants that the program needs to read during execution.

- `w`: This attribute permits data to be written to the memory. It is important for memory areas where the program stores data dynamically during execution.

- `x`: This attribute allows the execution of code from the specified memory region. It is typically assigned to flash memory where the program code resides.

- `rw`: This is a combination of read and write permissions, allowing both operations in the specified memory region. It's commonly used for sections such as SRAM where temporary data and variables are stored and modified.

- rx: This combines read and execute permissions. It's often used for flash memory to indicate that the region contains executable code that the processor can read and execute.

- rwx: This attribute combines all three permissions, making the memory region fully accessible for reading, writing, and executing. This is less commonly used due to security and system stability considerations but might be applicable in certain development or debug scenarios.

- empty: If no attribute is specified, the memory region does not grant any access permissions by default. This might be used in special cases where permissions are controlled or modified by other means within the firmware.

Now, let's examine the ENTRY directive.

The entry directive (ENTRY)

This directive specifies the entry point of the program, which is the first piece of code to execute upon reset.

Here is the usage template:

```
ENTRY(SymbolName)
```

Here is a usage example:

```
ENTRY(Reset_Handler)
```

In this example, Reset_Handler is designated as the entry point of the program, meaning, the first function to execute. In firmware development, Reset_Handler takes care of initializing the system and jumping to the main program.

Next, we have the SECTIONS directive.

The sections directive (SECTIONS)

This directive defines the mapping and ordering of sections from input files into the output file.

Usage example

Let's see a template of it:

```
SECTIONS
{
  .output_section_name address :
  {
    input_section_information
  } >memory_region [AT>load_address]  [ALIGN(expression)]  [:phdr_
    expression]  [=fill_expression]
}
```

The parameters are as follows:

- `output_section_name`: This is the name given to the output section being defined. Common names include `.text` for executable code, `.data` for initialized data, and `.bss` for uninitialized data.

- `address`: This is optional and specifies the start address of the section in memory. This is often left to the linker to determine, based on the order of sections and memory regions defined in the script.

- `input_section_information`: This determines which input sections (from the compiled object files) should be included in this output section. Wildcards such as `*(.text)` can be used to include all `.text` sections from all input files.

- `>memory_region`: This assigns the section to a specific memory region defined in the MEMORY block of the linker script. We use this to tell the linker where in the target's memory map this section should reside, for example, FLASH or SRAM.

- `[AT>load_address]`: This is optional and specifies the load address of the section. This is used in scenarios where the execution address differs from the load address.

- `[ALIGN(expression)]`: This is optional and aligns the start of the section to an address that is a multiple of the value specified by `expression`. This is particularly useful for ensuring that sections begin at addresses that meet specific alignment requirements, which can enhance access speed and compatibility.

- `[:phdr_expression]`: This is optional and associates the section with a program header. Program headers are part of the **Executable and Linkable Format** (**ELF**) file structure; they provide the system loader with information about how to load and run different segments of a program.

- `[=fill_expression]`: This is optional and specifies a byte value to fill gaps between sections or at the end of sections to reach a certain alignment. This can be useful for initializing memory regions to a known state.

Usage example

Let's see an example of the SECTIONS directive in action:

```
SECTIONS
{
  .text 0x08000000 :
  {
    *(.text)
  } >FLASH
}
```

In this example, we have the following:

- SECTIONS: This keyword begins the section of the linker script where output sections are defined. Output sections are areas of memory that hold the code and data from the input files being linked.

- .text 0x08000000: This line defines an output section named .text and sets its starting address to 0x08000000. The .text section typically contains executable code.

- { *(.text) }: This line specifies what goes into the .text output section. The *(.text) syntax means all .text sections from all input files.

- >FLASH: This directive tells the linker to place the .text section in a memory region named FLASH. The FLASH region will be defined in the MEMORY directive block.

To understand the importance of the *(.text) syntax, let's examine the process of merging sections.

Sections merging

As we learned earlier, the assembler generates an object file for each source file, with each containing its .text, .data, .bss, and other sections. These sections from all object files are then merged by the linker into unified .text, .data, and .bss sections for the final executable.

Consider a firmware project with two source files, main.c and delay.c. The assembly process yields main.o and delay.o, each with its own sections. The linker's task is to consolidate these into a single set of sections for the final executable.

The following figure depicts this process. Note that merging is *not* performed through an addition process; this is merely a visual aid to enhance your understanding.

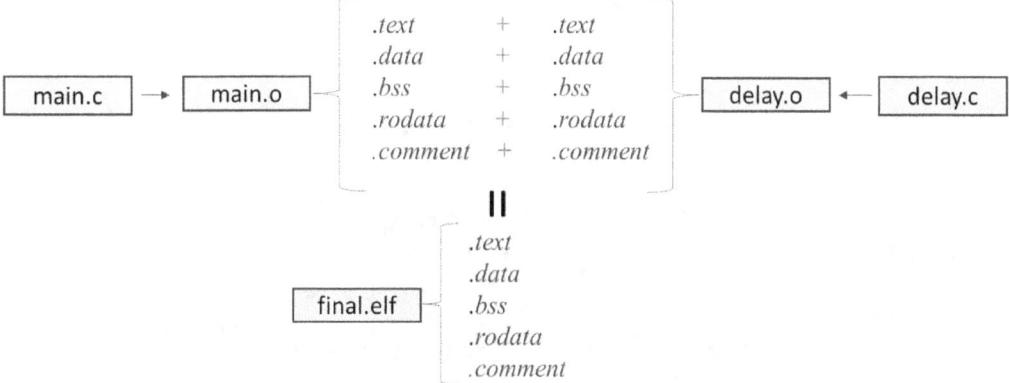

Figure 4.4: The merging process involving two source files: main.c and delay.c, resulting in the production of the final executable, final.elf

Now, let's explore the purpose of the AT > directive. To do this, it is essential to revisit the concepts of the LMA and VMA.

A closer look at the LMA and VMA

As we learned earlier, each allocatable and loadable output section in a binary output file is associated with two types of addresses: the LMA and the VMA. These addresses are crucial for defining how and where a section of the binary is processed during the system's startup and its subsequent runtime operations:

- **LMA**: This is the physical address in the binary image where the section is stored before program execution begins. It determines from where the system will load the section into memory when the program starts.

- **VMA**: In contrast, the VMA is the address where the section is intended to be accessed during the program's execution. This is the "runtime" address used by the system to refer to data or instructions in that section. For systems, particularly microcontrollers, that do not employ a **memory management unit (MMU)**, the VMA usually matches the section's physical memory address directly.

Why are LMA and VMA important?

The distinction between LMA and VMA allows for a flexible memory management approach where data can be stored in one location (such as flash memory) but run from another (such as SRAM). For example, initialized global and static variables (typically placed in the .data section) can be stored in flash memory but need to be copied to SRAM for faster access and to allow modification at runtime.

To fully understand this, let's consider the following snippet from a linker script generated by the STM32CubeIDE:

```
.data :
{
  . = ALIGN(4);
  _sdata = .;   /* create a global symbol at data start */
  *(.data)            /* .data sections */
  *(.data*)           /* .data* sections */
  *(.RamFunc)         /* .RamFunc sections */
  *(.RamFunc*)        /* .RamFunc* sections */

  . = ALIGN(4);
  _edata = .;   /* define a global symbol at data end */

} >SRAM AT> FLASH
```

In this script, the last line, >SRAM AT> FLASH, incorporates two important directives:

- >SRAM indicates that the output .data section is placed in the SRAM section of the memory during program execution (VMA).

- AT> FLASH specifies that although the section resides in SRAM when executed, it should initially be loaded into memory (**LMA**) at the corresponding address in FLASH. This is common for initialized data, which is stored in flash memory and then copied to SRAM at startup by the microcontroller's initialization code.

This detailed management of memory addresses highlights the critical role of LMA and VMA in maximizing the efficiency of resource-constrained microcontrollers. Through the effective use of LMA and VMA, we can ensure that even with limited memory resources, our microcontrollers operate reliably and efficiently, optimizing both storage and execution efficiency.

Before moving to explore the other features of the linker script, let's familiarize ourselves with some other commonly used directives.

Other commonly used directives

Some other commonly used directives include the KEEP, ALIGN, PROVIDE, >region, and AT directives. Let's examine them.

The KEEP directive

The KEEP directive ensures that specified sections or symbols are not eliminated by the linker during the optimization process, even if they appear unused. This is crucial for interrupt vector tables and initialization functions that must be present in the final binary.

Here is the usage template:

```
KEEP(section)
```

Here is a usage example:

```
KEEP(*(.isr_vector))
```

In this example, we are *keeping* the interrupt vector section. Next, let's see the region placement directive.

The >region directive

The (>region) region placement directive tells the linker to place a particular section into a specific memory region. The available memory regions must be defined in the MEMORY directive block of the linker script.

Here is the usage template:

```
section >region
```

Here is a usage example:

```
.data :
{
  *(.data)
} >SRAM
```

In this example, we are placing the `.data` section in the `SRAM` memory region.

The ALIGN directive

The `ALIGN` directive plays a crucial role in the linker script by adjusting the location counter to align with specified memory boundaries. **The location counter** tracks the current memory address allocated by the linker for placing sections or parts of the output file during linking.

The location counter is a built-in variable that represents the current address in memory where the linker is placing sections or parts of the output file during the linking process. It is denoted by a dot (`.`) in linker scripts.

As the linker processes the script, it assigns memory addresses to code and data sections according to the script's directives, with the location counter monitoring the progress. To ensure efficient memory access and adherence to hardware architecture requirements, sections and variables often need to be aligned to specific boundaries. The `ALIGN` directive allows us to achieve this by rounding up the location counter to the nearest address that matches the specified alignment, which must be a power of two.

Here is the usage template:

```
. = ALIGN(expression);
```

Here is a usage example:

```
. = ALIGN(4);
```

In this example, we are aligning the current location to a 4-byte boundary.

Next, let's see the `PROVIDE` directive.

The PROVIDE directive

The `PROVIDE` directive allows us to define symbols that the linker will include in the output file if they are not already defined. This can be used to set default values for symbols that may be optionally overridden by other modules.

Here is the usage template:

```
PROVIDE(symbol = expression);
```

Here is a usage example:

```
PROVIDE(_stack_end = ORIGIN(RAM) + LENGTH(RAM));
```

In this example, we are *providing* a default stack end address.

Next, we have the AT directive.

AT Directive

The AT directive specifies LMA for a section when it needs to be different from the section's VMA. This is commonly used for sections that need to be loaded into a different memory area during initialization before being moved to their runtime location.

Here is the usage template:

```
section AT> lma_region
```

Here is a usage example:

```
.data : AT> FLASH
{
    *(.data)
} >SRAM
```

In this example, the .data section is intended to reside in SRAM during the program's execution. However, it is initially loaded from FLASH, as indicated by AT> FLASH.

In the next section, we will explore another key aspect of linker scripts: the expression of numerical constants.

Understanding constants in linker scripts

When writing our linker script, we must keep in mind the interpretation of numerical prefixes and suffixes by the linker.

Firstly, let's clarify how the linker perceives integers with specific prefixes. An integer prefixed with 0 is read as an octal number by the linker. On the other hand, an integer starting with 0x is recognized as a hexadecimal value. This distinction is important for accurately defining memory addresses and sizes.

The use of the K and M suffixes introduces another layer of convenience, allowing us to denote large numbers succinctly. The K suffix multiplies the preceding number by 1024, while M expands the number by 1,024 twice over. Therefore, 4K translates to 4 times 1024, and 4M expands to 4 times 1,024 squared.

To put these principles into practice, let's explore an example that showcases the versatility of these notations. Imagine you need to specify a memory size of 4K. You could straightforwardly use 4K, or opt for its decimal equivalent, 4096, which results from multiplying 1,024 by 4. Alternatively, this quantity can be expressed in hexadecimal form as 0x1000.

Table 4.2 summarizes the key points to remember about using constants in linker scripts. It highlights the prefixes and suffixes that modify the base value, which offers a clear reference for interpreting and using these notations effectively in your linker script.

Notation	Meaning	Example	Equivalent Decimal	Hexadecimal Notation
0	Octal prefix	010	8	-
0x	Hexadecimal prefix	0x10	16	-
K	Multiplies by 1,024	4K	4096	0x1000
M	Multiplies by 1,024 twice (squared)	4M	4194304	0x400000

Table 4.2: Examples of linker script numerical prefixes and suffixes

In the next section, we shall learn about linker script symbols, further enhancing our understanding of linker scripts.

Linker script symbols

Linker symbols, also known simply as **symbols**, are fundamental elements in the process of converting source code into executable programs. At its core, a linker symbol comprises two essential components: a name and a value. These symbols are assigned integer values, representing memory addresses where variables, functions, or other program elements are stored in the microcontroller's memory.

Previously, we learned that after the assembly stage, the source code is transformed into object files. These object files contain machine code and unresolved references to variables and functions. The linker's primary task is to merge these object files, resolve these unresolved symbols, and generate a complete executable file ready for execution.

In the context of linker symbols, the value assigned to a symbol represents the memory address where the corresponding variable or function resides.

> **For example: X = 3500 means the memory address of X is 3500**
>
> A symbol named X might be assigned a value of 3500, indicating its memory address. It's crucial to note that in contrast to the variable's value in the source code, the X linker symbol represents its memory address.

Name	Type	Value (Memory Address)	Description
X	Symbol	3500	Represents the memory address where an X variable is stored.
Y	Symbol	0x3000	Represents the memory address where a Y variable is stored.
foo()	Symbol	0x4000	Represents the memory address where a foo() function is located.
bar()	Symbol	0x5000	Represents the memory address where a bar() function is located.
x	Variable	3500	Represents the value of a C variable named x.
y	Variable	4500	Represents the value of a C variable named y.

Table 4.3: Comparison of linker symbols and C source code variables assignments

Linker symbols can undergo various operations, such as those we use in C assignments. These operations include straightforward assignment (=), addition (+=), and subtraction (-=), among others.

```
symbol  =  expression ;
symbol +=  expression ;
symbol -=  expression ;
symbol *=  expression ;
symbol /=  expression ;
symbol <<= expression ;
symbol >>= expression ;
symbol &=  expression ;
symbol |=  expression ;
```

Figure 4.5: Examples of linker symbol operations

During the linking process, a symbol table is created, mapping each symbol to its corresponding address in memory. This table serves as a crucial reference for the linker to resolve symbol references and ensure proper linking of program components.

Let's consider a scenario where we have a `main.c` file and at the top of this file; we declare a variable named X, assigning it a value of 568. Additionally, within this file, there's a function named `blink`. Inside the `blink` function, there are operations to turn on an LED, wait, and then turn it off. This is depicted in *Figure 4.6*.

Now, let's take this `main.c` file and pass it through the build process to generate the `main.o` object file. During this process, a symbol table is generated. Each symbol in this table is associated with an address.

For instance, the X symbol would be assigned an address of `0x20000000`, and similarly, the `blink` function would also be assigned its address. In the following figure, the `blink` function is assigned the `0x08000000` address.

Essentially, just like in the C programming language, each variable has its value. In the object file, each symbol has its value, which essentially represents the address of the corresponding variable or function in C.

So, when referring to X in the object file, it wouldn't give us 568; rather, it would provide the address of X. This process of assigning values to symbols and associating them with addresses constructs the symbol table.

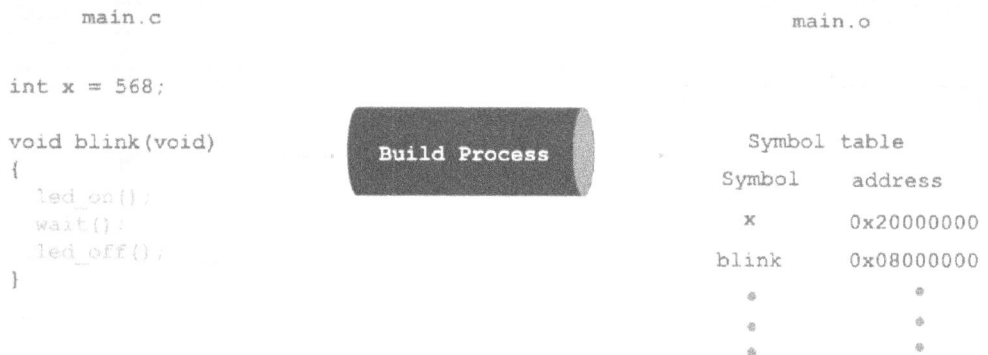

Figure 4.6: Representation of functions and variables from the source
file in the symbol table of the output object file

In this section, we delved deep into linker scripts, highlighting key components and directives. We carefully explored each directive, providing practical usage examples. Additionally, we distinguished between the LMA and VMA and also emphasized their important roles in guiding the linker on how to place sections. In the next section, we will learn how to write our own linker script and startup file from scratch, equipping you with another important skill in bare-metal firmware development.

Writing the linker script and startup file

Now that we have a good understanding of linker scripts and their essential components, we're prepared to write our own. However, before diving into writing the script, it's crucial to revisit the memory map of the microcontroller and gain insight into the `positions` load memory for various sections within the object file.

Understanding the load memory of different sections

As discussed earlier in this chapter, the output object file is structured into sections such as `.data`, `.rodata`, `.text`, and `.bss`. Together with the sections created by the assembler, we must define our own section to accommodate the vector table for **interrupt service routines (ISRs)**. We will name this section `.isr_vector_tbl`.

Each of these sections plays an important role in organizing the memory layout of the microcontroller, contributing to the functionality and efficiency of the final executable.

Figure 4.7 shows a zoomed-in view of the flash memory area, showing the required order for placing the different sections within the flash memory. Each section represents the combination of identical sections from all input files. For instance, the `.text` section depicted is a unified `.text` section, formed by merging all `.text` sections from the input files.

The diagram indicates that the placement must start with the `.isr_vector_tbl` section at the beginning of the flash memory. Following this, we must place the `.text` section, then the `.rodata` section, and finally, the `.data` section. The diagram does not show the placement of the `.bss` section, as we will place the `.bss` section directly in the SRAM. Additionally, during the startup code implementation, we must copy the content of the `.data` section from the flash memory to the SRAM.

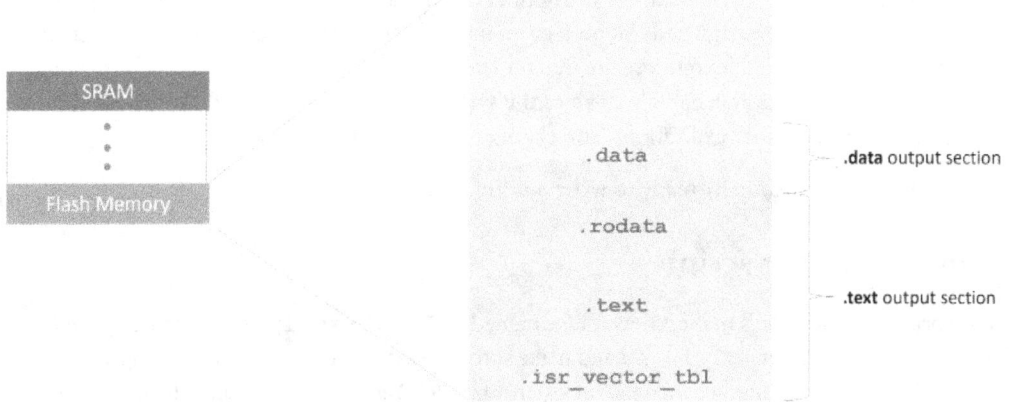

Figure 4.7: The flash memory area showing the order in which sections should be placed

Before we proceed, however, let's understand the concept of interrupts and the vector table.

Interrupts and the vector table

Interrupts are a fundamental concept in computing. They act as a powerful mechanism for managing how a computer or a microcontroller handles tasks and responds to external and internal events.

At its core, an **interrupt** is a signal to the processor from a hardware device or an internal software condition that temporarily halts the current operations. This signal indicates that immediate attention is required. When the processor receives an interrupt, it pauses its current tasks, saves its state, and executes a function known as an ISR to address the interrupt. Upon completing the ISR, the processor resumes its previous tasks, ensuring that critical signals receive prompt and efficient handling.

What are the types of interrupts?

We can broadly classify interrupts into two categories: hardware interrupts and software interrupts:

- **Hardware interrupts**: These originate from external devices, such as switches, network adapters, or any peripheral that needs to communicate with the processor. For example, pressing a push button may trigger a hardware interrupt that informs the processor to start a motor.

- **Software interrupts**: Unlike hardware interrupts, software interrupts are triggered by software instructions. These are used by programs to interrupt the current process flow and execute a specific routine.

What is the role of the interrupt vector table?

The interrupt vector table serves as an essential lookup table, guiding the processor to the correct ISR for each interrupt. An ISR is simply a function designed to address and manage the specific needs triggered by an interrupt. The table itself is **organized as an array of pointers**, with each pointer directing the system to the designated ISR for a given interrupt. Upon the occurrence of an interrupt, the system references this table to locate the exact memory address of the ISR required for handling the interrupt. This efficient mechanism enables the system to promptly respond to various events, such as external inputs, timer expirations, and changes in the internal state.

With this in mind, we are finally ready to write our linker script.

Writing the linker script

In our workspace folder, let's make a new folder named `3_LinkerscriptAndStartup`. In this folder, create a file called `stm32_ls.ld` and make sure its extension is `.ld`. If you're using Windows and it asks if you really want to change the file extension, click **Yes**. Then, right-click the file and open it with a basic text editor such as Notepad++.

Our objectives with the linker script can be summarized as follows:

- Specifying the firmware's entry point
- Detailing the available memory
- Specifying the necessary heap and stack sizes
- Defining output sections

This is our complete linker script, the contents of the `stm32_ls.ld` file:

```
/*Specifying the firmware's entry point*/
ENTRY(Reset_Handler)

/*Detailing the available memory*/
MEMORY
{
    FLASH(rx):ORIGIN =0x08000000,LENGTH =512K
    SRAM(rwx):ORIGIN =0x20000000,LENGTH =128K
}

_estack = ORIGIN(SRAM)+LENGTH(SRAM);

/*Specifying the necessary heap and stack sizes*/

__max_heap_size = 0x200;
__max_stack_size = 0x400;

/*Defining output sections*/
SECTIONS
{
    .text :
    {
    . = ALIGN(4);
    *(.isr_vector_tbl)
    *(.text)
    *(.rodata)
    . = ALIGN(4);
    _etext = .;
    }>FLASH
    .data :
    {
    . = ALIGN(4);
    _sdata = .;
    *(.data)
```

```
    . = ALIGN(4);
    _edata = .;
    } > SRAM AT> FLASH  /*>(vma) AT> (lma)*/
        .bss :
    {
    . = ALIGN(4);
    _sbss = .;
    *(.bss)
    . = ALIGN(4);
    _ebss = .;
    }> SRAM
}
```

Let's break it down.

Specifying the firmware's entry point

```
ENTRY(Reset_Handler)
```

As we learned earlier, the ENTRY directive specifies the entry point of the firmware, which is the first piece of code that gets executed when the firmware starts. In this case, the entry point is the function named Reset_Handler. We shall implement this function in the startup file.

Detailing the available memory

```
MEMORY
{
    FLASH(rx):ORIGIN =0x08000000,LENGTH =512K
    SRAM(rwx):ORIGIN =0x20000000,LENGTH =128K
}
```

Our script specifies two memory regions: FLASH and SRAM. The FLASH memory, with read and execute permissions (rx), starts at the 0x08000000 address and has a length of 512K. The SRAM memory, with read, write, and execute permissions (rwx), starts at the 0x20000000 address and has a length of 128K.

Symbol creation

```
_estack = ORIGIN(SRAM) +LENGTH(SRAM);
```

Over here, we create a symbol called _estack and we set it to the end of the SRAM memory region. We will use this symbol to initialize the stack pointer.

The **stack pointer** is an important register within our microcontroller that keeps track of the top of the stack. When a function is called, the address of the next instruction (return address) and the function's local variables are pushed onto the stack. The stack pointer is then adjusted to reflect the addition of these new items. Conversely, when a function returns, the stack pointer helps to pop the return address and the local variables off the stack, reverting it to its state before the function is called. The stack grows downward (from high memory to low memory addresses in most architectures), so setting the stack pointer to the end of SRAM ensures that it starts at the maximum available address, utilizing the SRAM space efficiently for stack operations.

The next lines of code in our linker script specify the heap and stack sizes.

Specifying the necessary heap and stack sizes

```
__max_heap_size = 0x200;
__max_stack_size = 0x400;
```

These lines define the maximum sizes for the heap (0x200 bytes) and stack (0x400 bytes). These sizes are important for dynamic memory allocation and function call management, respectively.

The next segment defines the output sections.

Defining output sections

In this section, we will go through the output sections.

The .text output section

This segment of our linker script shows the .text output section:

```
.text :
{
    . = ALIGN(4);
    *(.isr_vector_tbl)  /*merge all .isr_vector_tbl sections of input
    files*/
    *(.text)   /*merge all .text sections of input files*/
    *(.rodata) /*merge all .rodata sections of input files*/
    . = ALIGN(4);
    _etext = .;  /*Create a global symbol to hold end of text section*/
}>FLASH
```

Let's break it down:

- . = ALIGN(4);::

    This directive aligns the start of the .text section on a 4-byte boundary. This enhances memory access efficiency, which is a critical consideration for processors fetching instructions in word-sized chunks.

- `*(.isr_vector_tbl):`

  This directive pulls in all sections named `.isr_vector_tbl` from the input files into the current location in the `.text` section.

- `*(.text):`

  This directive pulls in all sections named `.text` from the input files into the current location in the `.text` section.

- `*(.rodata):`

  This directive pulls in all sections named `.rodata` from the input files into the current location in the `.text` section.

- `. = ALIGN(4);:`

  Again, this line ensures that the end of the section is aligned to a 4-byte boundary. Over here, we use it to align the end of a section, ensuring that the next section starts on an aligned boundary.

- `_etext = .;:`

  Over here, we define a symbol called `_etext` at the current location. This symbol marks the end of the `.text` section. We will use this symbol as a pointer to the end of the `.text` section in our startup file.

- `}>FLASH:`

  This directive specifies that the `.text` section should be placed in the FLASH memory segment as defined earlier in the MEMORY block of the linker script.

This segment shows the `.data` output section:

```
.data :
{
 . = ALIGN(4);
_sdata = .;    /*Create a global symbol to hold start of data
section*/
   *(.data)
 . = ALIGN(4);
_edata = .;    /*Create a global symbol to hold end of data
section*/
} > SRAM AT> FLASH  /*>(VMA) AT> (LMA)*/
```

Let's break it down:

- `= ALIGN(4);:`

  This directive aligns the start of the `.data` section on a 4-byte boundary.

- `_sdata = .;:`

  Over here, we create a symbol named `_sdata` to represent the start of the `.data` section by setting it to the current location counter. We will use this symbol as a pointer to the start of the `.data` section in our startup file.

- `*(.data):`

  This directive pulls in all sections named `.data` from the input files into the current location in the `.data` section.

- `. = ALIGN(4);:`

  This line ensures that the end of the section is aligned to a 4-byte boundary.

- `_edata= .;:`

  Similar to what we have done previously, we create a symbol named `_edata` to represent the end of the `.data` section by setting it to the current location counter. We will use this symbol in our startup file.

- `> SRAM AT> FLASH:`

  This directive specifies the LMA and the VMA of the `.data` section.

  `> SRAM` indicates that the section should be located in SRAM, allowing read and write access at runtime.

  `AT> FLASH` tells the linker that although the section is placed in SRAM for execution, its initial values should be stored in FLASH.

This segment shows the `.bss` output section:

```
.bss :
{
  . = ALIGN(4);
_sbss = .;
*(.bss)
  . = ALIGN(4);
_ebss = .;
}> SRAM
```

This is the final output section of our linker script.

As we learned earlier, the .bss section holds uninitialized global and static variables that we will be initialized to zero in our startup file. This zero-initialization ensures that all variables in this section begin with a known state, contributing to our firmware's stability and predictability.

Similar to other sections, we begin by aligning the section to a 4-byte boundary for efficient memory access, and then we define the _sbss and _ebss symbols to mark the start and end of the section, respectively. These symbols facilitate the calculation of the section's size and its initialization process. Finally, we place the section in the SRAM, emphasizing that, although it doesn't occupy space in the binary file on disk, it requires runtime allocation in memory.

With our linker script finalized, we'll move on to implementing the startup file. This shall be the focus of the next section.

Writing the startup file

The startup file is essential for initializing the firmware and it performs several critical tasks to ensure the system operates correctly from the moment it is powered on.

These tasks include the following:

- **Implementing the vector table**: This involves defining the vector table that maps interrupts to their handlers, ensuring the system can respond to various events efficiently.

- **Creating interrupt handlers**: For each interrupt listed in the vector table, an interrupt handler must be implemented to define how the system responds to that particular event.

- **Establishing the firmware's entry point**: This refers to implementing Reset_Handler, as specified in the linker script, which acts as the initial entry point of the firmware. This function is executed immediately after reset and is responsible for setting up the environment for the main application.

- **Transferring the .data section**: This involves copying the .data section from FLASH to SRAM.

- **Zeroing the .bss section**: This involves initializing the .bss section to zero, ensuring that all uninitialized global and static variables start with a known state for reliable operation.

In the current folder containing the linker script, create a file called stm32f411_startup.c and make sure its extension is .c. If you're using Windows and it asks if you really want to change the file extension, click **Yes**. Then, right-click the file and open it with a basic text editor such as Notepad++.

Let's analyze the complete startup code.

The following is our complete startup code written in C, the contents of the `stm32f411_startup.c` file. In the following snippet, we are not showing all the function prototypes of all the interrupts in the vector table. The complete source code can be found in the resources accompanying the book:

```c
extern uint32_t _estack;
extern uint32_t _etext;
extern uint32_t _sdata;
extern uint32_t _edata;
extern uint32_t _sbss;
extern uint32_t _ebss;

void Reset_Handler(void);
int main(void);
void NMI_Handler(void)__attribute__((weak,
alias("Default_Handler")));
void HardFault_Handler (void) __attribute__ ((weak, alias("Default_
Handler")));
void MemManage_Handler (void) __attribute__ ((weak, alias("Default_
Handler")));
.
.
.

uint32_t vector_tbl[] __attribute__((section(".isr_vector_tbl"))) = {
    (uint32_t)&_estack,
    (uint32_t)&Reset_Handler,
    (uint32_t)&NMI_Handler,
    (uint32_t)&HardFault_Handler,
    (uint32_t)&MemManage_Handler,
.


.
.
};
void Default_Handler(void) {
    while(1) {
    }
}
void Reset_Handler(void)
{
    // Calculate the sizes of the .data and .bss sections
    uint32_t data_mem_size = (uint32_t)&_edata - (uint32_t)&_sdata;
    uint32_t bss_mem_size = (uint32_t)&_ebss - (uint32_t)&_sbss;

    // Initialize pointers to the source and destination of the .data
    // section
```

```
    uint32_t *p_src_mem =   (uint32_t *)&_etext;
    uint32_t *p_dest_mem = (uint32_t *)&_sdata;

    /*Copy .data section from FLASH to SRAM*/
    for(uint32_t i = 0; i < data_mem_size; i++  )
    {
        *p_dest_mem++ = *p_src_mem++;
    }
    // Initialize the .bss section to zero in SRAM
    p_dest_mem =   (uint32_t *)&_sbss;

    for(uint32_t i = 0; i < bss_mem_size; i++)
    {
        /*Set bss section to zero*/
        *p_dest_mem++ = 0;
    }

        // Call the application's main function.
    main();
}
```

Let's break it down.

We start by declaring external symbols.

External symbol declarations

```
extern uint32_t _estack;
extern uint32_t _etext;
extern uint32_t _sdata;
extern uint32_t _edata;
extern uint32_t _sbss;
extern uint32_t _ebss;
```

These lines declare the external symbols that we defined in the linker script. Each symbol represents an important memory address used during the startup process:

- _estack: This is the initial top of the stack. This value is loaded into the main stack pointer register early in the startup process.

- _etext : This marks the end of the executable code section and the beginning of the data sections stored in flash memory. We use this as a reference point for copying initialized data from FLASH to SRAM.

- `_sdata` and `_edata` represent the start and end addresses of the initialized data section in SRAM, respectively. We use them to determine the size and destination for data copying from FLASH to RAM.

- `_sbss` and `_ebss` mark the start and end of the uninitialized data section (BSS section) in SRAM. We use these symbols to clear this section, setting it to zero.

Next in our snippet, we have the function prototypes and their attributes.

Function prototypes and attributes

```
void Reset_Handler(void);
int main(void);

void NMI_Handler(void)__attribute__((weak,
alias("Default_Handler")));
void HardFault_Handler (void) __attribute__ ((weak, alias("Default_
Handler")));
void MemManage_Handler (void) __attribute__ ((weak, alias("Default_
Handler")));
.
.
.
```

At this part of the startup file, we declare the prototype for the `Reset_Handler` function, the application's `main` function, and several interrupt handlers with specific attributes:

`__attribute__((weak, alias("Default_Handler")))`: This attribute makes each handler weakly linked and aliases it to a function named `Default_Handler`. It allows these handlers to be overridden by explicitly defined handlers with the same name elsewhere in the application.

Let's break down the statement further to understand its significance:

- `__attribute__`:

  We use this keyword to tell the compiler that the declaration it's applied to has certain properties that affect how it's treated by the linker and, potentially, at runtime. Attributes can be used to control optimizations, code generation, alignment, and, relevant to our discussion, linkage characteristics.

- `weak`:

  Declaring a function or variable as `weak` means that it does not prevent the linker from using another symbol of the same name with a stronger linkage. We use this to specify default implementations that can be overridden.

  In the context of our interrupt handlers, marking them as `weak` allows us to define default handlers in our startup file, which application-specific handlers can override without modifying the startup file.

- `alias("Default_Handler"):`

  This part of the attribute creates an alias for another symbol, in this case, `Default_Handler`. It means that the symbol (e.g., `NMI_Handler`) is not just weak, but it is also an alias for the `Default_Handler` function.

  This means that when an interrupt occurs, and a specific handler (such as `NMI_Handler`) has not been defined elsewhere in the application with stronger linkage (non-weak), the program will use `Default_Handler` in its place. This ensures that all interrupts have a handler, preventing the system from crashing due to unhandled events.

Next, we have the vector table array.

Vector table definition

```
uint32_t vector_tbl[] __attribute__((section(".isr_vector_tbl"))) = {
    (uint32_t)&_estack,
    (uint32_t)&Reset_Handler,
    (uint32_t)&NMI_Handler,
    (uint32_t)&HardFault_Handler,
    (uint32_t)&MemManage_Handler,
...

};
```

This array defines the microcontroller's interrupt vector table, placed in the `.isr_vector_tbl` section we defined in the linker script.

We set the `&_estack` symbol as the first element of the vector table to define the initial top of the stack in memory. In ARM Cortex microcontrollers, such as our STM32F411, the first word (32 bits) of the vector table must contain the initial value of the **main stack pointer** (**MSP**). Upon reset, the processor loads this value into the MSP register to set up the stack pointer correctly before executing any code.

Following this, we specify the address of `Reset_Handler`, then we proceed to list the addresses for `NMI_Handler` and other subsequent interrupt handlers in sequence. The precise placement of these handlers is crucial, as each must reside in a specific memory location to ensure correct functionality. This arrangement is detailed on *page 201* of the RM0383 document. Within the fully defined vector table in our `stm32f411_startup.c` file, you'll notice that there are zeros strategically placed among the interrupt handler addresses. These zeros act as placeholders for the positions corresponding to interrupts not supported by our specific microcontroller variant (STM32F411). The ARM Cortex-M core architecture is designed to support a comprehensive set of interrupts, yet not all interrupts are implemented across every microcontroller variant. By inserting zeros for these unsupported interrupts in the vector table, we maintain the required alignment with the architecture's specifications, ensuring the system operates correctly.

Let's take a closer look at the array declaration:

```
uint32_t vector_tbl[] __attribute__((section(".isr_vector_tbl")))={…}
```

- `uint32_t vector_tbl[]`:

   This specifies that each element of the `vector_tbl` array is an unsigned 32-bit integer. We chose this type because addresses in ARM Cortex-M microcontrollers are 32 bits in length, and the vector table consists of memory addresses pointing to the start of ISR handlers.

- `__attribute__((section(".isr_vector_tbl")))`:

   This attribute instructs the linker to place the `vector_tbl` array in a specific section of the output file named `.isr_vector_tbl`.

Next, we have our default handler function.

Default dandler

```
void Default_Handler(void) {
    while(1) {
        // Infinite loop
    }
}
```

This function serves as a universal fallback for any interrupt request for which a specific handler has not been implemented. Engaging in an infinite loop effectively prevents the program from proceeding into an undefined state following such an event.

It is linked to all interrupt handlers marked as `weak` and aliased to `Default_Handler` within the application. This strategy ensures a uniform and secure response throughout the system to any interrupt requests that lack a dedicated handler, thus upholding system stability and integrity.

Finally, we have our `Reset_Handler` function.

Reset handler implementation

```
void Reset_Handler(void) {
    uint32_t data_mem_size =  (uint32_t)&_edata - (uint32_t)&_sdata;
    uint32_t bss_mem_size  =  (uint32_t)&_ebss - (uint32_t)&_sbss;

    uint32_t *p_src_mem =  (uint32_t *)&_etext;
    uint32_t *p_dest_mem = (uint32_t *)&_sdata;

    for(uint32_t i = 0; i < data_mem_size; i++  ) {
        *p_dest_mem++ = *p_src_mem++;
```

```
    }
    p_dest_mem =   (uint32_t *)&_sbss;

    for(uint32_t i = 0; i < bss_mem_size; i++) {
        *p_dest_mem++ = 0;
    }

    main();
}
```

The job of Reset_Handler is to prepare the system before executing the main application.

In the function, we start by calculating the sizes of the .data and .bss sections:

```
uint32_t data_mem_size =  (uint32_t)&_edata - (uint32_t)&_sdata;
uint32_t bss_mem_size  =   (uint32_t)&_ebss - (uint32_t)&_sbss;
```

Here is a breakdown:

- Calculate the size of the .data section by subtracting the address of the start of the section (_sdata) from the address of the end (_edata). This size is used to copy initialized data from FLASH to SRAM.

- Calculate the size of the .bss section in a similar manner, using the start (_sbss) and end (_ebss) addresses. This size is used to zero out the .bss section in SRAM.

  Next, in the function, we initialize pointers for copying the .data section:

```
uint32_t *p_src_mem =   (uint32_t *)&_etext;
uint32_t *p_dest_mem = (uint32_t *)&_sdata;
```

Here is the breakdown:

- Initialize a source pointer (p_src_mem) to the address where initialized data is stored in flash memory, marked by _etext.

- Initialize a destination pointer (p_dest_mem) to the start of the .data section in SRAM (_sdata).

  Then, we copy the .data section from FLASH to SRAM:

```
for(uint32_t i = 0; i < data_mem_size; i++  ) {
    *p_dest_mem++ = *p_src_mem++;
}
```

The breakdown:

- Copy the .data section from FLASH to SRAM word (32-bit) by word. For each iteration, the content pointed to by p_src_mem is copied to the location pointed to by p_dest_mem, and then both pointers are incremented to the next word.

Next, we initialize the pointer for the .bss section zeroing:

```
p_dest_mem =  (uint32_t *)&_sbss;
```

We simply reset the destination pointer (p_dest_mem) to the start of the .bss section in SRAM (_sbss), preparing it for zeroing.

We then zero out the .bss section:

```
for(uint32_t i = 0; i < bss_mem_size; i++) {
    *p_dest_mem++ = 0;
}
```

This block zeroes out the .bss section in SRAM word by word. For each iteration, the location pointed to by p_dest_mem is set to 0, and then p_dest_mem is incremented to the next word.

Finally, we call the main() function located in the main.c file of our source code:

```
main();
```

After initializing the .data and .bss sections, this line calls the main function, transferring control to the main application code. This marks the end of the system initialization process and the beginning of the application execution.

Now that we have completed both our linker script and startup file, it is time to test our implementation by building the firmware using just our main.c source file, the stm32_ls.ld linker script, and the stm32f411_startup.c startup file.

Testing our linker script and startup file

Before diving into the command line, it's important to ensure that our linker script and startup file are correctly placed. Let's set up our project directory and add some modification to our main.c file:

1. **Create a new directory**: Create a new folder named 3_LinkerAndStartup in your workspace.

2. **Transfer required files**:

   - Locate the main.c file from the previous project (2_RegisterManipulation), which includes the foundational application code.

- Additionally, locate `stm32_ls.ld` (linker script) and `stm32f411_startup.c` (startup file).

- Copy and paste these files (`stm32_ls.ld`, `stm32f411_startup.c`, and `main.c`) into the `3_LinkerAndStartup` folder.

3. **Modify the main application code**: To visually confirm the successful execution of our latest firmware, let's adjust the LED blink rate within the `main.c` file from fast to slow:

   - **Open the main.c file for editing**: Right-click on the `main.c` file within the `3_LinkerAndStartup` folder and select the option to open it with a simple text editor, such as Notepad++.

   - **Change the blink rate**: Search for the code segment that controls the toggling of PA5 (`LED_PIN`). Adjust the delay intervals within this section to change the LED's blink rate from its current rapid pace to a slower one. The current one should look like this:

     ```
     //  22: Toggle PA5(LED_PIN)
        GPIOA_OD_R ^= LED_PIN;
     for(int i = 0; i < 100000; i++){}
     ```

   Replace the current code with the following snippet to toggle the state of PA5 at a slower rate:

     ```
     //  22: Toggle PA5(LED_PIN)
        GPIOA_OD_R ^= LED_PIN;
     for(int i = 0; i < 5000000; i++){}
     ```

   We are simply changing the loop iteration from 100,000 to 5,000,000.

   - **Save the changes**: After updating the code, save the `main.c` file.

Now, let's access our new folder through the Command Prompt following the steps we learned in *Chapter 3*. My favorite method for Windows users is the context menu method:

Navigate to the `3_LinkerAndStartup` folder in Windows Explorer. Once there, hold down the *Shift* key, *right-click* in a space within the folder, and select **Open Command Window Here**. This action will open a Command Prompt window directly in the `3_LinkerAndStartup` folder.

In the Command Prompt, we start by compiling the `main.c` file, and we do this by executing the following:

```
arm-none-eabi-gcc -c -mcpu=cortex-m4 -mthumb -std=gnu11 main.c -o
main.o
```

Then, we execute our startup file:

```
arm-none-eabi-gcc -c -mcpu=cortex-m4 -mthumb -std=gnu11 stm32f411_
startup.c -o stm32f411_startup.o
```

Once our `main.o` and `stm32f411_startup.o` object files are ready, we go ahead and link all object files (`*.o`) using our linker script:

```
arm-none-eabi-gcc -nostdlib -T stm32_ls.ld *.o -o 3_LinkerAndStartup.
elf.elf
```

This process produces the `3_LinkerAndStartup.elf` executable.

Next, we launch `openocd` to begin the uploading process:

```
openocd -f board/st_nucleo_f4.cfg
```

With OpenOCD running, the next step involves using the **GNU Debugger** (**GDB**) to upload the firmware to the microcontroller. Let's access another Command Prompt window (as OpenOCD should keep running in the first one) and enter the following command to start the GDB:

```
arm-none-eabi-gdb
```

Once GDB is open, we establish a connection to our microcontroller by running:

```
target remote localhost:3333
```

Let's reset and initialize the board as we learned in *Chapter 3* using the following command:

```
monitor reset init
```

Next, we load the firmware onto the microcontroller using the following command:

```
monitor flash write_image erase 3_LinkerAndStartup.elf
```

After successfully loading the firmware, we reset the board again with the same reset command:

```
monitor reset init
```

Finally, we resume the execution of the firmware on the microcontroller with the following:

```
monitor resume
```

There you have it; you should see the LED blinking at a slower rate, indicating the successful upload and execution of our new firmware.

Summary

In this chapter, we deeply explored the core components of embedded bare-metal programming, focusing on the microcontroller's memory model, writing linker scripts, and startup files. We began by exploring the STM32 microcontroller's memory layout, emphasizing the importance of flash memory and SRAM for storing executable code and runtime data.

We then dedicated a significant portion of the chapter to constructing and understanding linker scripts. Through this, we understood these scripts' critical role in the firmware build process by mapping the compiled firmware sections to the microcontroller's specific memory regions to ensure the executable operates correctly. We learned about the various directives within a linker script, such as MEMORY and SECTIONS. These directives are crucial for defining the memory layout and specifying where and how program sections are placed in memory.

Our discussion on linker scripts extended to the practicalities of defining memory regions, aligning sections, and managing section attributes for optimal memory utilization. We gave special attention to the LMA and VMA, which are essential for efficient program loading and execution.

Transitioning to the startup file, we meticulously outlined the startup file's role, covering the initialization of the vector table, setting up Reset_Handler, and preparing the system for the execution of the main application. We learned the procedures for copying the .data section from FLASH to SRAM and zeroing the .bss section, ensuring a predictable start for our firmware.

In the next chapter, we will explore build systems, highlighting the essential role of the Make tool. This knowledge will enable us to streamline our build process by automating it, instead of manually entering each command in the command line.

Unlock this book's exclusive benefits now

This book comes with additional benefits designed to elevate your learning experience.

Note: Have your purchase invoice ready before you begin.

https://www.packtpub.com/
unlock/9781835460818

5

The "Make" Build System

In this chapter, we will learn how to automate our entire build process using build systems, specifically focusing on the **make** build system – an indispensable tool for automating the compilation and linking processes in software development. We start by defining what a build system is and then exploring its fundamental purpose, which primarily involves automatically transforming source code into deployable software, such as executables or libraries.

Throughout the chapter, we will systematically uncover the components of the make build system, starting with the essential elements of a **Makefile**, including *targets*, *prerequisites*, and *recipes*. In the latter part of the chapter, I will provide a step-by-step guide on writing a Makefile, highlighting the syntax and structure necessary to execute builds effectively.

In this chapter, we're going to cover the following main topics:

- An introduction to build systems
- The Make build system
- Writing Makefiles for firmware projects

By the end of this chapter, you will have a solid understanding of how to leverage the make build system to streamline your development process, improve build times, and reduce manual errors in building and deploying your firmware.

Technical requirements

All the code examples for this chapter can be found on GitHub at https://github.com/ PacktPublishing/Bare-Metal-Embedded-C-Programming.

An introduction to build systems

In the world of software development, build systems are pivotal tools that enable the transformation of source code into executable programs or other usable software formats. These systems automate the process of compiling and linking code, managing dependencies, and ensuring that software builds are reproducible and efficient. Simply put, a build system refers to a set of tools that automate the processes of compiling source code into binary code, linking binaries with libraries, and packaging the results into deployable software units. These systems are designed to handle complex dependency chains by tracking which parts of a software project need recompilation, thereby optimizing the build process. Build systems are responsible for a range of tasks. These include the following:

- **Dependency management**: This involves identifying and resolving interdependencies among various components or libraries that software requires.

- **Code compilation**: Converting source code, whether it's written in C, C++, or another programming language, into machine-readable object code.

- **Linking**: This process integrates the compiled object files and necessary libraries into a unified executable or library file.

- **Packaging**: This step prepares the software for deployment, which might include creating installer packages or compressing software into distributable archives.

- **Testing and validation**: Executing automated tests to confirm that software adheres to the predefined quality benchmarks before its release.

- **Documentation generation**: Build systems can also automate the creation of documentation. This is achieved by integrating with tools such as Doxygen for C/C++, Javadoc for Java, or Sphinx for Python, which extract annotated comments and metadata from source code to produce structured documentation. This automation ensures that documentation stays synchronized with changes in the source code, thereby maintaining consistency and reducing manual errors.

By incorporating these diverse functions, build systems significantly boost the efficiency and reliability of the software development process. Modern software projects often involve complex configurations, including thousands of source files and a wide array of external dependencies. Build systems provide a crucial framework to manage these complexities efficiently. They automate repetitive tasks, minimize the likelihood of human errors, and guarantee consistent builds across various environments. This streamlining not only enhances productivity but also supports the adoption of continuous integration and continuous delivery practices, which are essential for timely and effective software delivery.

Choosing a build system depends on various factors, including the programming language used in a project, the platform compatibility required, and the development team's familiarity with the tool. Some of the commonly used build systems include `make` and `maven`.

Make

Make is one of the oldest and most fundamental build systems available. It is primarily used for C and C++ projects. Make uses Makefiles to specify how to compile and link source files. Its primary advantage lies in its simplicity and broad support across different platforms.

Its key features include the following:

- **Flexibility**: Make allows us to define explicit rules on how files should be compiled and linked.

- **Wide support**: It is available on most Unix-like systems as well as Windows. Also, make can be used with a variety of compilers and programming languages. On Windows, make can be used in environments such as **Minimalist GNU for Windows (MinGW)** or Cygwin.

Next, let's look at maven.

Maven

Maven is primarily used for Java projects. It is designed to provide a comprehensive and standard framework for building projects, handling documentation, reporting, dependencies, **Source Control Management (SCM)** systems, releases, and distribution. The key features include the following:

- **Convention over configuration**: Maven uses a standard directory layout and a default build life cycle to decrease the time spent on configuring projects

- **Dependency management**: It can automatically download libraries and plugins from repositories and incorporate them into the build process

- **Project information management**: Maven can generate project documentation, reports, and other information from a project's metadata

- **Build and release management**: Maven supports the entire build life cycle, from compilation, packaging, and testing to deployment and release management

- **Extensibility**: Maven's plugin-based architecture allows it to be extended with custom plugins to support additional tasks

Other notable build systems include Apache Ant, which is a Java-based build system, and Gradle, which supports multiple programming languages but is especially favored within the Java ecosystem.

Before exploring the specifics of the make build system, it is important to familiarize ourselves with the fundamental components of build systems. These components form the backbone of the build process and include the following:

- **Source code**: The raw, human-readable code written in programming languages such as C, Java, and Python.

- **Compiler**: The tool that converts source code into machine-readable object files. Common compilers include GCC for C/C++ and javac for Java.

- **Linker**: The tool that combines object files into a single executable or library file.

- **Build scripts**: Scripts that describe the build process. They define what commands need to be run and their order.

- **Dependencies**: External code libraries or tools required by a project that need to be integrated during the build process.

- **Artifacts**: The output of build systems, which can include executables, libraries, or other formats needed to deploy or run software.

In the upcoming section, we will explore the fundamentals of the make build system and learn how to write Makefiles that automate the build process for firmware projects.

The Make build system

In this section, we will explore the Make build system, from its basic concepts to practical usage in firmware development.

The basics of Make

The primary component of the make build system is the Makefile, which contains a set of directives used by the tool to generate a *target*. At its core, a Makefile consists of **rules**. Each rule begins with a *target*, followed by *prerequisites*, and then a *recipe*:

- **Target**: This is typically the name of the file that the rule generates – in other words, the output file that needs to be generated, such as main.o or app.exe. The target can also be the name of the action to carry out.

- **Prerequisites**: These are the source files needed to create the target (e.g., main.c and adc.c).

- **Recipes**: The recipe is a series of commands that make executes in order to build the target.

The following diagram illustrates a simple make rule:

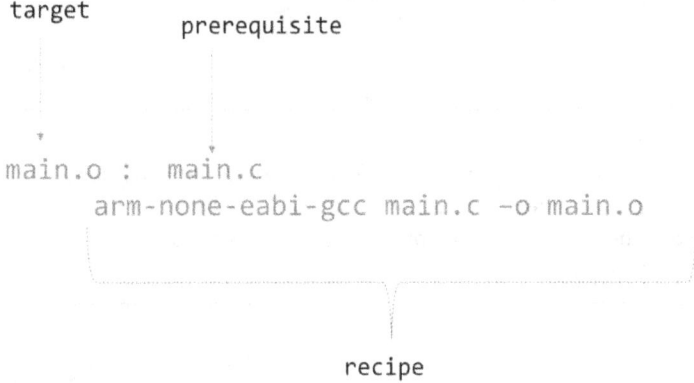

Figure 5.1: A Make rule, with main.o as the target file to be generated from the prerequisite, main.c, using the arm-none-eabi-gcc main.c –o main.o recipe

> **Note**
> The line of the recipe must start with a tab.

Makefiles also allow us to use variables to simplify and manage complex build commands and configurations. For instance, we can define a variable name, CC, to represent the compiler command, as shown in *Figure 5.2*:

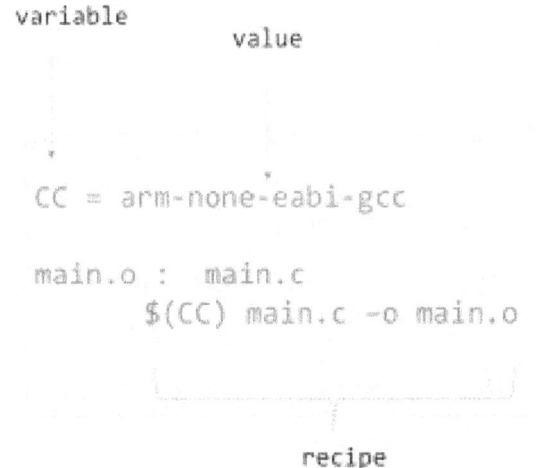

Figure 5.2: A Make rule, with main.o as the target file to be generated from the prerequisite, main.c, where the compiler command in the recipe is replaced by a variable

Variables in Makefiles allow us to store text strings that can be reused throughout a file. The most basic way to define a variable is by simple assignment (=):

```
CC = arm-none-eabi-gcc
```

💡 **Quick tip**: Enhance your coding experience with the **AI Code Explainer** and **Quick Copy** features. Open this book in the next-gen Packt Reader. Click the **Copy** button (**1**) to quickly copy code into your coding environment, or click the **Explain** button (**2**) to get the AI assistant to explain a block of code to you.

Copy   Explain

```
function calculate(a, b) {
  return {sum: a + b};
};
```

1   2

🔒 **The next-gen Packt Reader** is included for free with the purchase of this book. Unlock it by scanning the QR code below or visiting `https://www.packtpub.com/unlock/9781835460818`.

This line sets the CC variable to the `arm-none-eabi-gcc` cross-compiler.

Once defined, variables can be used throughout the Makefile to simplify commands and definitions. To use a variable, enclose its name in $(...) or ${...}:

```
$(CC) main.c -o main.o
```

The recipe uses the $(CC) variable to refer to the compiler set earlier (`arm-none-eabi-gcc`).

Apart from user-defined variables, there are special variables related to targets and prerequisites that come in handy when writing Makefiles.

In make, special variables related to targets help streamline the process of specifying filenames and file paths, making the rules within a Makefile more general and reusable. One of the most commonly used special variables for targets is '$@' - Target Name. This variable represents the name of the target for the rule. It is particularly useful when the target name is repeated multiple times within a rule, which is common in link and compile commands.

Let's see an example:

```
main.o :  main.c
        arm-none-eabi-gcc main.c -o $@
```

In this example, $@ is replaced by main.o, which is the target of the rule.

Make also provides special variables to reference prerequisites. One of the most commonly used is $^. This variable lists all the prerequisites of a target, with spaces between them (if more than one). Let's see an example:

```
main.o :  main.c
        arm-none-eabi-gcc $^ -o main.o
```

When the preceding snippet of the Makefile is executed, $^ is replaced with main.c, effectively running the arm-none-eabi-gcc main.c -o main.o command.

Special variables in Makefiles are very useful in improving efficiency and flexibility when defining build rules. By effectively using these variables, we can create more robust and maintainable build systems.

We will conclude this section with *Figure 5.3*. This figure illustrates the revised rule from *Figure 5.2*, now incorporating our user-defined variable along with the special variables related to targets and prerequisites.

```
CC = arm-none-eabi-gcc

main.o :  main.c
        $(CC) $^ -o $@


                recipe
```

Figure 5.3: A Make rule using a user-defined variable and two special variables

In the next section, I will guide you through the process of setting up a make build system on your development computer.

Installing and configuring Make

In this section, we will go through the process of downloading, installing, and configuring the make build system on a Windows environment. Let's begin:

1. **Download make**: We begin by navigating to the appropriate website to download GNU Make for Windows. For this example, we'll use SourceForge, a popular repository for open source projects. Go to `https://gnuwin32.sourceforge.net/packages/make.htm`.

   Under the **Complete package, except sources** option description, click **Setup** under the **Download** column to start downloading.

2. **Install make**: Once the download is complete, follow the installation wizard to install make on your computer. When you reach the step titled **Select Destination Location**, make sure the path is set to `C:\Program Files (x86)\GnuWin32`.

3. **Set up the environment variable**: To use make from any command line or script, we need to add its executable to our system environment variables, following the same process we used in *Chapter 1* to add OpenOCD to the environment variables.

   We do this by navigating to the `bin` folder where make was installed (`C:\Program Files (x86)\GnuWin32\bin`) and then copying the path.

   Then, we do the following:

   I.   Right-click on **This PC**, and then choose **Properties**.

   II.   Search for and select **Edit the system environment variables**.

   III.   Click the **Environment Variables** button in the **System Properties** pop-up window.

   IV.   Under the **User Variables** section of the **Environment Variables** pop-up, double-click the **Path** entry.

   V.   In the **Edit environment variable** popup, click on **New** to create a row for a new path entry.

   VI.   Paste the previously copied make path into this new row.

   VII.   Confirm your changes by clicking **OK** on the various pop-up windows.

To confirm that the make is properly set up, open command prompt and simply type make, as shown here:

```
make
```

It should return the following:

```
make: *** No targets specified and no makefile found.  Stop.
```

This confirms that the make build system is properly configured on the Windows machine.

On many Linux distributions, make is readily available through the distribution's package manager. For instance, on Ubuntu and other Debian-based distributions, we can install make (along with other build essentials, such as the GCC compiler) by running the following:

```
sudo apt install build-essential
```

On macOS, make is part of the Command Line Tools package that comes with Xcode, Apple's suite of development tools. This means that if you have installed Xcode, you will have make already installed. We can also install the standalone Command Line Tools package by running the following command in the terminal:

```
xcode-select --install
```

In this section, we successfully set up the make build system on our development machine. In the following section, we will apply the concepts covered in this chapter to write our own Makefile.

Writing Makefiles for firmware projects

The focus of this section is to write a Makefile and successfully test it. Let's begin.

In our workspace folder, let's make a new folder named 4_Makefiles. In this folder, create a file called Makefile. This file must start with a capital **M** and should have **no extension**.

If you're using Windows and it asks whether you really want to change the file extension, click **Yes**. Then, right-click the file and open it with a basic text editor, such as Notepad++.

Our objectives with the Makefile can be summarized as follows:

1. **The compilation of source code**: We want to compile source files (main.c and stm32f411_startup.c) into object files (main.o and stm32f411_startup.o).

2. **Linking object files into an executable**: Then, link the compiled object files, along with setting a specific memory layout using the linker script, (stm32_ls.ld) to create a final executable (4_makefile_project.elf).

3. **Loading and cleaning**:

   I.   Invoke OpenOCD to start the process of uploading the final executable onto the target hardware

   II. Conclude by providing a way to clean the build environment by removing all generated files (*.o, *.elf., and *.map), allowing for a fresh start with no leftover artifacts from previous builds

This is our complete Makefile:

```
final : 4_makefile_project.elf

main.o : main.c
    arm-none-eabi-gcc -c -mcpu=cortex-m4 -mthumb -std=gnu11 main.c -o
    main.o

stm32f411_startup.o : stm32f411_startup.c
    arm-none-eabi-gcc -c -mcpu=cortex-m4 -mthumb -std=gnu11 stm32f411_
    startup.c -o stm32f411_startup.o

4_makefile_project.elf : main.o stm32f411_startup.o
    arm-none-eabi-gcc -nostdlib -T stm32_ls.ld *.o -o 4_makefile_
    project.elf -Wl,-Map=4_makefile_project.map

load :
    openocd -f board/st_nucleo_f4.cfg
clean:
del    -f *.o *.elf *.map
```

Let's break it down:

1. **Create the final target**: `final : 4_makefile_project.elf`

   This line deals with the creation of the final target.

   When we execute `make final`, make will check whether the `4_makefile_project.elf` target needs to be updated before executing it. This is a dependency relationship where `final` acts purely as an aggregation point to invoke all the build processes leading up to `4_makefile_project.elf`.

2. **Create an object file for main.c**:

   `main.o : main.c`

   `arm-none-eabi-gcc -c -mcpu=cortex-m4 -mthumb -std=gnu11 main.c -o main.o`

   The make rule compiles the `main.c` source file into an object file named `main.o`. Upon close inspection, you can see that the command employed here is identical to the one we used at the command prompt to manually compile `main.c` in the previous chapter.

3. **Create an object file for stm32f411_startup.c:**

   ```
   stm32f411_startup.o : stm32f411_startup.c
   arm-none-eabi-gcc -c -mcpu=cortex-m4 -mthumb -std=gnu11 stm32f411_
   startup.c -o stm32f411_startup.o
   ```

   This compiles the stm32f411_startup.c source file into an object file, named stm32f411_startup.o.

4. **Link to create the final executable:**

   ```
   4_makefile_project.elf : main.o stm32f411_startup.o
   arm-none-eabi-gcc -nostdlib -T stm32_ls.ld *.o -o 4_makefile_
   project.elf -Wl,-Map=4_makefile_project.map
   ```

   This rule links the main.o and stm32f411_startup.o object files to produce the final executable, 4_makefile_project.elf. Additionally, it generates a map file named 4_makefile_project.map that shows where each part of the code and data is loaded in memory, which is useful for debugging.

5. **Load the target:**

   ```
   load:
   openocd -f board/st_nucleo_f4.cfg
   ```

   This rule initiates OpenOCD to begin the process of loading the final executable onto the target hardware. It executes OpenOCD, using a configuration file tailored for the STM32 Nucleo F4 board, specifically st_nucleo_f4.cfg.

6. **Clean the target:**

   ```
   clean:
   del -f *.o *.elf *.map
   ```

   This command cleans the build directory by removing all generated files, ensuring a clean environment for subsequent builds. The del -f command forcefully deletes files, preventing prompts that ask for deletion confirmation. The *.o, *.elf, and *.map patterns specify that all object files, ELF executables, and map files, respectively, should be deleted.

With the Makefile ready, it is time to test it out.

Testing our Makefile

Before proceeding to the command line, let's ensure that the linker script, startup file, and all source files are placed in the correct directory. Additionally, we will slightly update the main.c file, which will allow us to validate that the most recent version of the firmware executes correctly:

1. **Transfer the required files**:

    I. Locate the main.c file from the previous project (3_LinkerscriptAndStartup), which includes the foundational application code.

    II. Additionally, locate stm32_ls.ld (the linker script) and stm32f411_startup.c (the startup file).

    III. Copy and paste these files (stm32_ls.ld, stm32f411_startup.c, and main.c) into the 4_Makefiles folder.

2. **Modify the main application code**: To visually confirm the successful execution of our latest firmware, let's adjust the LED blink rate within the main.c file from slow to fast:

    I. **Open the main.c file for editing**: Right-click on the main.c file within the 4_Makefiles folder and select the option to open it with a simple text editor, such as Notepad++.

    II. **Change the blink rate**: Search for the code segment that controls the toggling of PA5 (LED_PIN). Adjust the delay intervals within this section to change the LED's blink rate from its current slower pace to a rapid one. The current one should look like this:

    ```
    //  22: Toggle PA5(LED_PIN)
    GPIOA_OD_R ^= LED_PIN;
    for(int i = 0; i < 5000000; i++){}
    ```

    Replace the current code with the following snippet to toggle the state of PA5 to a faster rate:

    ```
    //  22: Toggle PA5(LED_PIN)
    GPIOA_OD_R ^= LED_PIN;
    for(int i = 0; i < 100000; i++){}
    ```

    We simply change the loop iteration from 5 million to 100,000.

    III. **Save the changes**: After updating the code, save the main.c file.

Now, let's access our new folder through the command prompt, following the steps we used in the previous chapter.

In the command prompt, simply execute the following:

```
make final
```

This will create our final executable, 4_makefile_project.elf.

Next, we will begin uploading the final executable onto our microcontroller by executing the following:

```
make load
```

This will launch OpenOCD. The next step involves using the **GNU Debugger** (**GDB**) to upload the firmware to the microcontroller, as we did in the previous chapter. Let's access another command prompt window (as OpenOCD should keep running in the first one) and enter the following command to start the GDB:

```
arm-none-eabi-gdb
```

Once GDB is open, we establish a connection to our microcontroller by running the following:

```
target remote localhost:3333
```

Let's reset and initialize the board, as we learned in *Chapter 3*, using the following command:

```
monitor reset init
```

Next, we load the firmware onto the microcontroller using the following command:

```
monitor flash write_image erase 4_makefile_project.elf
```

After successfully loading the firmware, we reset the board again with the same reset command:

```
monitor reset init
```

Finally, we resume the execution of the firmware on the microcontroller with the following:

```
monitor resume
```

You should see the LED blinking at a rapid rate, indicating the successful upload and execution of our new firmware.

We can stop the GDB by executing the following:

```
quit
```

And then, we execute the following when asked if we want to quit anyway:

```
y
```

To clean our build directory, we will open the command prompt in the build directory and execute the following command:

```
make clean
```

This will delete all the .o, .elf, and .map files in the build directory.

Before concluding this chapter, let's explore how our Makefile appears when we incorporate special and user-defined variables.

Applying special and user-defined variables

Let's apply special variables and user-defined variables to our makefile:

```
CC = arm-none-eabi-gcc
CFLAGS = -c -mcpu=cortex-m4 -mthumb -std=gnu11
LDFLAGS = -nostdlib -T stm32_ls.ld -Wl,-Map= 5_makefile_project_v2.map


final : 5_makefile_project_v2.elf

main.o : main.c
    $(CC) $(CFLAGS) $^ -o $@

stm32f411_startup.o : stm32f411_startup.c
    $(CC) $(CFLAGS) $^ -o $@

5_makefile_project_v2.elf : main.o stm32f411_startup.o
    $(CC) $(LDFLAGS) $^ -o $@

load :
    openocd -f board/st_nucleo_f4.cfg
clean:
    del    -f *.o *.elf *.map
```

In this version, we've defined three essential variables to streamline our Makefile:

- CC: This variable represents the compiler used to compile the source files. It simplifies the Makefile by centralizing the compiler definition, making it easier to update or change if needed.

- CFLAGS: This holds the compilation flags necessary to build the source files.

- LDFLAGS: This contains the linker flags that dictate how an executable is linked from the object files.

Additionally, we've used special variables for target names ($@) and prerequisite lists ($^) to replace explicit mentions of these components in the make recipes, further simplifying the Makefile structure.

Update your current makefile to this new version and upload the new executable, named 5_makefile_project_v2.elf, onto your microcontroller. This updated version should function seamlessly, just like the previous one.

Summary

In this chapter, we embarked on an exploration of the `make` build system, a cornerstone tool for automating the build process in software development. The journey began with an introduction to what build systems are and their critical role in converting source code into deployable software, such as executables and libraries.

We then delved into the specific mechanics of the `make` build system, starting with the foundational elements of a Makefile, which include *targets*, *prerequisites*, and *recipes*. These components were thoroughly discussed to provide a clear understanding of how they interact within `make` to manage and streamline the compilation and linking of software projects.

This chapter wrapped up with a practical demonstration of writing Makefiles, effectively consolidating the theoretical concepts discussed throughout. This hands-on experience ensures that you are well-equipped to apply these strategies to your own firmware projects.

In the next chapter, we will transition to another critical aspect of firmware development – the development of peripheral drivers, beginning with **General Purpose Input/Output (GPI/O)** drivers. This shift will introduce you to the fundamentals of interfacing with hardware components, a pivotal skill in embedded systems development.

Unlock this book's exclusive benefits now

This book comes with additional benefits designed to elevate your learning experience.

Note: Have your purchase invoice ready before you begin.

```
https://www.packtpub.com/
unlock/9781835460818
```

6

The Common Microcontroller Software Interface Standard (CMSIS)

In this chapter, we will delve into the **Common Microcontroller Software Interface Standard** (CMSIS), a critical framework for Cortex-M and some Cortex-A processors. We will begin by learning how to define hardware registers using C structures. This foundational knowledge will enable us to read and understand CMSIS-compliant header files provided by microcontroller manufacturers.

Next, we will explore CMSIS itself, discussing its components and how it facilitates efficient software development. Finally, we will set up the necessary header files from our silicon manufacturer, demonstrating how CMSIS compliance can streamline production and improve code portability.

In this chapter, we're going to cover the following main topics:

- Defining peripheral registers with C structures
- Understanding CMSIS
- Setting up the required CMSIS files

By the end of this chapter, you will be equipped with a solid understanding of CMSIS and how to use it to enhance code portability in your Arm Cortex-M projects.

Technical requirements

All the code examples for this chapter can be found on GitHub at https://github.com/PacktPublishing/Bare-Metal-Embedded-C-Programming.

Defining peripheral registers with C structures

In embedded systems development, defining hardware registers using C structures is a fundamental technique that enhances code readability and maintainability. In this section, we will explore how to use C structures to represent peripherals and their registers, drawing on practical examples and analogies to simplify the concept.

In previous chapters, we configured a **General Purpose Input/Output** (**GPIO**) pin (PA5) to turn on an LED by manually defining the address of each required register. We learned how to find the correct addresses from documentation, define registers, and define register bits. This method, while effective, can become cumbersome as projects grow in complexity.

To streamline this process, we can use C structures to represent peripherals and their registers. This approach groups related registers into a cohesive unit to match the hardware architecture and memory map of our microcontroller, making the code more intuitive.

Let's create a structure to represent the GPIO peripherals and their associated registers.

To achieve this, we need to get the base address of each GPIO port and the offset of each register within these ports. Here, offset refers to the register's address relative to the peripheral's base address.

Before diving into the details of creating the structure, it's important to understand how to obtain the necessary base addresses and offsets.

Getting the base address and offsets of registers

In *Chapter 2*, we learned how to locate the base addresses of peripherals in our datasheet. Specifically, we examined pages 54 to 56 of the STM32F411 datasheet, which lists the base addresses for the microcontroller's peripherals. Here are the extracted base addresses for the GPIO and **Reset and Clock Control** (**RCC**) peripherals:

Peripheral	Base Address
GPIOA	0x4002 0000
GPIOB	0x4002 0400
GPIOC	0x4002 0800
GPIOD	0x4002 0C00
GPIOE	0x4002 1000
GPIOH	0x4002 1C00
RCC	0x4002 3800

Table 6.1: Base addresses of GPIO and RCC

Also, in *Chapter 2*, we covered how to extract the register offsets from the reference manual (**RM383**). Here are the extracted offsets for all the GPIO registers:

Register	Offset
GPIOx_MODER	0x00
GPIOx_OTYPER	0x04
GPIOx_OSPEEDR	0x08
GPIOx_PUPDR	0x0C
GPIOx_IDR	0x10
GPIOx_ODR	0x14
GPIOx_BSRR	0x18
GPIOx_LCKR	0x1C
GPIOx_AFRL	0x20
GPIOx_AFRH	0x24

Table 6.2: Offsets of GPIO register

Table 6.1 shows all the GPIO registers of our STM32F411 microcontroller along with their offsets, arranged in the same order they appear in memory. Almost all registers in our microcontroller are 32 bits (4 bytes) in size. As illustrated in *Table 6.1*, each register is offset by 4 bytes from the previous one. For instance, the GPIOx_OTYPER register at 0x04 is 4 bytes from the GPIOx_MODER register at 0x00 (0x04 - 0x00 = 4). Similarly, the GPIOx_PUPDR register at 0x0C is 4 bytes from the GPIOx_OSPEEDR register at 0x08.

This tells us that the registers are contiguously arranged in that memory region, since we know that each register is 4 bytes in size.

However, this contiguous arrangement is not always the case. There are instances where gaps of a few bytes are left between registers within a peripheral.

Now, what's the relationship between the offset and the base address?

Imagine your microcontroller as a large apartment complex. Each apartment represents a peripheral, such as GPIO or RCC, and the entrance to each apartment is the peripheral's base address. Inside each apartment, there are several rooms, which represent the registers. Each room has a specific purpose and is located at a certain distance from the entrance, known as the **offset**.

For example, when you enter the GPIO apartment (peripheral), the living room might be the GPIOx_MODER register located right at the entrance (offset 0x00). The kitchen could be the GPIOx_OTYPER register, located a bit further down the hallway (offset 0x04). The bedroom might be the GPIOx_OSPEEDR register, located even further down the hallway (offset 0x08), and so on.

This arrangement shows that each room (register) is placed at a fixed distance (offset) from the entrance (base address). In our case, since each room is 4 bytes in size, every subsequent room is 4 bytes away from the previous one. However, in some apartments (peripherals), there might be extra space between rooms, indicating non-contiguous placement of registers. This is similar to having a small hallway between rooms, which you'll notice when you examine peripherals such as the RCC peripheral in the reference manual.

Now, let's go ahead with implementing the peripheral structures using what we have learned so far.

Implementing the peripheral structures

The following is our GPIO_TypeDef structure, representing the GPIO peripherals:

```
typedef struct
{
    volatile uint32_t MODER;    /*offset: 0x00    */
    volatile uint32_t OTYPER;   /*offset: 0x04    */
    volatile uint32_t OSPEEDR;  /*offset: 0x08    */
    volatile uint32_t PUPDR;    /*offset: 0x0C    */
    volatile uint32_t IDR;      /*offset: 0x10    */
    volatile uint32_t ODR;      /*offset: 0x14    */
    volatile uint32_t BSRR;     /*offset: 0x18    */
    volatile uint32_t LCKR;     /*offset: 0x1C    */
    volatile uint32_t AFRL;     /*offset: 0x20  */
    volatile uint32_t AFRH;     /*offset: 0x24    */

} GPIO_TypeDef;
```

Let's break down the syntax:

The line typedef struct begins the definition of a new structure type. typedef is used to create an alias for the structure, allowing us to use GPIO_TypeDef as a type name later in the code.

Each member of the structure is declared as volatile uint32_t. Here's the breakdown:

- volatile: This keyword indicates that the value of the variable can change at any time, often due to hardware changes. The compiler should not optimize accesses to this variable.

- `uint32_t`: This indicates that each member of the structure is a 32-bit (4-byte) unsigned integer. This is important because the registers we are working with are also 32 bits in size. To ensure that the structure members accurately represent these registers, they must match this size. This alignment guarantees that each member corresponds correctly to its respective register in the memory map.

Also note that the structure members are arranged in the same order and have the same size as the registers, as specified in the reference manual.

Now, as we discussed in *Chapter 2*, to use any peripheral in the microcontroller, we first need to enable clock access to that peripheral. This is done through the RCC peripheral. Let's create the structure for the RCC peripheral.

This is our RCC structure:

```
typedef struct
{
    volatile uint32_t DUMMY[12];
    volatile uint32_t AHB1ENR;          /*offset: 0x30*/

} RCC_TypeDef;
```

The RCC peripheral of the STM32F411 microcontroller has about 24 registers, which are not contiguous, leaving gaps in the memory region. The register we are interested in for the purposes of GPIO peripherals is the `AHB1ENR` register, which has an offset of 0x30.

In our `RCC_TypeDef`, we have added padding to the structure with the number of `uint32_t` (4 bytes) items required to reach the offset 0x30. In this case, it is 12 items. This is *because 4 bytes multiplied by 12 equals 48 bytes*, which corresponds to *0x30 in hexadecimal notation*.

At this point, we have defined two important structures (`GPIO_TypeDef` and `RCC_TypeDef`) required to configure and control our GPIO pins. The next step involves creating pointers to the base addresses of the GPIO and RCC peripherals using these structures. This allows us to access and manipulate the peripheral registers in a structured and readable manner. Here is the code snippet that accomplishes this:

```
#define     RCC_BASE       0x40023800
#define     GPIOA_BASE     0x40020000
#define     RCC            ((RCC_TypeDef*) RCC_BASE)
#define     GPIOA          ((GPIO_TypeDef*)GPIOA_BASE)
```

Let's break it down:

- `#define RCC_BASE 0x40023800`

    This line defines the base address of the RCC peripheral. The address value is taken from *Table 6.1*.

- `#define GPIOA_BASE 0x40020000`

    This line defines the base address of the GPIOA peripheral.

- `#define RCC ((RCC_TypeDef*) RCC_BASE)`

    This line defines a macro, RCC, that casts the RCC_BASE base address to a pointer of type RCC_TypeDef*.

    By doing this, RCC becomes a pointer to the RCC peripheral, allowing us to access its registers through the RCC_TypeDef structure.

- `#define GPIOA ((GPIO_TypeDef*) GPIOA_BASE)`

    Similarly, this line defines a macro, GPIOA, that casts the GPIOA_BASE base address to a pointer of type GPIO_TypeDef*.

    This makes GPIOA a pointer to the GPIOA peripheral, enabling access to its registers via the GPIO_TypeDef structure.

With this accomplished, we are now ready to test out our implementation. Let's do that in the next section.

Evaluating the structure-based register access method

Let's update our previous project to use the structure-based register access approach:

```c
// 0: Include standard integer types header for fixed-width //integer
types
#include <stdint.h>

// 1: GPIO_TypeDef structure definition
typedef struct
{
  volatile uint32_t MODER;    /*offset: 0x00    */
  volatile uint32_t OTYPER;   /*offset: 0x04    */
  volatile uint32_t OSPEEDR;  /*offset: 0x08    */
  volatile uint32_t PUPDR;    /*offset: 0x0C    */
  volatile uint32_t IDR;      /*offset: 0x10    */
  volatile uint32_t ODR;      /*offset: 0x14    */
  volatile uint32_t BSRR;     /*offset: 0x18    */
  volatile uint32_t LCKR;     /*offset: 0x1C    */
  volatile uint32_t AFRL;     /*offset: 0x20    */
```

```
    volatile uint32_t AFRH;        /*offset: 0x24        */

} GPIO_TypeDef;

// 2: RCC_TypeDef structure definition
typedef struct
{
    volatile uint32_t DUMMY[12];
    volatile uint32_t AHB1ENR;         /*offset: 0x30*/

} RCC_TypeDef;


// 3: Base address definitions
#define    RCC_BASE      0x40023800
#define    GPIOA_BASE    0x40020000

// 4: Peripheral pointer definitions
#define RCC            ((RCC_TypeDef*) RCC_BASE)
#define GPIOA          ((GPIO_TypeDef*)GPIOA_BASE)

//5: Bit mask for enabling GPIOA (bit 0)
#define GPIOAEN        (1U<<0)
//6: Bit mask for GPIOA pin 5
#define PIN5           (1U<<5)
//7: Alias for PIN5 representing LED pin
#define LED_PIN        PIN5

//  8: Start of main function
int main(void)
{
    //  9: Enable clock access to GPIOA
    RCC->AHB1ENR |=  GPIOAEN;

    GPIOA->MODER |= (1U<<10);  //  10: Set bit 10 to 1
    GPIOA->MODER &= ~(1U<<11); //  11: Set bit 11 to 0

    //  21: Start of infinite loop
    while(1)
    {
        //  12: Set PA5(LED_PIN) high
        GPIOA->ODR^= LED_PIN;
```

```
            // 13: Simple delay
                for(int i=0;i<100000;i++){}


    }

  }
```

In this new implementation, we access the required registers using the C structure pointer operator (`->`). Here's a breakdown of `RCC->AHB1ENR`:

- `RCC`: This is the pointer to a structure of type `RCC_TypeDef`. This pointer allows us to access the RCC registers using the structure's members.

- `->`: This is the structure pointer operator in C. It is used to access a member of a structure through a pointer.

- `AHB1ENR`: This is a member of the `RCC_TypeDef` structure.

Similarly, we use the same approach to access GPIOA registers.

Here's a breakdown of `GPIOA->MODER` and `GPIOA->ODR`:

- `GPIOA`: This is a pointer to the structure of type `GPIO_TypeDef`, allowing access to GPIOA registers

- `MODER`: A member of the `GPIO_TypeDef` structure, representing the GPIO port mode register

- `ODR`: Another member of the `GPIO_TypeDef` structure, representing the GPIO port output data register

Build the project and run it on your development board. It should work the same way as the previous project.

In this section, we learned how to define hardware registers using C structures. This technique is an important step toward understanding CMSIS-compliant code, which will be covered in the next sections.

Understanding CMSIS

In this section, we will explore CMSIS, an important framework designed for Arm Cortex-M and some Cortex-A processors. **CMSIS** is a vendor-independent hardware abstraction layer that standardizes software interfaces across various Arm Cortex-based microcontroller platforms, promoting software portability and reusability. Let's start by understanding CMSIS and its key benefits.

What is CMSIS?

CMSIS, pronounced *See-M-Sys*, is a standard developed by Arm to provide a consistent and efficient way to interface with Cortex-based microcontrollers. It includes a comprehensive set of APIs, software components, tools, and workflows designed to simplify and streamline the development process for embedded systems.

Its key benefits include the following:

- **Standardization**: CMSIS standardizes the interface for all Cortex-based microcontrollers, making it easier for you to switch between different microcontrollers and tools without having to relearn or reconfigure your code base

- **Portability**: By providing a consistent API, CMSIS allows software developed for one microcontroller to be easily ported to another, enhancing code reuse and reducing development time

- **Efficiency**: CMSIS includes optimized libraries and functions, such as **Digital Signal Processing** (**DSP**) and neural network kernels, which improve performance and reduce the memory footprint

CMSIS has several components. Let's explore some of the commonly used ones.

Key components of CMSIS

CMSIS comprises several components, each serving a unique purpose:

- **CMSIS-Core (M)**: This component is designed for all Cortex-M and SecurCore processors. It provides standardized APIs for configuring the Cortex-M processor core and peripherals. It also standardizes the naming of device peripheral registers, which helps reduce the learning curve when switching between different microcontrollers.

- **CMSIS-Driver**: This provides generic peripheral driver interfaces for middleware, facilitating the connection between microcontroller peripherals and middleware such as communication stacks, filesystems, or graphical user interfaces.

- **CMSIS-DSP**: This component offers a library of over 60 functions for various data types, optimized for the **Single Instruction Multiple Data** (**SIMD**) instruction sets available on Cortex-M4, M7, M33, and M35P processors.

- **CMSIS-NN**: This stands for **neural network**. It includes a collection of efficient neural network kernels optimized to maximize the performance and minimize the memory footprint on Cortex-M processor cores.

- **CMSIS-RTOS**: There are two versions, RTOS v1 and RTOS v2. RTOS v1 supports Cortex-M0, M0+, M3, M4, and M7 processors, providing a common API for real-time operating systems. RTOS v2 extends support to Cortex-A5, A7, and A9 processors, including features such as dynamic object creation and multi-core system support.

- **CMSIS-Pack**: Pack outlines a delivery system for software components, device parameters, and evaluation board support. It streamlines software reuse and facilitates effective product life cycle management.

- **CMSIS-SVD**: **SVD** stands for **System Viewer Description**. It defines device description files maintained by the silicon vendor, containing comprehensive descriptions of the microcontroller peripherals and registers in XML format. Development tools import these files to automatically construct peripheral debug windows.

Another crucial aspect of CMSIS is its coding standard. Let's take a closer look to familiarize ourselves with these guidelines.

The CMSIS coding rules

Here are the essential coding rules and conventions used in CMSIS:

- **Compliance with ANSI C (C99) and C++ (C++03)**: This ensures compatibility with widely accepted programming standards.

- **Standard data types**: Uses ANSI C standard data types defined in `<stdint.h>`, ensuring consistency in data representation.

- **Complete data types**: Variables and parameters are defined with complete data types, avoiding ambiguity.

- **MISRA 2012 conformance**: While CMSIS conforms to the MISRA 2012 guidelines, it does not claim full MISRA compliance. Any rule violations are documented for transparency.

Additionally, CMSIS uses specific qualifiers:

- `__I` for read-only variables (equivalent to **volatile const** in ANSI C)

- `__O` for write-only variables

- `__IO` for read and write variables (equivalent to **volatile** in ANSI C)

These qualifiers in CMSIS provide a convenient way to specify the intended access mode for variables, particularly for memory-mapped peripheral registers.

With this background knowledge, let's go ahead and learn how to use CMSIS in our embedded projects.

The CMSIS-Core files

The CMSIS-Core files are categorized into two primary groups, each serving a specific purpose in the development process. These groups are the CMSIS-Core standard files and the CMSIS-Core device files. *Figure 6.1* provides a comprehensive overview of the CMSIS-Core file structure, illustrating the different types of files and their roles in a project.

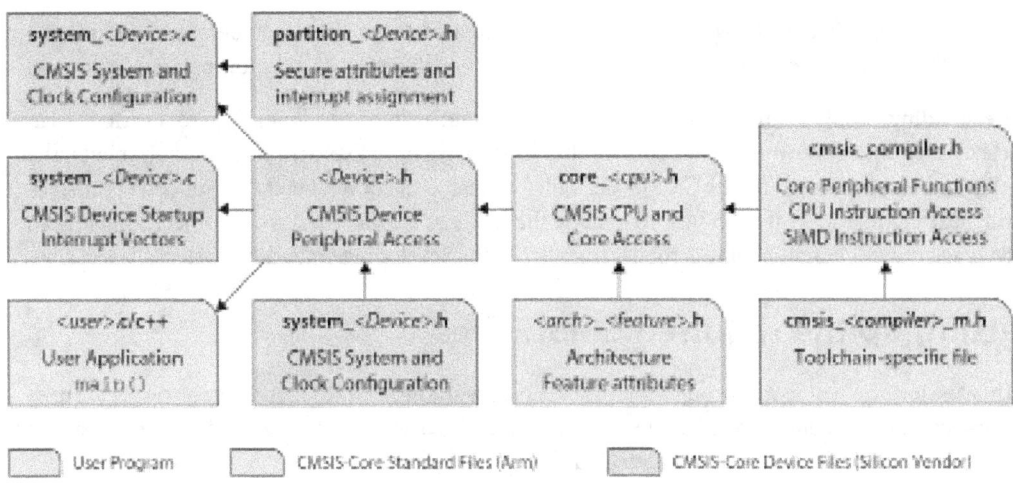

Figure 6.1: The CMSIS-Core files

Let's analyze the diagram.

The files are divided into three categories: **CMSIS-Core Standard Files**, **CMSIS-Core Device Files**, and **User Program Files**.

The **CMSIS-Core Standard Files** category of files is provided by Arm and typically does not require modifications. They include the following:

- `core_<cpu>.h`: This provides access to the CPU and core-specific functionalities
- `cmsis_compiler.h`: This contains core peripheral functions, CPU instruction access, and SIMD instruction access
- `<arch>_<feature>.h`: This defines architecture-specific attributes and features
- `cmsis_<compiler>_m.h`: This is a toolchain-specific file that aids in compiler compatibility and optimization

The next category is **CMSIS-Core Device Files**. These files are provided by silicon vendors (such as STMicroelectronics) and may require application-specific modifications. They include the following:

- `system_<Device>.c`: This file handles system and clock configuration
- `partition_<Device>.h`: This manages secure attributes and interrupt assignments
- `startup_<Device>.c`: This contains the device startup interrupt vectors
- `<Device>.h`: This provides access to the CMSIS device peripheral functionalities
- `system_<Device>.h`: This assists in system and clock configuration

The third category is **User Program Files**. These are created by us, the developers, and include the main application code along with other user-defined functionalities essential for the project's operation.

Understanding CMSIS is fundamental for efficient and standardized development across various Arm Cortex-based microcontroller platforms. In this section, we have explored its key components, the CMSIS coding standards, and the CMSIS-Core files crucial to the development of embedded projects. In the next section, we will learn how to include the required CMSIS files in our project, enabling us to leverage the full potential of this robust framework for developing embedded systems.

Setting up the required CMSIS files

In this section, we will work through the process of integrating CMSIS files into our project. These files also contain the definitions of all the registers and their respective bits, making it easier to manage and configure peripherals without manually defining each register.

Getting the right header files

Let's start by downloading the package for our microcontroller from the STMicroelectronics website:

1. Open your browser and go to `https://www.st.com/content/st_com/en.html`.
2. Search for `STM32CubeF4` to locate the package for the STM32F4 microcontroller series.
3. Next, download the STM32CubeF4 package:

    I.   Locate the STM32CubeF4 package and download the latest version. Make sure not to download the patch version.

    II. Once the download is complete, unzip the package. You will find several subfolders inside, including the `Drivers` folder.

Our next steps involve organizing the files:

1. Inside your project workspace, create a new folder named `chip_headers`.
2. Within the `chip_headers` folder, create another folder named `CMSIS`.
3. Navigate to `Drivers/CMSIS` in the extracted package.
4. Copy the entire `Include` folder from `Drivers/CMSIS` into `chip_headers/CMSIS`.
5. Next, copy the `Device` folder from `Drivers/CMSIS` into `chip_headers/CMSIS`.

Finally, we clean up the `Device` folder:

1. Navigate to `chip_headers/CMSIS/Device/ST/STM32F4xx`.

2. Delete all files and folders, *except the Include folder*. This ensures you keep only the necessary header files for your specific microcontroller.

Moving forward, we will consistently include the `chip_headers` folder in our project directories. This practice ensures that anyone running our projects on a different computer won't encounter errors due to missing header files.

In the next section, we'll create a new project in STM32CubeIDE. After setting up the project, we'll copy the `chip_headers` folder and paste it into the project directory. Subsequently, we'll add the subfolders within `chip_headers` to our project's `include` paths, ensuring seamless integration of the necessary CMSIS files.

Working with CMSIS files

In this section, we will integrate the CMSIS files into our project by adding the relevant folders to our project's `include` paths. We will then test our setup by updating our `main.c` file to use these CMSIS files instead of our manually defined peripheral structures.

Let's begin by creating a new project following the steps outlined in *Chapter 1*. I will name my project `header_files`. Once the project is created, I'll copy and paste the `chip_headers` folder into the project folder.

Our next task involves adding the paths of the subfolders within the `chip_headers` folder to our project's `include` paths:

1. Open STM32CubeIDE, right-click on your project, and select **Properties**.

2. Once the **Properties** window opens, expand the **C/C++ General** options.

3. Select **Paths and Symbols**.

4. Under the **Includes** tab, click **Add** to add a new directory.

5. Enter `${ProjDirPath}\chip_headers\CMSIS\Include` to add the `Include` folder located in our `CMSIS` folder, and then click **OK** to save.

6. Click **Add** again to add another directory.

7. Enter `${ProjDirPath}\chip_headers\CMSIS\Device\ST\STM32F4xx\Include` to add the `Include` folder located in the `STM32F4xx` subdirectory and click **OK** to save.

Figure 6.2 illustrates the Project Properties window after adding these directories to the project's `include` paths.

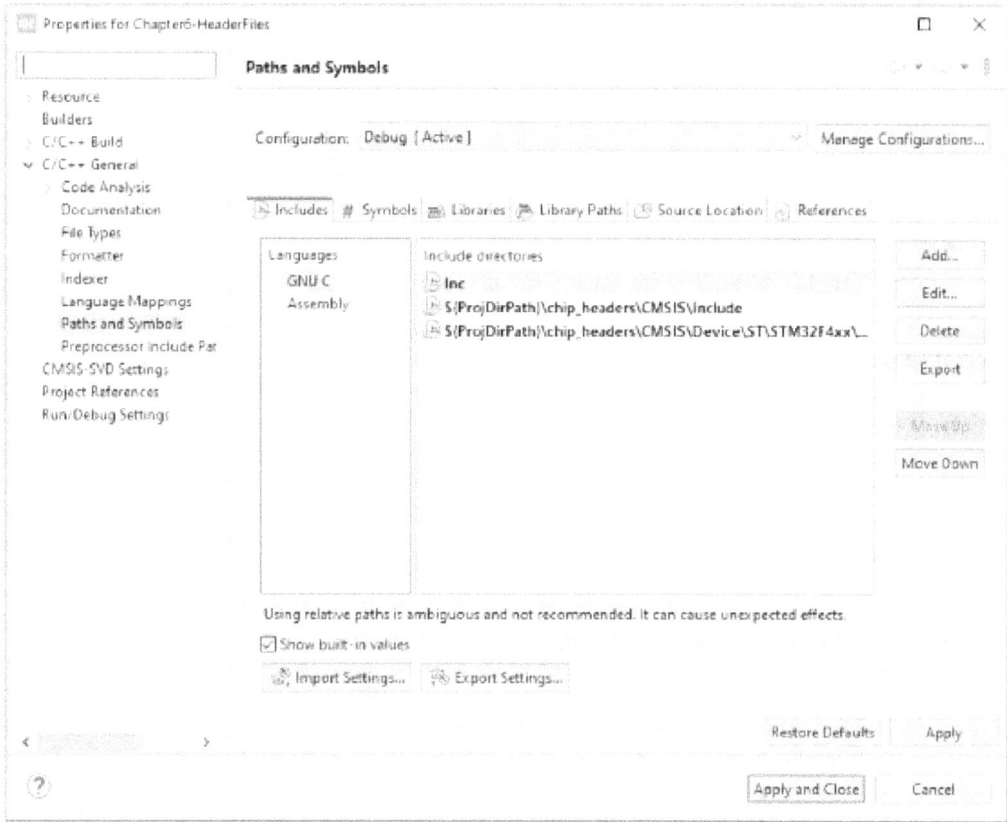

Figure 6.2: The Includes tab in the project properties window

Let's break down the two lines we have just added:

- `${ProjDirPath}\chip_headers\CMSIS\Include`:

  - `${ProjDirPath}`: This is a macro in STM32CubeIDE that represents the root directory of your current project. It's a placeholder that dynamically points to the directory where your project is located.

- `\chip_headers\CMSIS\Include`: This specifies the path relative to the project's root directory. It points to the `Include` folder within the `CMSIS` directory inside the `chip_headers` folder. This folder contains general CMSIS include files, which provide core functionalities and definitions for the Cortex-M processor.

- `${ProjDirPath}\chip_headers\CMSIS\Device\ST\STM32F4xx\Include`:

  - `${ProjDirPath}`: As mentioned earlier, this macro represents the root directory of your current project.

  - `\chip_headers\CMSIS\Device\ST\STM32F4xx\Include`: This specifies another path relative to the project's root directory. It points to the `Include` folder within the `STM32F4xx` subdirectory inside the `Device` directory, which is located in the `CMSIS` directory in the `chip_headers` folder. This folder contains device-specific include files for the STM32F4xx series microcontrollers, providing definitions and configurations specific to this family of microcontrollers.

Before closing the project properties dialog box, we need to specify the exact version of the STM32F4 microcontroller we are using. This ensures that the appropriate header file for our specific microcontroller is enabled in our project. As we can see, the `STM32F4xx\Include` subfolder contains header files for various microcontrollers within the STM32F4 family. The NUCLEO-F411 development board has the STM32F411 microcontroller.

To configure our project for the STM32F411 microcontroller:

1. Click on the **# Symbols** tab.
2. Click **Add…** to add a new symbol.
3. In the **Add Symbol** dialog box, enter `STM32F411xE` in the **Name** field and then click **OK**.
4. Click **Apply and Close** to save all changes and close the Project Properties window.

Figure 6.3 illustrates the **Symbols** tab of the project properties window after adding the STM32F411xE symbol.

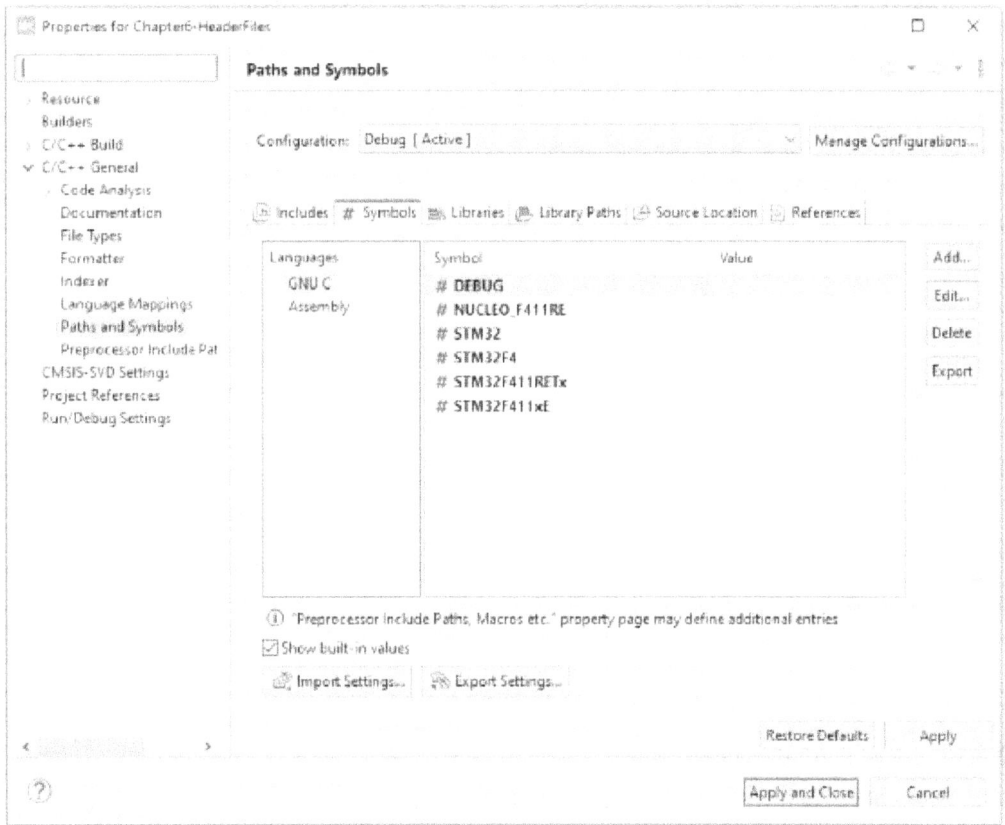

Figure 6.3: The Symbols tab in the project properties window

To test our setup, let's follow these steps:

1. Copy the entire content of the main.c file from our previous project that used the structure-based access approach.

2. Open the main.c file in the current project.

3. Delete all the existing content in the main.c file.

4. Paste the copied content into the main.c file.

5. In the main.c file, delete all the code related to manually defined addresses and structures, as we will now use the register definitions provided in our header files.

6. Include the stm32f4xx.h header file in your project to access these definitions.

7. Build the project in the IDE and run it on the development board.

The following snippet shows the updated main.c file:

```c
//1: Include the stm32f4 header file
#include "stm32f4xx.h"
//2: Bit mask for enabling GPIOA (bit 0)
#define GPIOAEN        (1U<<0)
//3: Bit mask for GPIOA pin 5
#define PIN5           (1U<<5)
//4: Alias for PIN5 representing LED pin
#define LED_PIN        PIN5
int main(void)
{
    //  5: Enable clock access to GPIOA
    RCC->AHB1ENR |=  GPIOAEN;
    //  6: Set PA5 to output mode
    GPIOA->MODER |= (1U<<10);
    GPIOA->MODER &= ~(1U<<11);
    while(1)
    {
        //  7: Set PA5(LED_PIN) high
        GPIOA->ODR^= LED_PIN;
        // 8: Simple delay
        for(int i=0;i<100000;i++){}
    }
}
```

You'll observe that this project builds without any errors and works in the same way as our previous one. This implementation accesses the microcontroller registers defined in chip_headers/CMSIS/ Device/ST/STM32F4xx/Include/stm32f411xe.h.

Upon inspecting this file, you will find that it contains a typedef structure for each peripheral of our microcontroller, similar to the one we manually created a few sections ago. This means that moving forward, we don't need to manually extract the base addresses and register offsets from the documentation. Instead, we can simply include the stm32f4xx.h header file in our project. This header file will in turn include the stm32f411xe.h file because we specified in the **Symbols** tab of the project properties window that we are using the *STM32F411xE* microcontroller.

Setting up the required header files significantly simplifies configuring and using peripherals on our STM32 microcontroller. This approach allows us to leverage pre-defined register addresses and bit definitions, making our code more readable and maintainable while also reducing development time.

Summary

In this chapter, we explored CMSIS, a critical framework for Cortex-M and some Cortex-A processors. This chapter gave us the foundational knowledge to enhance code portability and efficiency in our Arm Cortex-M projects.

We began by learning how to define hardware registers using C structures, a fundamental technique for improving code readability and maintainability. This knowledge allowed us to understand how the CMSIS-compliant header files provided by microcontroller manufacturers give us access to the register definitions.

Next, we explored CMSIS itself, discussing its components and how it facilitates efficient software development. We examined the key benefits of CMSIS, such as standardization, portability, and efficiency, and introduced its main components, including CMSIS-Core, CMSIS-Driver, CMSIS-DSP, CMSIS-NN, CMSIS-RTOS, CMSIS-Pack, and CMSIS-SVD.

We then moved on to setting up the necessary CMSIS files from our silicon manufacturer. This process involved downloading the relevant packages, organizing the files, and integrating them into our project.

Finally, we tested our setup by updating our previous project to use CMSIS files instead of our manually defined peripheral structures. This practical application showcased how CMSIS simplifies accessing microcontroller registers, making the code more readable, maintainable, and efficient.

In the next chapter, we will learn about the GPIO peripheral. This chapter will provide a comprehensive understanding of how to configure and use GPIO for input/output applications in embedded systems.

7

The General-Purpose Input/ Output (GPIO) Peripheral

In this chapter, we will explore the **General-Purpose Input/Output** (**GPIO**) peripheral, an essential component in microcontrollers. This peripheral is crucial for interfacing with microcontrollers, making it fundamental to embedded systems development.

We will start by exploring the organization of GPIO ports and pins, covering both the general-purpose and alternate functions of these pins. Next, we will examine the key registers associated with the GPIO peripheral in STM32 microcontrollers. Finally, we will apply this knowledge to develop input and output drivers using the detailed register information we learn in this chapter.

In this chapter, we will cover the following main topics:

- Understanding the GPIO peripheral
- The STM32 GPIO registers
- Developing input and output drivers

By the end of this chapter, you will be able to use the GPIO peripheral to interface effectively with microcontrollers, which will enable you to handle various input and output tasks with confidence.

Technical requirements

All the code examples for this chapter can be found on GitHub at `https://github.com/PacktPublishing/Bare-Metal-Embedded-C-Programming`.

Understanding the GPIO peripheral

Since we introduced the GPIO peripheral in *Chapter 2*, this section will reiterate the key points to remember regarding GPIOs.

Microcontroller pins are grouped into ports. For instance, a microcontroller might have ports named **GPIOA**, **GPIOB**, and **GPIOC**. See *Figure 2.10* in *Chapter 2*. Each port is composed of individual pins, which are referred to by their port name, followed by their pin number. The following are examples of this naming convention:

- PA1 refers to port A, pin 1
- PD7 refers to port D, pin 7

This naming convention helps in identifying and configuring specific pins for various functions.

The STM32F411xC/E microcontroller series features six ports: **PORTA**, **PORTB**, **PORTC**, **PORTD**, **PORTE**, and **PORTH**. Each port is equipped with a comprehensive set of registers to manage configuration, data handling, and functionality.

These ports offer a variety of features designed for versatility and performance. The features offered include the following:

- **I/O control**: They allow us to manage up to 16 input/output pins per port.
- **Output states**: Pins can be configured for push-pull or open-drain modes, with optional pull-up or pull-down resistors.
- **Output data**: Output data is driven by the GPIOx_ODR register when the pin is configured as a general-purpose output. For alternate function configurations, the associated peripheral drives the output data.
- **Speed selection**: The operating speed for each I/O pin can be set.
- **Input states**: Pins can be configured as floating, pull-up, pull-down, or analog inputs.
- **Input data**: Data can be read from the GPIOx_IDR register or an associated peripheral when configured for alternate function input.
- **Configuration locking**: The GPIOx_LCKR register can be used to lock the I/O configuration, preventing accidental changes.
- **Alternate function selection**: Up to 16 alternate functions per I/O pin can be configured, providing flexibility in pin usage.

In the next section, we shall explore some of the GPIO registers of the STM32F411 microcontroller.

The STM32 GPIO registers

In this section, we will explore the characteristics and functions of some of the common registers within the GPIO peripheral.

Each GPIO port includes a set of 32-bit registers essential for configuration and control. The configuration registers comprise the following:

- `GPIOx_MODER` (mode register)
- `GPIOx_OTYPER` (output type register)
- `GPIOx_OSPEEDR` (output speed register)
- `GPIOx_PUPDR` (pull-up/pull-down register)

The data registers include the following:

- `GPIOx_IDR` (input data register)
- `GPIOx_ODR` (output data register)

`GPIOx_BSRR` (the bit-set/reset register) and `GPIOx_LCKR` (the locking register) are used to control pin states and access. Additionally, the alternate function selection registers, `GPIOx_AFRH` and `GPIOx_AFRL`, manage the alternate function assignments for the pins within the GPIO port.

Let's start with the GPIO mode register.

The GPIO mode register (GPIOx_MODER)

The GPIO port mode register (`GPIOx_MODER`) is an important register for configuring the mode of each pin in the GPIO port. This register allows us to set each pin in different modes, such as **input**, **output**, **alternate function**, or **analog**.

It is a 32-bit register divided into 16 pairs of bits. Each pair of bits corresponds to a specific pin in the GPIO port, allowing the individual configuration of each pin. See *Figure 2.17* in *Chapter 2*.

The possible configurations for these bits are the following:

- **00**: Input mode (reset state)

  In this mode, the pin is configured as an input, which can be used to read signals from external devices such as a push button.

- **01**: General-purpose output mode

  This mode configures the pin as an output, which can be used to drive external signals or components such as an LED.

- **10**: Alternate function mode

  This mode sets the pin to an alternate function, allowing it to interface with various peripherals such as UART, SPI, or I2C.

- **11**: Analog mode

  This mode configures the pin for analog input, which is useful for **analog-to-digital converter (ADC)** operations.

Let's see a practical example.

Consider configuring a pin (e.g., **PA5**) on port A:

1. To set **PA5** as a general-purpose output (**01**), we can follow these steps:

    - Locate the bit pair corresponding to **PA5** (**MODER5[1:0]**); these are **bit11** and **bit10**
    - Write **0** to **bit11** and **1** to **bit10**

2. To set **PA5** as an alternate function (**10**), we should write **1** to **bit11** and **0** to **bit10**.

3. To set **PA5** as an analog input (**11**), we should write **1** to **bit11** and **1** to **bit10**.

This is all there is to know about the GPIOx_MODER register.

Let's move on to examine two other important registers: the output data register (GPIOx_ODR) and the input data register (GPIOx_IDR).

The GPIO output data register (GPIOx_ODR) and the GPIO input data register (GPIOx_IDR)

The GPIO output data register (GPIOx_ODR) and GPIO input data register (GPIOx_IDR) are essential for managing the data flow through the GPIO pins. These registers allow for reading the state of pins and setting the state of pins, enabling our microcontroller to interact with external devices effectively.

GPIOx_ODR is a 32-bit register, but only the lower 16 bits are used to control the output state of the pins. Each bit in the register corresponds to a pin in the GPIO port.

By writing to this register, we can set the logic level (high or low) of each pin configured as an output. *Figure 7.1* shows the structure of the GPIO **output data register** (**ODR**) extracted from the reference manual.

31	30	29	28	27	26	25	24	23	22	21	20	19	18	17	16
							Reserved								

15	14	13	12	11	10	9	8	7	6	5	4	3	2	1	0
ODR15	ODR14	ODR13	ODR12	ODR11	ODR10	ODR9	ODR8	ODR7	ODR6	ODR5	ODR4	ODR3	ODR2	ODR1	ODR0
rw	rw	rw	rw	rw	rw	rw	rw	rw	rw	rw	rw	rw	rw	rw	rw

Figure 7.1: GPIO ODR

Take the following examples:

- Writing **1** to `bit5` (ODR5) sets **PA5** to a high state
- Writing **0** to `bit5` (ODR5) sets **PA5** to a low state

How about the GPIO input data register?

The GPIO input data register (`GPIOx_IDR`) is used to read the current state of the GPIO pins configured as inputs. By reading from this register, we can determine whether each input is at a high or low logic level.

It is a 32-bit register, but similar to the ODR, only the lower 16 bits are used to read the state of the pins. Each bit in the register corresponds to a pin in the GPIO port.

A bit value of **1** indicates that the corresponding pin is at a high logic level, while a bit value of **0** indicates that it is at a low logic level.

Figure 7.2 shows the structure of the GPIO input data register:

31	30	29	28	27	26	25	24	23	22	21	20	19	18	17	16
							Reserved								

15	14	13	12	11	10	9	8	7	6	5	4	3	2	1	0
IDR15	IDR14	IDR13	IDR12	IDR11	IDR10	IDR9	IDR8	IDR7	IDR6	IDR5	IDR4	IDR3	IDR2	IDR1	IDR0
r	r	r	r	r	r	r	r	r	r	r	r	r	r	r	r

Figure 7.2: GPIO input data register

Take the following examples:

- If `bit5` (IDR5) reads **1**, **PA5** is at a high state
- If `bit5` (IDR5) reads **0**, **PA5** is at a low state

Another commonly used register is the GPIO bit set/reset register (`GPIOx_BSRR`). Let's examine this register in the next section.

The GPIO bit-set/reset register (GPIOx_BSRR)

The GPIO bit-set/reset register (GPIOx_BSRR) is a crucial register for controlling the state of GPIO pins. It provides atomic bitwise operations to set or reset individual bits, which ensures that no interrupts can disrupt the operation, maintaining data integrity during modifications.

GPIOx_BSRR is a 32-bit register divided into two 16-bit sections:

- **Bits 15:0 (BSy)**: These bits are used to set the corresponding GPIO pin

- **Bits 31:16 (BRy)**: These bits are used to reset the corresponding GPIO pin

Figure 7.3 shows the structure of the GPIO **bit-set/reset register (BSRR)**.

31	30	29	28	27	26	25	24	23	22	21	20	19	18	17	16
BR15	BR14	BR13	BR12	BR11	BR10	BR9	BR8	BR7	BR6	BR5	BR4	BR3	BR2	BR1	BR0
w	w	w	w	w	w	w	w	w	w	w	w	w	w	w	w
15	14	13	12	11	10	9	8	7	6	5	4	3	2	1	0
BS15	BS14	BS13	BS12	BS11	BS10	BS9	BS8	BS7	BS6	BS5	BS4	BS3	BS2	BS1	BS0
w	w	w	w	w	w	w	w	w	w	w	w	w	w	w	w

Figure 7.3: BSRR

🔍 **Quick tip**: Need to see a high-resolution version of this image? Open this book in the next-gen Packt Reader or view it in the PDF/ePub copy.

📱 **The next-gen Packt Reader** and a **free PDF/ePub copy** of this book are included with your purchase. Unlock them by scanning the QR code below or visiting https://www.packtpub.com/unlock/9781835460818.

Let's analyze the bits in the register:

- **Bits 31:16 (BRy)**: Writing **1** to any of the upper 16 bits resets the corresponding pin to a low state

- **Bits 15:0 (BSy)**: Writing **1** to any of the lower 16 bits sets the corresponding pin to a high state

For example, let's see how to set PA5 as high and low using the BSRR:

- To set PA5 as **high**: We write **1** to **bit5(BS5)** in the GPIOA_BSRR register

- To set PA5 as **low**: We write **1** to **bit21(BR21)** in the GPIOA_BSRR register

The GPIO BSRR (GPIOx_BSRR) provides a robust mechanism for controlling the state of GPIO pins. By understanding its structure and functionality, we can perform efficient and atomic operations to set or reset individual pins.

Another pair of commonly used GPIO registers are the GPIO alternate function high and alternate function low registers. Let's explore them in the next section.

The GPIO alternate function registers (GPIOx_AFRL and GPIOx_AFRH)

The GPIO **alternate function registers (AFRs)** are important for configuring the alternate functions of the GPIO pins in STM32 microcontrollers. These registers allow each pin to be assigned a specific peripheral function, enhancing the versatility and functionality of the microcontroller.

Each GPIO port has two AFRs:

- **GPIOx_AFRL: Alternate function low register (AFRL)**, for pins 0 to 7

- **GPIOx_AFRH: Alternate function high register (AFRH)**, for pins 8 to 15

These registers enable the selection of alternate functions for each pin, facilitating the use of the pins for various peripheral interfaces such as UART, I2C, and SPI.

GPIOx_AFRL is a 32-bit register, divided into eight 4-bit fields. Each 4-bit field corresponds to one pin in the range of pins 0 to 7 in the GPIO port.

GPIOx_AFRH is also a 32-bit register, divided into eight 4-bit fields. Here, each 4-bit field corresponds to one pin in the range of pins 8 to 15 in the GPIO port.

31	30	29	28	27	26	25	24	23	22	21	20	19	18	17	16
AFRL7[3:0]				AFRL6[3:0]				AFRL5[3:0]				AFRL4[3:0]			
rw	rw	rw	rw	rw	rw	rw	rw	rw	rw	rw	rw	rw	rw	rw	rw
15	14	13	12	11	10	9	8	7	6	5	4	3	2	1	0
AFRL3[3:0]				AFRL2[3:0]				AFRL1[3:0]				AFRL0[3:0]			
rw	rw	rw	rw	rw	rw	rw	rw	rw	rw	rw	rw	rw	rw	rw	rw

Figure 7.4: Alternate function low register

Figure 7.4 illustrates the AFRL. To understand the various alternate functions each pin can assume based on the values of the corresponding 4-bit fields, we will refer to *Figure 7.5* as our guide:

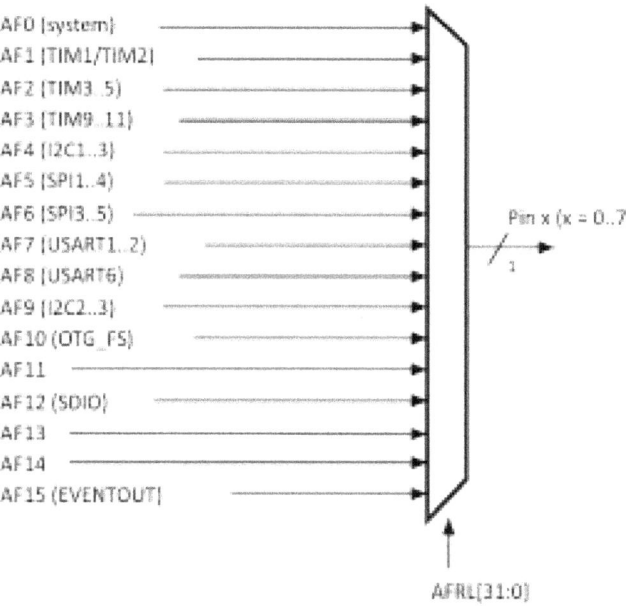

For pins 8 to 15, the GPIOx_AFRH[31:0] register selects the dedicated alternate function

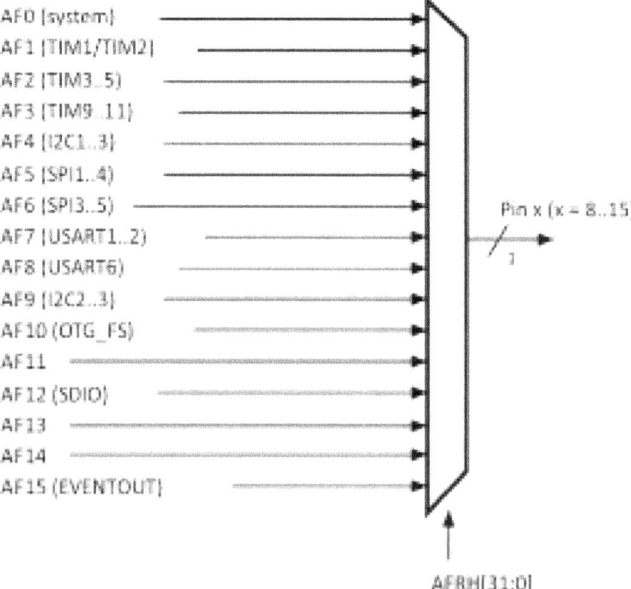

Figure 7.5: Alternate function selection

Figure 7.5 shows a diagram with two multiplexers. The first multiplexer represents the selector for the AFRL, while the second represents the selector for the AFRH. This diagram is sourced from page 150 of the reference manual. It effectively demonstrates how to configure GPIO pins for alternate functions on STM32F411 microcontrollers using these registers.

Let's break down what we see.

For both `GPIOx_AFRL` and `GPIOx_AFRH`, the diagram provides a list of the possible alternate functions that can be selected for each pin. The alternate functions are designated by **AF0** through **AF15**, each associated with specific peripheral functions, as follows:

Alternate function description	Binary value
AF0: System functions	0000
AF1: TIM1/TIM2	0001
AF2: TIM3/TIM4/TIM5	0010
AF3: TIM9/TIM10/TIM11	0011
AF4: I2C1/I2C2/I2C3	0100
AF5: SPI1/SPI2/SPI3/SPI4	0101
AF6: SPI3/SPI4/SPI5	0110
AF7: USART1/USART2	0111
AF8: USART6	1000
AF9: I2C2/I2C3	1001
AF10: OTG_FS	1010
AF11: Reserved	1011
AF12: SDIO	1100
AF13: Reserved	1101
AF14: Reserved	1110
AF15: EVENTOUT	1111

Table 7.1: Alternate function selection

For example, to configure pin 5 (**PA5**) to be used by TIM9 (**AF3**), we write the value **0011** to bits [23:20] of the `GPIOA_AFRL` register. We use `GPIOA_AFRL` because PA5 falls within the range of pins 0 to 7, which are managed by this register. To configure pin 10 (**PA10**) to be used by the USART1 peripheral (**AF7**), we write the value **0111** to bits [7:4] of the GPIOA_AFRH register. This is because **PA10** falls within the range of pins 8 to 15, which are managed by the GPIOA_AFRH register.

In this section, we explored the characteristics and functions of several essential registers within the STM32 GPIO peripheral. We began with the GPIO mode register (GPIOx_MODER), which configures the mode of each GPIO pin, allowing settings for the input, output, alternate function, or analog mode. We then examined the GPIO output data register (GPIOx_ODR) and input data register (GPIOx_IDR), which manage the data flow through the GPIO pins by setting and reading pin states. Next, we looked at the GPIO BSRR (GPIOx_BSRR), which provides atomic operations for setting and resetting individual pin states. Finally, we covered the GPIO alternate function registers (GPIOx_AFRL and GPIOx_AFRH), which assign specific peripheral functions to each pin, enhancing the microcontroller's versatility.

In the next section, we will develop GPIO drivers using the knowledge gained in this section. Specifically, we will focus on creating input and output drivers. We will explore the practical usage of the alternate function registers in the chapters dedicated to communication peripherals.

Developing input and output drivers

In this section, we will apply the knowledge gained about the GPIO peripheral to develop practical input and output drivers for STM32 microcontrollers. Since we are already familiar with developing the output driver to toggle an LED using the ODR, this section will focus on developing the output driver using the BSRR.

The GPIO output driver using the BSRR

Let's start by making a copy of our last project in our IDE:

1. Right-click on the last project and select **Copy**.
2. Right-click in the **Project Explorer** pane and select **Paste**.
3. Rename the copied project to GpioInput-Output.

Next, we will modularize our code by creating dedicated files for the GPIO driver code:

1. Right-click on the Src folder in the project and select **New | File**.
2. In the **File Name** field, type gpio.c.

Then, we will create the corresponding header file:

1. Right-click on the Inc folder in the project and select **New | File**.
2. In the **File Name** field, type gpio.h.

Our next task is to implement the driver code in the gpio.c file and declare the public functions in the gpio.h file.

Populate your gpio.c with the following code:

```
#include "gpio.h"

#define GPIOAEN            (1U<<0)
#define LED_BS5            (1U<<5)   /*Bit Set Pin 5*/
#define LED_BR5            (1U<<21)  /*Bit Reset Pin 5*/

void led_init(void)
{
    /*Enable clock access to GPIOA*/
    RCC->AHB1ENR |= GPIOAEN;
    /*Set PA5 mode to output mode*/
    GPIOA->MODER |=(1U<<10);
    GPIOA->MODER &=~(1U<<11);
}
void led_on(void)
{
    /*Set PA5 high*/
    GPIOA->BSRR |=LED_BS5;
}

void led_off(void)
{
    /*Set PA5 high*/
    GPIOA->BSRR |=LED_BR5;

}
```

Let's break down the unique elements of the gpio.c file, focusing on the usage of the BSRR. We start with the header file inclusion:

```
#include "gpio.h"
```

This line includes the gpio.h header file, which in turn includes stm32fxx.h to provide access to the register definitions.

Next, we define all the macros we need:

```
#define GPIOAEN            (1U<<0)
#define LED_BS5            (1U<<5)   /* Bit Set Pin 5 */
#define LED_BR5            (1U<<21)  /* Bit Reset Pin 5 */
```

Let's break these macros down:

- GPIOAEN: This macro is used to enable the clock for GPIOA by writing to the AHB1ENR register.

- LED_BS5: This macro represents the bit-set operation for pin PA5. Writing this value to the BSRR sets PA5 to high, turning the LED on.

- LED_BR5: This macro represents the bit-reset operation for pin PA5. Writing this value to the BSRR resets PA5 to low, turning the LED off.

Next, we implement the function for turning on the LED:

```
void led_on(void)
{
    /* Set PA5 high */
    GPIOA->BSRR |= LED_BS5;
}
```

The GPIOA->BSRR |= LED_BS5 line uses the BSRR to set PA5 to high. Writing **1** to bit 5 of the BSRR sets the corresponding pin (PA5) to high, turning the LED on.

And then we implement the function to turn the LED off:

```
void led_off(void)
{
    /* Set PA5 low */
    GPIOA->BSRR |= LED_BR5;
}
```

Similarly, GPIOA->BSRR |= LED_BR5 uses the BSRR to reset PA5 to low. Writing **1** to bit 21 of the BSRR resets the corresponding pin (PA5) to low, turning the LED off.

Here is the content for the gpio.h file:

```
#ifndef GPIO_H_
#define GPIO_H_

#include "stm32f4xx.h"

void led_init(void);
void led_on(void);
void led_off(void);

#endif /* GPIO_H_ */
```

Let's break it down, starting with the header guard:

```
#ifndef GPIO_H_
#define GPIO_H_
...
#endif /* GPIO_H_ */
```

The header guards prevent multiple inclusions of the same header file, which can lead to errors and redundant declarations. The #ifndef GPIO_H_ directive checks whether GPIO_H_ has been defined yet. If it hasn't, it proceeds to define GPIO_H_ and includes the rest of the file. The #endif directive at the end closes the conditional directive that began with #ifndef.

Next, we have the include directive:

```
#include "stm32f4xx.h"
```

This directive includes the stm32f4xx.h header file, which in turn includes the stm32f411xe.h header file, which provides definitions and declarations for all the registers in our microcontroller.

And then we have the function declarations:

```
void led_init(void);
void led_on(void);
void led_off(void);
```

These declarations allow us to access the functions defined in the gpio.c file from other files, such as main.c.

Now that our GPIO output driver for PA5 is complete, let's test it by updating the main.c file to call the functions defined in the gpio.c file. Here is the updated main.c code:

```
#include "gpio.h"
int main(void)
{
    /*Initialize LED*/
    led_init();
    while(1)
    {
        led_on();
        for(int i = 0; i < 100000; i++){}
        led_off();
        for(int i = 0; i < 100000; i++){}
    }
}
```

The code begins by including the `gpio.h` header file to access the GPIO functions defined in `gpio.c`. Within the `main` function, it first calls `led_init()` to initialize PA5 as an output pin. Then, it enters an infinite loop where it alternately turns the LED on and off by calling `led_on()` and `led_off()`, respectively. We use simple delay loops between these calls to keep the LED on and off for visible durations, effectively making the LED blink continuously.

Proceed to build the project and run it on the development board. You should see the green LED blinking.

In the next section, we shall develop the GPIO input driver using **PC13**. We are using **PC13** because the blue push button of the development board is connected to this pin.

The GPIO input driver

Let's start by analyzing the initialization function. Add this function to the `gpio.c` file:

```
#define GPIOAEN            (1U<<0)
#define GPIOCEN            (1U<<2)
#define BTN_PIN            (1U<<13)

void button_init(void)
{
    /*Enable clock access to PORTC*/
    RCC->AHB1ENR |=GPIOCEN;

    /*Set PC13 as an input pin*/
    GPIOC->MODER &=~(1U<<26);
    GPIOC->MODER &=~(1U<<27);
}
```

This function enables clock access to GPIO port C and configures pin PC13 as an input pin:

- **GPIOC->MODER**: This is the GPIO port mode register for port C. Each pair of bits in this register corresponds to the mode configuration of a specific pin.

- **Clearing bits 26 and 27**: The bits corresponding to pin 13 in the **GPIOC->MODER** register are bits **26** and **27**.

  The bitwise AND operator combined with the bitwise NOT operator (&=~) clears these bits, setting them to **00**. As we learned earlier, configuring these bits to **00** sets PC13 as an input pin.

Next, add the function for reading the state of the pin:

```
bool get_btn_state(void)
{
```

```
    /*Note : BTN is active low*/

    /*Check if button is pressed*/
    if(GPIOC->IDR & BTN_PIN)
    {
        return false;
    }
    else
    {
        return true;
    }

}
```

This function reads the state of the button. The button is internally connected as an active-low input. This means the pin reads at a low logic level (0) when the button is pressed and a high logic level (1) when it is not pressed.

- **GPIOC->IDR**: This is the input data register for GPIO port C. It holds the current state of all the pins in the port.

- **Bitwise AND Operator (&)**: This checks whether the bit corresponding to **BTN_PIN** in the IDR register is set to high.

- **Return values**:

  - `false`: If the bit is set, the button is not pressed

  - `true`: If the bit is not set, the button is pressed

To be able to access these new functions from other files, such as main.c, we need to add their prototypes to the gpio.h file.

Add the following lines to gpio.h:

```
#include <stdbool.h>
void button_init(void);
bool get_btn_state(void);
```

Now, let's test the new functions by updating the main.c file to call them.

The following is the updated main.c code:

```
#include "gpio.h"

bool btn_state;
```

```
int main(void)
{
    /*Initialize LED*/
    led_init();

    /*Initialize Pushbutton*/
    button_init();

    while(1)
    {
        /*Get Pushbutton State*/
        btn_state = get_btn_state();

        if(btn_state)
        {
            led_on();
        }
        else
        {
            led_off();
        }

    }
}
```

Let's break it down:

- `bool btn_state`: This variable holds the state of the push button
- `led_init()`: Configures PA5 as an output pin
- `button_init()`: Configures PC13 as an input pin
- `while(1) {…}`:
  - Continuously reads the state of the push button using `get_btn_state()`
  - Turns the LED on if the button is pressed (`btn_state` is true)
  - Turns the LED off if the button is not pressed (`btn_state` is false)

Build the project and run it on the development board. You should see the green LED light up only when you press the push button.

Summary

In this chapter, we explored the GPIO peripheral, a critical peripheral in microcontrollers that is essential for interfacing with various external components. We began by understanding the organization of GPIO ports and pins, covering both general-purpose and alternate functions.

The STM32F411 microcontroller series features several ports, each equipped with registers to manage configuration, data handling, and functionality.

We introduced the registers associated with the GPIO peripheral, including configuration registers such as `GPIOx_MODER`, `GPIOx_OTYPER`, `GPIOx_OSPEEDR`, and `GPIOx_PUPDR`, as well as data registers such as `GPIOx_IDR` and `GPIOx_ODR`. We also covered `GPIOx_BSRR` (bit-set/reset register) for atomic pin state control and `GPIOx_LCKR` (locking register) for preventing accidental configuration changes. Additionally, we explored the GPIO alternate function registers (`GPIOx_AFRL` and `GPIOx_AFRH`), which enable versatile pin usage by assigning specific peripheral functions.

In practical terms, we developed both output and input drivers. We first created an output driver using the `GPIOx_BSRR` register to control the LED connected to pin PA5. This involved setting up the necessary macros, implementing initialization and control functions, and testing the driver by making the LED blink. We then developed an input driver for reading the state of the push button connected to pin PC13. This included configuring PC13 as an input pin, implementing a function to read the button state, and testing the driver by making the LED respond to button presses.

In the next chapter, we shall explore another important peripheral: the **system tick (SysTick)** timer.

Unlock this book's exclusive benefits now

This book comes with additional benefits designed to elevate your learning experience.

Note: Have your purchase invoice ready before you begin.

`https://www.packtpub.com/ unlock/9781835460818`

System Tick (SysTick) Timer

In this chapter, we will learn about the **System Tick (SysTick)** timer, an important core peripheral in all Arm Cortex microcontrollers. We will begin by introducing the SysTick timer and discussing its most common use cases. Following this, we will explore the SysTick timer registers in detail. Finally, we will develop a driver for the SysTick timer.

In this chapter, we're going to cover the following main topics:

- Introduction to the SysTick timer
- Developing a driver for the SysTick timer

By the end of this chapter, you will have a good understanding of the SysTick timer and be able to effectively implement and utilize it in your Arm Cortex-M projects.

Technical requirements

All the code examples for this chapter can be found on GitHub at

`https://github.com/PacktPublishing/Bare-Metal-Embedded-C-Programming`.

Introduction to the SysTick timer

The **System Tick** timer, commonly known as **SysTick**, is a fundamental component of all Arm Cortex microcontrollers. Regardless of the processor core—whether it's Cortex-M0, Cortex-M1, or Cortex-M7—and the silicon manufacturer—be it STMicroelectronics, Texas Instruments, or any other—every Arm Cortex microcontroller includes a SysTick timer. In this section, we will learn about this essential peripheral and explore its registers in detail.

Overview of the SysTick timer

The SysTick timer is a **24-bit down counter** integral to all Arm Cortex-M processors. It is designed to offer a configurable time base that can be used for various purposes, such as **task scheduling**, **system monitoring**, and **time tracking**. This timer provides us with a simple and efficient means of **generating periodic interrupts** and serves as a cornerstone for implementing system timing functions, including **operating system** (**OS**) tick generation for **real-time operating systems** (**RTOSs**). Using SysTick makes our code more portable since it is part of the core and not a vendor-specific peripheral.

The key features of the SysTick timer include the following:

- **24-bit Reloadable Counter**: The counter decrements from a specified value to zero, then reloads automatically to provide a continuous timing operation

- **Core Integration**: Being part of the core, it requires minimal configuration and offers low-latency interrupt handling

- **Configurable Clock Source**: SysTick can operate either from the core clock or an external reference clock, providing flexibility in timing accuracy and power consumption

- **Interrupt Generation**: When the counter reaches zero, it can trigger an interrupt

The SysTick timer typically serves three primary use cases:

- **OS Tick Generation**: In an RTOS environment, SysTick is commonly used to generate the system tick interrupt, which drives the OS scheduler

- **Periodic Task Execution**: It can be used to trigger regular tasks, such as sensor sampling or communication checks

- **Time Delay Functions**: SysTick can provide precise delays for various timing functions within the firmware

Now let's explore the registers in the SysTick timer.

SysTick timer registers

The SysTick timer consists of four primary registers:

- SysTick Control and Status Register (`SYST_CSR`)
- SysTick Reload Value Register (`SYST_RVR`)
- SysTick Current Value Register (`SYST_CVR`)
- SysTick Calibration Value Register (`SYST_CALIB`)

Let's analyze them one by one, starting with the Control and Status Register.

The SysTick Control and Status Register (SYST_CSR)

The SYST_CSR register controls the SysTick timer's operation and provides status information. It has the following bits:

- **ENABLE (Bit 0)**: Enables or disables the SysTick counter

- **TICKINT (Bit 1)**: Enables or disables the SysTick interrupt

- **CLKSOURCE (Bit 2)**: Selects the clock source (0 = external reference clock, 1 = processor clock)

- **COUNTFLAG (Bit 16)**: Indicates whether the counter has reached zero since the last read (1 = yes, 0 = no)

This is the structure of the SysTick Control and Status Register:

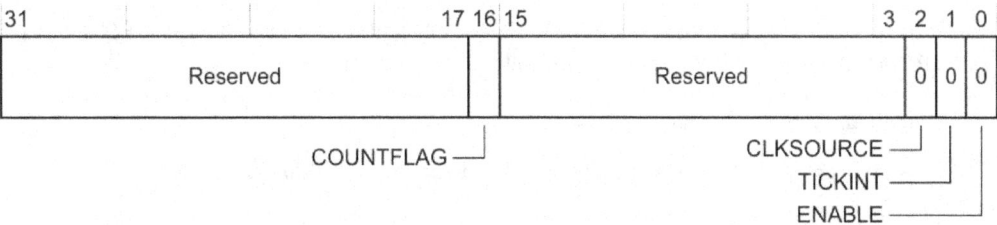

Figure 8.1: The SysTick Control and Status Register

The next register is the SysTick Reload Value Register (SYST_RVR).

The SysTick Reload Value Register (SYST_RVR)

This register specifies the start value to load into the SysTick Current Value Register. It is crucial for setting the timer's period and understanding its bit assignments and calculations is essential for effective SysTick configuration.

It has the following fields:

- **Bits [31:24] Reserved**: These bits are reserved

- **Bits [23:0] RELOAD**: This field specifies the value that the SysTick timer will load into the SYST_CVR register when the counter is enabled and when it reaches zero

This is the structure of the SysTick Reload Value Register:

Figure 8.2: The SysTick Reload Value Register

Since SysTick is a 24-bit timer, the RELOAD value can be any value in the range **0x00000001** to **0x00FFFFFF**.

To calculate the RELOAD value based on the desired timer period, we determine the number of clock cycles for the desired period, and then subtract 1 from this number to get the RELOAD value.

For example, if the **core clock frequency** is **16 MHz** and we want the SysTick timer **to trigger every 1 ms**, the RELOAD value would be calculated as follows:

1. Calculate the number of clock cycles in 1 ms:

    Clock cycles = 16,000,000 cycles/second * 0.001 second = 16,000 cycles

    Note: 1ms = 0.001 second

2. Subtract 1 from the calculated number of clock cycles:

    RELOAD = 16,000 - 1 = 15,999 since counting from 0 to 15,999 will give us 16000 ticks.

Meaning, to configure the SysTick timer for a 1 ms period with a 16 MHz clock, we would set the RELOAD value to **15,999**. The next register is the SysTick Current Value Register (SYST_CVR).

The SysTick Current Value Register

The SysTick Current Value Register (SYST_CVR) holds the current value of the SysTick counter. We can use this register to monitor the countdown process and to reset the counter when necessary.

It has the following fields:

*   **Bits [31:24] Reserved**: These bits are reserved and should not be modified. They must be written as zero.

*   **Bits [23:0] CURRENT**: This field contains the current value of the SysTick counter. Reading this field returns the current counter value. Writing any value to this field clears it to zero and also clears the COUNTFLAG bit in the SysTick Control and Status Register (SYST_CSR). This is the SysTick Current Value Register:

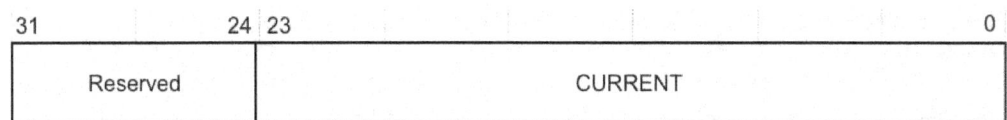

Figure 8.3: The SysTick Current Value Register

The SysTick Calibration Value Register

The final register of the SysTick timer is the SysTick Calibration Value Register (SYST_CALIB). This register provides us with the calibration properties of the SysTick timer.

The names of these registers are slightly different in the STM32 header files. *Table 8.1* provides a clear correspondence between the register names used in the *Arm Generic User Guide* documentation and those in the STM32-specific header files. This correspondence will help us understand and reference them correctly in our code.

Function	Arm Generic User Guide	STM32 Header File
Control and Status	SYST_CSR	SysTick->CTRL
Reload Value	SYST_RVR	SysTick->LOAD
Current Value	SYST_CVR	SysTick->VAL
Calibration Value	SYST_CALIB	SysTick->CALIB

Table 8.1: Correspondence of SysTick register names

In the next section, we will use the information we have learned to develop a driver for the SysTick timer.

Developing a driver for the SysTick timer

In this section, we will develop a driver for the SysTick timer to generate precise delays.

Let's start by making a copy of our last project in our IDE, following the steps we learned in *Chapter 7*. Rename the copied project to `SysTick`. Next, create a new file named `systick.c` in the `Src` folder and another file named `systick.h` in the `Inc` folder, just like we did for the GPIO drivers in the previous lesson.

Populate your `systick.c` file with the following code:

```
#include "systick.h"

#define CTRL_ENABLE         (1U<<0)
```

```c
#define CTRL_CLCKSRC      (1U<<2)
#define CTRL_COUNTFLAG    (1U<<16)

/*By default, the frequency of the MCU is 16Mhz*/
#define ONE_MSEC_LOAD      16000


void systick_msec_delay(uint32_t delay)
{

    /*Load number of clock cycles per millisecond*/
    SysTick->LOAD =  ONE_MSEC_LOAD - 1;

    /*Clear systick current value register*/
    SysTick->VAL = 0;

    /*Select internal clock source*/
    SysTick->CTRL = CTRL_CLCKSRC;

    /*Enable systick*/
    SysTick->CTRL |=CTRL_ENABLE;

    for(int i = 0; i < delay; i++)
    {
        while((SysTick->CTRL & CTRL_COUNTFLAG) == 0){}
    }

    /*Disable systick*/
    SysTick->CTRL = 0;


}
```

Let's break it down.

We start with the header file inclusion:

```c
#include "systick.h"
```

This line includes the header file, systick.h, which in turn includes stm32fxx.h to provide access to the register definitions.

Next, we define all the macros we need:

- `#define CTRL_ENABLE (1U << 0)`: Macro to enable the SysTick timer.
- `#define CTRL_CLKSRC (1U << 2)`: Macro to select the internal clock source for the SysTick timer.
- `#define CTRL_COUNTFLAG (1U << 16)`: Macro to check the `COUNTFLAG` bit, which indicates when the timer has counted to zero.
- `#define ONE_MSEC_LOAD 16000`: Macro to define the number of clock cycles in 1 millisecond. This assumes the microcontroller's clock frequency is 16 MHz. This is the default configuration of the NUCLEO-F411 development board.

Next, we move on to the function implementation.

First, we have the following:

```
SysTick->LOAD = ONE_MSEC_LOAD - 1;
```

This line loads the SysTick timer with the number of clock cycles for 1 millisecond.

Then, we clear the Current Value register with the following to reset the timer:

```
SysTick->VAL = 0;
```

Next, we select the internal clock source:

```
SysTick->CTRL = CTRL_CLKSRC;
```

To enable the SysTick timer, we use the following:

```
SysTick->CTRL |= CTRL_ENABLE;
```

Now, we enter the loop that handles the delay:

```
for (int i = 0; i < delay; i++)
{
    while ((SysTick->CTRL & CTRL_COUNTFLAG) == 0) {}
}
```

This loop runs for the specified delay duration. Inside each iteration, it waits for the `COUNTFLAG` bit to be set, which indicates the timer has counted down to zero.

Finally, we disable the SysTick timer:

```
SysTick->CTRL = 0;
```

And that's it! With these steps, we've successfully implemented a delay function using the SysTick timer.

Our next task is to populate the `systick.h` file.

Here is the code:

```
#ifndef SYSTICK_H_
#define SYSTICK_H_

#include <stdint.h>
#include "stm32f4xx.h"

void systick_msec_delay(uint32_t delay);

#endif
```

Over here, the `#include <stdint.h>` directive is needed to ensure that we have access to standard integer type definitions provided by the C standard library. These definitions include fixed-width integer types such as `uint32_t`, `int32_t`, `uint16_t`, and so on, which are essential for writing portable and clear code, especially in embedded systems programming.

With the driver files complete, we are now ready to test inside `main.c`.

First, let's enhance our `gpio.c` file by adding a new function that toggles the LED. This will simplify our code by allowing us to toggle the LED with a single function call instead of calling `led_on()` and `led_off()` separately.

Add the following function to your `gpio.c` file:

```
#define LED_PIN               (1U<<5)
void led_toggle(void)
{
    /*Toggle PA5*/
    GPIOA->ODR ^=LED_PIN;
}
```

This function toggles the state of the LED connected to pin PA5 by using the bitwise XOR operation on the **Output Data Register (ODR)**.

Next, declare this function in the `gpio.h` file by adding the following line:

```
void led_toggle(void)
```

Finally, update your `main.c` file as shown here to call the SysTick delay and LED toggle functions:

```
#include "gpio.h"
#include "systick.h"

int main(void)
```

```
{
    /*Initialize LED*/
    led_init();


    while(1){
        /*Delay for 500ms*/
        systick_msec_delay(500);
            /* Toggle the LED */
        led_toggle();
    }
}
```

In this example, we are **toggling the LED at a 500ms interval**. Build the project and run it on your development board. You should see the green LED blinking. To experiment further, you can modify the delay value and observe how the blinking rate of the LED changes.

Summary

In this chapter, we explored the SysTick timer, a core peripheral of all Arm Cortex microcontrollers. We began with an introduction to the SysTick timer, discussing its significance and common applications, such as generating OS ticks in real-time operating systems, executing periodic tasks, and providing precise time delays.

We then examined the SysTick timer's registers in detail. These included the Control and Status Register (SYST_CSR), which manages the timer's operation and status; the Reload Value Register (SYST_RVR), which sets the timer's countdown period; the Current Value Register (SYST_CVR), which holds the current value of the countdown; and the Calibration Value Register (SYST_CALIB), which provides essential calibration properties for accurate timing. We also provided a comparison between the register names used in the Arm Generic User Guide and those in the STM32 header files to ensure clear correspondence for accurate coding.

The chapter concluded with the development of a SysTick timer driver. We walked through the creation and implementation of the systick_msec_delay function, which introduces millisecond delays using the SysTick timer. To test the driver, we integrated it with GPIO functions to toggle our green LED, demonstrating how to achieve precise timing and control in embedded systems.

In the next chapter, we shall learn about another timer peripheral. Unlike the SysTick timer, the configuration of this timer peripheral is specific to STM32 microcontrollers.

**Unlock this book's exclusive
benefits now**

This book comes with additional benefits designed
to elevate your learning experience.

Note: Have your purchase invoice ready before you begin.

https://www.packtpub.com/
unlock/9781835460818

9

General-Purpose Timers (TIM)

In this chapter, we will delve into the **general-purpose timers** (**TIM**) found in the STM32F411 microcontroller. Unlike the SysTick timer, these timer peripherals are unique to STM32 microcontrollers and offer a range of capabilities essential for various applications.

We will begin by discussing the common uses of timers, providing a foundation for understanding their importance in embedded systems. Following this, we will explore the specific characteristics of the timers on STM32 microcontrollers. This includes understanding the timer clock source, the mechanics of prescaling, and a detailed look at the commonly used timer registers.

In the latter part of the chapter, we will put theory into practice by developing a timer driver and applying the knowledge we've gained to create functional and efficient code.

In this chapter, we're going to cover the following main topics:

- Introduction to timers and their uses

- Common use cases of timers

- STM32 timers

- Developing the timer driver

By the end of this chapter, you will have a good understanding of STM32 timers and how to develop drivers for them, enhancing your ability to utilize these peripherals effectively in your projects.

Technical requirements

All the code examples for this chapter can be found on GitHub at https://github.com/PacktPublishing/Bare-Metal-Embedded-C-Programming.

Introduction to timers and their uses

Timers are crucial components in embedded systems, serving essential functions in a wide array of applications. In this section, we will explore the concept of timers, their types, and their various applications.

The question is, *What are timers?*

Timers are hardware peripherals found in microcontrollers that **count clock pulses**. **These pulses can be used to measure time intervals, generate precise delays, or trigger events at specific intervals**. Timers can operate in several modes, including **counting up**, **counting down**, and generating **Pulse Width Modulation** (**PWM**) signals.

The choice between using the SysTick timer and general-purpose timers on STM32 microcontrollers comes down to the complexity and precision of your timing needs. While the SysTick timer is great for simple system timing tasks such as generating periodic interrupts or creating delays, it's limited in flexibility and features, as it's often dedicated to **real-time operating system** (**RTOS**) ticks. On the other hand, general-purpose timers, as we have just discussed, offer far more versatility, with the ability to handle complex tasks such as PWM generation, input capture, and output compare. They also support multiple channels for concurrent timing events and provide advanced functionalities such as precise frequency measurement and low-power operation.

Let's explore some typical use cases to illustrate the importance of timers.

Common use cases of timers

Let's start with timers for time interval measurement.

Time interval measurement

One common use of timers is in ultrasonic sensors for distance measurement, widely used in robotics, automotive parking systems, and obstacle detection.

Timers play a vital role in the operation of these sensors. Here's how they work:

1. The microcontroller sends a trigger signal to the ultrasonic sensor, causing it to emit an ultrasonic pulse.
2. The sensor waits for the echo of the pulse to return after bouncing off an object.
3. A timer starts counting when the pulse is emitted and stops when the echo is received.
4. The time interval measured by the timer is used to calculate the distance to the object based on the speed of sound.

By accurately measuring the time interval between sending the pulse and receiving the echo, the system can determine the distance to objects, enabling precise navigation and obstacle avoidance.

Timers are also essential for generating precise delays, such as in the refresh mechanism of **light-emitting diode** (**LED**) matrix displays.

Delay generation

In embedded systems, LED matrix displays are used for various applications, including digital signage, scoreboards, and simple graphical displays. To display images or text, the microcontroller needs to refresh the display at a precise rate to ensure smooth visuals.

Over here, the following happens:

- The microcontroller drives an LED matrix display and needs to refresh the display rows sequentially.

- A timer generates precise delays to control the time each row is activated before moving to the next row.

  For example, the timer might be set to generate a delay of 1 millisecond between switching rows.

  This ensures that each row is displayed for a consistent period, maintaining a stable and flicker-free image.

Now, let's see an example of how timers can be used for triggering events in embedded systems.

Event trigger

Periodic data sampling from sensors is important for applications such as environmental monitoring and **industrial process control (IPC)**. A timer can be used to trigger the **analog-to-digital converter (ADC)** at specific intervals to ensure consistent data sampling.

Over here, the following happens:

- The microcontroller is connected to an environmental sensor (for example, a temperature or humidity sensor) that outputs an analog signal

- A general-purpose timer is configured to trigger the ADC conversion periodically (for example, every 100 milliseconds)

- The ADC is set up to use the timer trigger option, automatically starting a conversion at each timer event

- The main loop of the microcontroller regularly checks for completed ADC conversions and processes the data

In the next section, we will focus on the features and characteristics of STM32 timers.

STM32 timers

The timers in STM32 microcontrollers are classified into two main categories: **general-purpose timers** and **advanced timers**.

Introduction to general-purpose timers and advanced timers

General-purpose timers are highly versatile and can be used for a variety of applications, whereas advanced timers offer more sophisticated features than general-purpose timers, making them suitable for high-precision and complex timing tasks.

In the STM32F411 microcontroller, timers **TIM2**, **TIM3**, **TIM4**, **TIM5**, **TIM9**, **TIM10**, and **TIM11** are general-purpose timers, whereas **TIM1** is an advanced timer.

Key features of the general-purpose timers include the following:

- **Counter size**: They feature 16-bit counters for TIM3 and TIM4 and 32-bit counters for TIM2 and TIM5, capable of operating in up, down, or up/down auto-reload modes

- **Prescaler**: They are equipped with a 16-bit programmable prescaler, which allows us to divide the counter-clock frequency by any factor ranging from 1 to 65,536

- **Channels**: They offer up to four independent channels

- **Synchronization**: The timers can synchronize with external signals and interconnect with multiple timers, enhancing flexibility and coordination in complex applications

- **Interrupt/direct memory access (DMA) generation**: The timers can generate interrupts or DMA requests based on various events, including counter overflow/underflow, counter initialization, trigger events, input capture, and output compare

- **Encoder support**: These timers also support incremental (quadrature) encoders and Hall-sensor circuitry, making them suitable for precise positioning applications

The advanced timer, TIM1, has a 16-bit counter size. Apart from the difference in counter size, it shares all the features of the general-purpose timers with the following additional features:

- **Complementary outputs**: Support for complementary outputs with programmable dead-time insertion

- **Repetition counter**: Allows updating the timer registers only after a specific number of counter cycles

- **Break input**: It has the ability to put the timer's output signals into a reset or known state upon activation

Before we start analyzing some of the key registers provided in the reference manual, let's first understand how STM32 timers work.

How STM32 timers work

The core of the timer is the 16-bit/32-bit counter and its associated auto-reload register. The counter can count upward or downward, and its clock can be divided by a prescaler. We can read from and write to the counter, auto-reload register, and prescaler register even while the counter is running.

The time-base unit includes the following registers:

- **Counter register** (TIMx_CNT)
- **Prescaler register** (TIMx_PSC)
- **Auto-reload register** (TIMx_ARR)

The prescaler register (TIMx_PSC) divides the timer's input clock frequency by a programmable value between 1 and 65,536. This allows for slower counting rates to accommodate longer timing intervals.

The auto-reload register (TIMx_ARR) defines the value at which the counter resets to zero in up-counting mode or to the auto-reload value in down-counting mode. We use it to set the period of the timer.

Apart from the registers in the time-base unit, the timer peripheral also includes control registers (TIMx_CR1, TIMx_CR2, and so on) for configuring the various operational parameters, such as enabling the counter, setting the counting direction, and configuring **update events** (**UEVs**). A **status register** (**SR**) (TIMx_SR) indicates the status of the timer, such as whether a UEV has occurred and other auxiliary registers for controlling the timer, among other things.

As we mentioned earlier, the timer has two counting modes: up-counting mode and down-counting mode. Let's break them down:

- **Up-counting mode**:

    The counter counts from 0 to the auto-reload value (TIMx_ARR) and generates a UEV on overflow

- **Down-counting mode**:

    The counter counts down from the auto-reload value to 0 and generates a UEV on underflow

The next question that arises is: *What exactly is a UEV?*

Let's break this down:

- In simple terms, it is when the timeout should occur
- When the counter reaches the auto-reload value or 0 (based on the count mode), a UEV occurs
- This event can trigger an interrupt or a DMA request and set the **update interrupt flag** (**UIF**) in the SR (TIMx_SR)
- UEVs can also be generated manually by setting the **UG** bit in the event generation register (TIMx_EGR)

Now that we know what a UEV is, let's understand how timer clock prescaling is achieved; this will enable us to accurately calculate the timing for UEVs.

Timer clock prescaling

We use prescaling to reduce the high-frequency system clock to a lower frequency suitable for driving the timer. *Figure 9.1* illustrates the process of timer clock pre-scaling in our STM32 microcontroller, showing how the system clock (SYSCLK) is divided down to drive the timer counters (TIMx_CNT) through a series of prescalers:

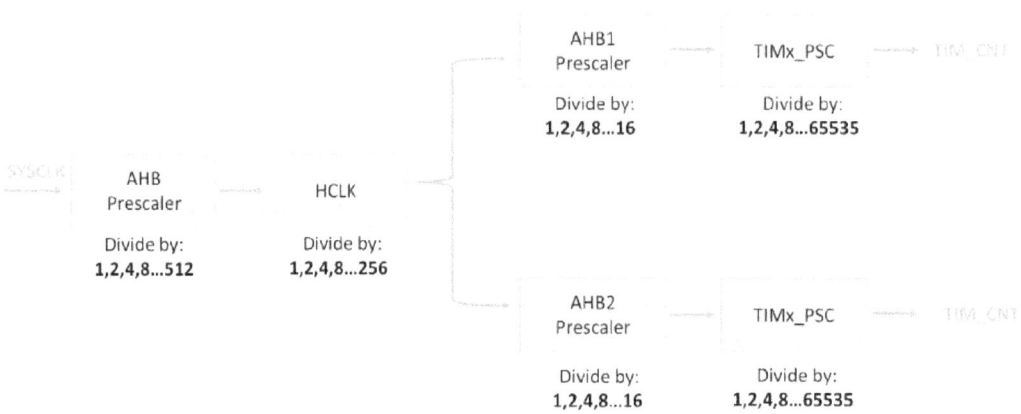

Figure 9.1: Timer clock prescaling

Let's explain each component and step involved in this process:

- **System clock (SYSCLK)**

  This is the main system clock that drives the microcontroller. It serves as the initial clock source for all subsequent operations.

- **Advanced High-performance Bus (AHB) prescaler**

  This is the first prescaler in the chain, which divides the SYSCLK frequency by a factor of 1, 2, 4, 8, 16, 64, 128, 256, or 512. The resulting clock signal is referred to as **HCLK** (AHB clock).

- **High-performance Bus Clock (HCLK)**

  This is the clock signal generated after the AHB prescaler. It is used to drive the core and the system bus.

- **Advanced Peripheral Bus (APB) prescalers**

    - **APB1 prescaler**: This further divides HCLK to generate the clock for the APB1 peripheral bus. The division factors available are 1, 2, 4, 8, and 16.

    - **APB2 prescaler**: Similar to the APB1 prescaler but used for the APB2 peripheral bus. It also divides HCLK by factors of 1, 2, 4, 8, and 16.

- **Timer prescaler (TIMx_PSC)**

    Each timer has its own prescaler register that can divide the APB1 or APB2 clock further by any value between 1 and 65,535. This flexibility allows for precise control over the timer's counting rate.

- **Timer counter (TIMx_CNT)**

    This is the final counter that uses the clock derived from `TIMx_PSC`. The rate at which this counter increments or decrements depends on the combined division effect of the previous prescalers.

The prescalers directly influence the rate at which the timer counter (`TIMx_CNT`) increments or decrements. While the other prescalers (AHB and APB) are shared among all peripherals on the same bus, **the TIMx_PSC prescaler is unique to each timer**. By adjusting the `TIMx_PSC` prescaler, we can precisely control the resolution and period of that specific timer without affecting other peripherals.

Now, let's learn how to compute a UEV.

Computing a UEV

A UEV occurs when the timer counter reaches the value set in the auto-reload register (`TIMx_ARR`) in up-counting mode. Calculating this event is crucial for determining the timer's period and ensuring it operates at the desired frequency.

We can derive the frequency of the UEV with the following formula:

$$\text{Update Event Frequency} = \frac{Timer\ Clock}{(Prescaler + 1) \times (Period + 1)}$$

Here, we have the following:

- **Timer clock**: The clock frequency driving the timer (APB1 or APB2). The APB1 and APB2 bus frequencies are equal to `SYSCLK` when using the default clock configuration (in a bare-metal setup) on the NUCLEO-F411 development board.

- **Prescaler**: The value set in the `TIMx_PSC` register.

- **Period**: The value set in the `TIMx_ARR` register.

Let's see an example. Let's say we have the following parameters:

- **Timer clock (APB1 clock)**: 16 MHz
- **Prescaler (TIMx_PSC value)**: 15999
- **Period (TIMx_ARR value)**: 499

This means we have the following values:

- The timer clock is 16,000,000 Hz
- The TIMx_PSC register is set to 15999, but since the prescaler divides the clock by the value plus one, we use 15999 + 1 = 16000
- The TIMx_ARR register is set to 499, but since the period counts to the value plus one, we use 499 + 1 = 500

Plugging these values into our formula, we get the following:

$$\text{Update Event Frequency} = \frac{16{,}000{,}000}{(15999 + 1) \times (499 + 1)}$$

$$= \frac{16{,}000{,}000}{8000000}$$

$$= 2\text{Hz}$$

This means the UEV occurs at a frequency of **2 Hz, or twice every second**. This concludes this section. In the next section, we will develop our timer driver.

Developing the timer driver

In this section, we will apply the knowledge gained about the TIM peripheral to develop a driver for generating delays.

First, create a copy of your previous project in your IDE, following the steps outlined in earlier chapters. Rename this copied project GTIM. Next, create a new file named tim.c in the Src folder and another file named tim.h in the Inc folder.

Our goal is to develop a driver that initializes TIM2 to generate a 1 Hz timeout. Populate your tim.c file with the following code:

```
#include "tim.h"
#define TIM2EN          (1U<<0)
#define CR1_CEN         (1U<<0)

void tim2_1hz_init(void)
{
    /*Enable clock access to tim2*/
```

```
    RCC->APB1ENR |=TIM2EN;

    /*Set prescaler value*/
    TIM2->PSC =  1600 - 1 ;
    /*Set auto-reload value*/
    TIM2->ARR =  10000 - 1;
    /*Clear counter*/
    TIM2->CNT = 0;

    /*Enable timer*/
    TIM2->CR1 = CR1_CEN;
}
```

💡 **Quick tip**: Enhance your coding experience with the **AI Code Explainer** and **Quick Copy** features. Open this book in the next-gen Packt Reader. Click the **Copy** button (**1**) to quickly copy code into your coding environment, or click the **Explain** button (**2**) to get the AI assistant to explain a block of code to you.

```
                                                    Copy      Explain
  function calculate(a, b) {
                                                     1          2
    return {sum: a + b};
  };
```

🔒 **The next-gen Packt Reader** is included for free with the purchase of this book. Unlock it by scanning the QR code below or visiting `https://www.packtpub.com/unlock/9781835460818`.

Let's break it down:

```
#define TIM2EN      (1U<<0)
#define CR1_CEN     (1U<<0)
```

The TIM2EN instance is defined as (1U<<0), which sets bit 0. This is used to enable the clock for TIM2.

The CR1_CEN instance is defined as (1U<<0), which also sets bit 0. This is used to enable the counter in the TIM2 control register 1 (CR1):

```
/* Enable clock access to TIM2 */
RCC->APB1ENR |= TIM2EN;
```

This line enables the clock for TIM2 by setting the appropriate bit in the APB1 peripheral clock enable register (RCC->APB1ENR):

```
/* Set prescaler value */
TIM2->PSC = 1600 - 1;  // 16,000,000 / 1,600 = 10,000
```

The prescaler value is set to 1599. The prescaler divides the input clock frequency (16 MHz) by (1599 + 1), resulting in a 10,000 Hz (10 kHz) timer clock:

```
/* Set auto-reload value */
TIM2->ARR = 10000 - 1;
```

The auto-reload value is set to 9999. This means the timer will count from 0 to 9999, creating a period of 10,000 ticks. Since the timer clock is 10 kHz, counting 10,000 ticks results in 1 second (10,000 / 10,000 Hz = 1 s):

```
/* Clear counter */
TIM2->CNT = 0;
```

This line resets the timer counter to 0. It ensures that the counting starts from 0 when the timer is enabled:

```
/* Enable timer */
TIM2->CR1 = CR1_CEN;
```

This line enables the timer by setting the CEN (Counter Enable) bit in the TIM2 control register 1 (CR1). This starts the timer, and it begins counting based on the configured prescaler and auto-reload values.

In summary, our code accomplishes the following:

1. Enables the clock for TIM2.
2. Sets a prescaler value to divide the input clock to 10 kHz.
3. Sets an auto-reload value to make the timer count up to 10,000, creating a 1-second period.
4. Clears the timer counter.
5. Enables the timer.

Our next task is to populate the tim.h file.

Here is the code:

```
#include "stm32f4xx.h"
#ifndef TIM_H_
#define TIM_H_

#define SR_UIF   (1U<<0)
void tim2_1hz_init(void);

#endif
```

The following line defines a SR_UIF macro that sets the 0th bit (least significant bit) to 1:

```
#define SR_UIF   (1U << 0)
```

This bit represents the UIF in the SR of the timer. When the timer overflows or reaches the auto-reload value, this flag is set, indicating a UEV. We will need to access this bit in the main.c file.

We are now ready to test inside main.c.

Update your main.c file as shown next:

```
#include "gpio.h"
#include "tim.h"

int main(void)
{
    /*Initialize LED*/
    led_init();
    /*Initialize timer*/
    tim2_1hz_init();
    while(1)
    {
            led_toggle();
             /*Wait for UIF */
            while(!(TIM2->SR & SR_UIF)){}
            /*Clear UIF*/
            TIM2->SR &=~SR_UIF;

    }
}
```

In this code, we initialize the LED and TIM2 to toggle the LED at a 1 Hz frequency. The code works by waiting for the timer's UIF to be set, indicating that 1 second has passed, then toggling the LED and clearing the UIF to repeat the process. This cycle continues indefinitely in the main loop, resulting in the LED toggling every second.

Summary

In this chapter, we explored the general-purpose timers (TIM) in the STM32F411 microcontroller, which are distinct from the SysTick timer and offer a variety of features crucial for embedded systems applications. We began by discussing the fundamental uses of timers, emphasizing their importance in tasks such as time interval measurement, delay generation, and event triggering.

We then learned about the specifics of STM32 timers, detailing their classification into general-purpose and advanced timers.

Next, we examined the workings of STM32 timers, focusing on key registers such as the counter register (`TIMx_CNT`), prescaler register (`TIMx_PSC`), and auto-reload register (`TIMx_ARR`). We explained how the timer's clock prescaling mechanism reduces the system clock to a suitable frequency for the timer and how this affects the timer's operation.

We also provided a practical example of computing UEV frequency, demonstrating how to calculate the period and frequency of the timer based on the prescaler and auto-reload values.

Finally, we applied the theoretical knowledge by developing a timer driver for TIM2 to generate a 1 Hz timeout. In the next chapter, we will learn about another useful peripheral.

The Universal Asynchronous Receiver/Transmitter Protocol

In this chapter, we will learn about the **universal asynchronous receiver/transmitter** (**UART**) protocol, an important communication method widely used in embedded systems. UART is fundamental for enabling communication between microcontrollers and various peripherals, making it an essential component in embedded systems development.

We will start by discussing the significance of communication protocols in embedded systems and highlight common use cases for UART alongside other protocols such as SPI and I2C. Following this, we will provide a comprehensive overview of the UART protocol, detailing its operational principles and features. Next, we will extract and examine the relevant registers for UART from the STM32 reference manual, providing the necessary foundational knowledge for driver development. Finally, we will apply this knowledge to develop a bare-metal UART driver, illustrating the practical aspects of initializing and transmitting data via UART.

In this chapter, we will cover the following main topics:

- Introduction to communication protocols
- Overview of the UART protocol
- The STM32F4 UART peripheral
- Developing the UART driver

By the end of this chapter, you will have a good understanding of the UART protocol and the skills needed to develop bare-metal drivers for UART communication.

Technical requirements

All the code examples for this chapter can be found on GitHub at `https://github.com/PacktPublishing/Bare-Metal-Embedded-C-Programming`.

Introduction to communication protocols

In the world of embedded systems, communication protocols are essential conduits that enable microcontrollers and peripheral devices to talk to each other seamlessly. Think of them as the languages that different devices use to understand and exchange information, ensuring that everything from your smartphone to your smart home devices works smoothly.

Let's dive into what communication protocols are, how they are grouped, their unique features and advantages, and explore some common use cases to see these protocols in action.

What are communication protocols?

Communication protocols are sets of rules and conventions that allow electronic devices to communicate with each other. These protocols define how data is formatted, transmitted, and received, ensuring that devices can exchange information accurately and reliably. Without these protocols, it would be like trying to have a conversation with someone who speaks a completely different language – communication would be chaotic and error-prone.

In embedded systems, these protocols are crucial because they facilitate the interaction between microcontrollers and peripherals such as sensors, actuators, displays, and other microcontrollers. Whether it's sending a simple temperature reading from a sensor to a microcontroller or streaming video data from a camera module, communication protocols make it happen.

Let's analyze the classification of communication protocols, starting with the big picture: what communication protocols can be broadly classified into – **serial** and **parallel** communication.

Serial versus parallel communication

Let's start with serial communication.

Serial communication

In this category, communication protocols can be further broken down into asynchronous and synchronous protocols:

- **Asynchronous**: This type of communication sends data one bit at a time without a clock signal to synchronize the sender and receiver. Think of it as sending letters through the mail without a scheduled delivery time. A common example is UART, which is simple and efficient for many applications.

- **Synchronous**: Unlike asynchronous communication, this form of communication uses a clock signal to coordinate the transmission of bits. It's like having a drumbeat to ensure everyone marches in step. Examples include **Serial Peripheral Interface** (**SPI**) and **Inter-Integrated Circuit** (**I2C**). These protocols ensure data integrity and timing, making them suitable for more complex tasks.

Parallel communication

This type involves transmitting multiple bits simultaneously over multiple channels. Imagine sending a whole fleet of cars instead of a single one – it's faster but requires more lanes (or pins, in our case). While parallel communication is faster, it's less common in embedded systems due to the higher pin count. Also, it's prone to crosstalk and signal integrity problems, especially over longer distances.

We can also classify communication protocols based on their architecture. In this classification system, we have point-to-point communication and multi-device communication.

Point-to-point versus multi-device communication

Let's look at the differences.

Point-to-point

This is a direct line of communication between two devices. **UART** is a classic example, where data flows directly between a microcontroller and a peripheral device. It's straightforward, reliable, and ideal for many embedded systems.

Multi-device (bus) communication

Here, multiple devices share the same communication lines, which can be either of the following:

- **Multi-master**: Multiple devices can control the communication bus. **I2C** is a great example as it allows multiple masters and slaves on the same bus. It's like a group of friends taking turns talking in a conversation.

- **Master-slave**: One master device controls the communication, directing traffic to and from multiple slave devices. **SPI** operates this way, with a single master communicating with multiple slaves through dedicated lines. **I2C** can also operate this way. It's akin to a teacher (master) calling on students one at a time to speak.

Lastly, communication protocols can be classified based on their data flow capabilities.

Full-duplex versus half-duplex

Let's see the differences between full-duplex and half-duplex:

- **Full-duplex**: This allows simultaneous two-way communication. Imagine a two-lane road where cars can travel in both directions at the same time. **UART** and **SPI** support full-duplex communication, making them highly efficient for real-time data exchange.

- **Half-duplex**: Here, communication can occur in both directions, but not at the same time – it's like a single-lane road where cars must take turns. **I2C** typically operates in half-duplex mode, which works well for its intended applications but can be a limitation in high-speed data scenarios.

Now, let's closely compare the three common communication protocols that are used in modern embedded systems.

Comparing UART, SPI, and I2C

Let's start with UART.

UART

Here are some key features of UART:

- **Asynchronous communication**: UART doesn't require a clock signal. Instead, it uses start and stop bits to synchronize data transmission.

- **Full-duplex**: UART can send and receive data simultaneously, which is ideal for many applications requiring real-time communication.

- **Simple and cost-effective**: With minimal hardware requirements, UART is easy to implement and cost-effective.

The following are some of its advantages:

- **Ease of use**: Setting up UART communication is straightforward, making it a popular choice for beginners and simple applications

- **Wide support**: UART is universally supported by most microcontrollers and peripheral devices

- **Low overhead**: The lack of a clock signal means fewer pins are used, reducing complexity

However, it also has some disadvantages:

- **Speed limitations**: UART is generally slower compared to SPI and I2C, making it less suitable for high-speed data transfer

- **Limited distance**: Susceptibility to noise over long distances can limit the range of reliable communication

- **Point-to-point only**: UART is designed for direct, point-to-point communication, which can be a limitation if multiple devices need to communicate

Next, we have SPI.

SPI

Here are some key features of SPI:

- **Synchronous communication**: SPI uses a clock signal along with data lines, ensuring synchronized data transfer
- **Full-duplex**: It allows data to be sent and received simultaneously
- **Master-slave architecture**: One master device controls multiple slave devices, with dedicated lines for each

The following are some of its advantages:

- **High speed**: SPI supports high-speed data transfer, making it ideal for applications requiring fast communication
- **Versatility**: SPI can connect multiple devices with different configurations, providing flexibility in design

However, it also has some disadvantages:

- **More pins required**: Each slave device needs a separate select line, which can increase the pin count significantly
- **No standardized acknowledgment**: Unlike I2C, SPI does not have a built-in acknowledgment mechanism, which can make error detection more challenging
- **Limited multi-master capability**: SPI is not designed for multi-master systems, which can be a limitation in some scenarios

The final common communication protocol we'll cover is I2C.

I2C

Here are some key features of I2C:

- **Synchronous communication**: I2C uses a clock signal for synchronized data transfer
- **Multi-master capability**: Multiple master devices can share the same bus, which is useful in more complex systems
- **Two-wire interface**: I2C requires only two lines (SDA and SCL) for communication, minimizing the pin count

The following are some of its advantages:

- **Simplicity in wiring**: The two-wire interface reduces the complexity and number of pins required

- **Multi-device support**: I2C easily connects multiple devices on the same bus, each with a unique address

- **Built-in addressing**: I2C has a built-in addressing mechanism, making communication with multiple devices straightforward

However, it does have some disadvantages:

- **Slower speed**: I2C is generally slower than SPI, which can be a limitation for high-speed applications

- **Complex protocol**: The protocol is more complex than UART and SPI, requiring more sophisticated handling of data transfers and addressing

- **Susceptible to noise**: Like UART, I2C can be susceptible to noise over longer distances, potentially affecting communication reliability

Choosing the right communication protocol depends on your specific application needs. If you need simple, straightforward communication and can tolerate slower speeds, **UART** is a great choice. For high-speed applications with a need for full-duplex communication, **SPI** is ideal, especially if you can manage the higher pin count. When you need to connect multiple devices with minimal wiring and have a complex communication setup, **I2C** is your go-to protocol. To help you better understand when to choose which protocol, let's explore some common use cases.

Common use cases for the UART, SPI, and I2C protocols

When designing embedded systems, selecting the right communication protocol is crucial for ensuring efficient and reliable data exchange. UART, SPI, and I2C each have unique strengths, making them suitable for different applications. Let's explore the practical use cases and compelling case studies for each protocol, highlighting their professional and real-world relevance.

UART

Let's look at some common use cases for the UART protocol:

- **Serial communication with PCs**: UART is often used for serial communication between microcontrollers and computers, particularly for debugging, firmware updates, and data logging

- **GPS modules**: UART can be used to transmit location data from a GPS module to a microcontroller

- **Bluetooth modules**: UART enables wireless communication with devices via Bluetooth

These use cases represent some of the most common applications of UART, but the protocol is versatile and can be used in many other scenarios that require simple serial communication.

Case study – GPS module integration for autonomous drones

Imagine you're developing an autonomous drone that requires precise navigation to perform tasks such as surveying and mapping. Integrating a GPS module using UART can provide real-time location data essential for navigation.

Setup: Connect the GPS module's transmit (TX) pin to the microcontroller's receive (RX) pin and vice versa. Configure the baud rate so that it matches the GPS module's output.

Operation: The GPS module continuously sends NMEA sentences (text strings) containing location data. The microcontroller reads these strings via UART, parses them, and uses the location information to navigate the drone accurately.

Advantage: UART's simplicity and widespread support make it straightforward to integrate the GPS module, providing reliable and continuous data flow without a complex setup.

Next, we'll look at SPI.

SPI

The following are some common use cases for the SPI protocol:

- **High-speed data transfer**: It's ideal for applications such as memory cards, **analog-to-digital converters (ADCs)**, **digital-to-analog converters (DACs)**, and displays

- **Display modules**: SPI can be used for communicating with high-resolution displays requiring fast refresh rates

- **Sensors and actuators**: SPI can handle high-frequency data outputs from various sensors

Like UART, these examples highlight some typical uses of SPI, but the protocol's high-speed capabilities make it suitable for a wide range of other applications requiring rapid data transfer.

Case study – SD card data logging for industrial equipment

Consider an industrial monitoring system that logs data from various sensors to an SD card for long-term analysis. SPI is the perfect protocol for this high-speed data transfer.

Setup: Connect the microcontroller to the SD card using SPI pins (MISO, MOSI, SCLK, and CS). Initialize the SPI bus and configure the SD card.

Operation: The microcontroller collects data from sensors (for example, temperature, pressure, and vibration) and writes this data to the SD card in real time.

Advantage: SPI's high-speed data transfer ensures that large amounts of data are logged quickly and efficiently, preventing any data loss and ensuring accurate monitoring.

Using SPI in this scenario allows the industrial system to maintain precise logs of critical parameters, which are essential for predictive maintenance and operational efficiency.

Finally, we have I2C.

I2C

Let's consider two common use cases related to I2C:

- **Multiple sensor integration systems**: This involves connecting several sensors with different addresses on the same I2C bus

- **Peripheral expansion**: This involves adding more GPIO pins to a microcontroller using I2C expanders

These use cases are just two examples of I2C's applications. Its ability to support multiple devices on a single bus makes it an excellent choice for many other scenarios where scalability is important.

Case study – environmental monitoring system for smart agriculture

Let's say you're developing a smart agriculture system that uses multiple sensors (temperature, humidity, and soil moisture) to optimize farming conditions. I2C is the ideal protocol for this multi-sensor integration.

Setup: Connect all sensors to the I2C bus (SDA and SCL lines). Assign each sensor a unique address.

Operation: The microcontroller queries each sensor in sequence, collects the data, and processes it to provide insights and control irrigation, ventilation, and lighting systems.

Advantage: I2C's ability to support multiple devices on the same bus with just two lines simplifies wiring, reduces costs, and saves GPIO pins, making it an efficient solution for complex sensor networks.

Starting with the next section, we'll focus exclusively on the UART protocol. We'll cover the I2C and SPI protocols in the following chapters.

Overview of the UART protocol

One of the most fundamental and widely used protocols is UART. Whether you're debugging hardware or enabling communication between a microcontroller and peripherals, understanding UART is crucial. Let's delve into the workings of this protocol.

What is UART?

UART is a hardware communication protocol that operates using asynchronous serial communication, allowing for adjustable data transmission speeds. The "asynchronous" nature of UART means it doesn't require a clock signal to align the transmission of bits between the sender and receiver. Instead, both devices must agree on a specific baud rate, which dictates the speed at which data is exchanged. Let's take a look at the interface.

The interface

The UART interface employs two wires for communication: TX and RX. To establish a connection between two devices, we simply connect the TX pin of the first device to the RX pin of the second device, and the RX pin of the first device to the TX pin of the second device. Additionally, it's crucial to connect the ground pins of both devices to ensure a common electrical reference. *Figure 10.1* shows the connection between two UART devices:

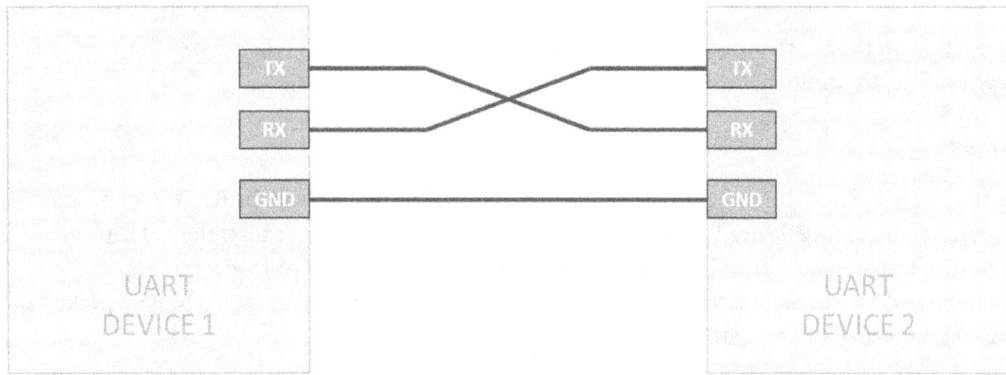

Figure 10.1: The UART interface

How UART works

Data in UART is transmitted as frames containing a **start bit**, **data bits**, an optional **parity bit**, and **stop bits**:

Start Bit	Data Frame	Parity Bits	Stop Bits
1 bit	5 to 9 bits	1bit	1 or 2 bits

Figure 10.2: The UART data packet

Here's a step-by-step breakdown of the process:

1. **Start bit**: The transmission line is normally held high. To start the data transfer, the transmitting UART pulls the line low for one clock cycle. This indicates the start of a new data frame.

2. **Data frame**: Following the start bit, the data frame typically consists of 5 to 9 bits and is sent from the **least significant bit** (**LSB**) to the **most significant bit** (**MSB**).

3. **Parity bit**: This is optional and is used for error checking. It ensures that the number of set bits (1s) in the data is even or odd.

4. **Stop bits**: This is one or two bits indicating the end of the data packet. The line is driven high during the stop bits.

Let's take a closer look at the start, stop, and parity bits.

The start, stop, and parity bits

These bits form the backbone of the UART protocol, allowing devices to synchronize and verify the integrity of the transmitted data.

Start bit

The start bit is the initial signal that marks the beginning of a data frame in UART communication. When the transmitting device is idle, the data line is held at a high voltage level (logic 1). To signal the start of transmission, the UART transmitter pulls the line to a low voltage level (logic 0) for a 1-bit duration. This transition from high to low alerts the receiving device that a new data packet is incoming, allowing it to synchronize and prepare for data reception.

Stop bit

After the data bits and optional parity bit are transmitted, the stop bit signals the end of the data frame. The transmitter drives the data line back to a high voltage level (logic 1) for 1 or 2-bit durations, depending on the configuration. The stop bit(s) ensure that the receiver has time to process the last data bit and prepare for the next start bit. In essence, the stop bit acts as a buffer, providing a clear demarcation between successive data frames and helping maintain synchronization between the communicating devices.

Parity bit

The parity bit is an optional feature that's used for basic error checking in UART communication. It provides a simple method to detect errors that may have occurred during data transmission. The parity bit can be configured for either even or odd parity:

- **Even parity**: The parity bit is set to 0 if the number of 1s in the data frame is even, and set to 1 if the number of 1s is odd. This ensures that the total number of 1s (including the parity bit) is even.

- **Odd parity**: The parity bit is set to 0 if the number of 1s in the data frame is odd, and set to 1 if the number of 1s is even. This ensures that the total number of 1s (including the parity bit) is odd.

When the receiver gets the data frame, it checks the parity bit against the received data bits. If there's a mismatch, it indicates that an error occurred during transmission. While parity doesn't correct errors, it helps in identifying them, prompting for retransmission if necessary.

The start, stop, and parity bits are essential components of UART communication, each playing a critical role in ensuring data integrity and synchronization. The start bit signals the beginning of transmission, the stop bit marks the end, and the parity bit provides a basic error-checking mechanism. Together, they create a robust framework for reliable and efficient serial communication between devices.

Before wrapping up this section, let's take a moment to understand the unit of speed that's used in UART communication.

Understanding the baud rate – the speed of communication in embedded systems

In the world of embedded systems, **baud rate** is a term you'll encounter frequently. Whether you're debugging a microcontroller, setting up a serial communication link, or working with various peripherals, understanding the baud rate is essential. But what exactly is the baud rate, and why is it so important? Let's break it down.

What is the baud rate?

The baud rate is essentially the speed at which data is transmitted over a communication channel. It's measured in **bits per second (bps)**. Think of it as the speed limit on a highway: the higher the baud rate, the more data can travel along the communication path in a given amount of time.

For example, a baud rate of *9,600* means *9,600 bits* of data are transmitted each *second*. In other words, it sets the pace for how fast data packets are sent and received.

However, it's important to distinguish between the baud rate and the **bit rate**. While the baud rate refers to the number of signal changes per second, the bit rate is the number of bits transmitted per second. In simple systems, *each signal change can represent one bit*, making the baud rate and bit rate the same. In more complex systems, each signal change can represent multiple bits, resulting in a bit rate higher than the baud rate.

Why does the baud rate matter?

Imagine trying to have a conversation with someone who speaks at a wildly different speed than you. It would be confusing and inefficient, right? The same principle applies to electronic devices communicating with each other. Both the transmitting and receiving devices need to agree on a common baud rate to understand each other correctly. If they don't, the data might get lost or garbled, leading to communication errors.

For successful communication, both the sender and receiver must have the same baud rate to synchronize correctly. If one device is set to 9,600 bps and the other to 115,200 bps, the communication will fail, similar to how a conversation fails if one person is speaking too fast or too slow for the other to understand.

There are standard baud rates that are commonly used in serial communication. Here are a few:

- **300 bps**: Very slow, often used for long-distance communication where bandwidth is limited
- **9,600 bps**: A widely used default rate for many devices, including microcontrollers
- **19,200 bps**: Faster, often used in more data-intensive applications
- **115,200 bps**: High-speed communication, common in applications requiring quick data transfer

This concludes our overview of the UART protocol. In the next section, we will explore the UART peripheral in the STM32F4 microcontroller.

The STM32F4 UART peripheral

STM32 microcontrollers often include several UART peripherals, though the number varies depending on the specific model. The STM32F411 microcontroller has three UART peripherals:

- USART1
- USART2
- USART6

> **USART versus UART**
>
> Our STM32 documentation refers to the UART peripheral as **USART** because it stands for universal **synchronous/asynchronous** receiver/transmitter. This name reflects the dual functionality of the peripheral:
>
> **Asynchronous mode** (**UART**): In this mode, the USART operates as a traditional UART. It transmits and receives data without needing a clock signal, which is typical for standard serial communication.
>
> **Synchronous mode** (**USART**): In this mode, the USART can also operate with a synchronous clock signal, allowing it to communicate with devices that require a clock line in addition to the data lines.

Let's analyze the key registers of this peripheral, starting with the USART Status Register.

USART Status Register (USART_SR)

The USART_SR register is one of the main registers used to monitor the status of the UART peripheral. It provides real-time information about various operational flags and errors.

Let's consider the key bits in this register:

- **Transmit data register empty (TXE)**: This bit is set when the data register is empty and ready for new data to be written. It indicates that the transmitter can send more data.

- **Read data register not empty (RXNE)**: This bit indicates that the data register contains data that has not been read yet. It signals that there is incoming data to be processed.

- **Transmission complete (TC)**: This bit is set when the last transmission has been completed, including all the stop bits. It shows that the data has been fully sent.

- **Overrun error (ORE)**: This bit indicates that the data was lost because the data register wasn't read before new data arrived. It flags an error condition.

You can find detailed information about this register on *page 547* of the *STM32F411 reference manual (RM0383)*. Next, we have the **USART Data Register** (USART_DR).

USART Data Register (USART_DR)

The USART_DR register is used for both transmitting and receiving data. It acts as the primary interface for data exchange through the UART peripheral.

The following are the key functions in this register:

- **Transmitting data**: Writing a byte to USART_DR sends the data through the TX line. The UART peripheral handles the conversion and transmission serially.

- **Receiving data**: Reading from USART_DR retrieves the data received on the RX line. This should be done promptly to avoid data overrun.

Next, we have the **USART Baud Rate Register** (USART_BRR).

USART Baud Rate Register (USART_BRR)

The USART_BRR register is used to set the baud rate for the UART communication, which is critical for synchronizing the data transfer speed between devices.

This register has two fields:

- **Mantissa**: The integer part of the division factor that sets the baud rate

- **Fraction**: The fractional part of the division factor that fine-tunes the baud rate

The final register we will examine is the **USART Control Register 1** (USART_CR1).

USART Control Register 1 (USART_CR1)

The USART_CR1 register is a comprehensive control register that enables various UART functionalities and configurations.

Let's consider the key bits in this register:

- **USART enable (UE)**: This bit enables or disables the UART peripheral. It must be set to activate UART communication.

- **Word length (M)**: This bit configures the word length, allowing 8-bit or 9-bit data frames.

- **Parity control enable (PCE)**: This bit enables parity checking for error detection.

- **Parity selection (PS)**: This bit selects even or odd parity.

- **Transmitter enable (TE)**: This bit enables the transmitter, allowing data to be sent.

- **Receiver enable (RE)**: This bit enables the receiver, allowing data to be received.

With these registers in mind, we're now ready to develop the UART driver. We will dive into that in the next section.

Developing the UART driver

In this section, we will apply everything we've learned about the UART peripheral to develop a driver for transmitting data using the USART2 peripheral.

Let's begin by identifying the GPIO pins connected to the UART2 peripheral. To do this, refer to the table on *page 39* of the *STM32F411RE datasheet*. This table lists all the GPIO pins of the microcontroller, along with their descriptions and additional functionalities. As shown in *Figure 10.3*, part of this table reveals that PA1 has an alternate function labeled as USART2_TX:

12	16	E5	25	K3	PA2		I/O	FT	-	TIM2_CH3, TIM5_CH3, TIM9_CH1, I2S2_CKIN, USART2_TX, EVENTOUT	ADC1_2

Figure 10.3: The USART2_TX pin

To use PA2 as the USART2_TX line, we need to configure PA2 as an alternate function pin in the GPIOA_MODER register and then specify the alternate function number for USART2_TX in the GPIOA_AFRL register. The STM32F4 microcontroller allows us to choose from 16 different alternate functions, numbered from AF00 to AF15. The alternate function mapping table, which you can find on *page 47* of the datasheet, outlines these functions and their corresponding numbers. As shown in *Figure 10.4*, sourced from the datasheet, configuring PA2 as AF07 will set it to function as the USART2_TX line:

Port	AF00	AF01	AF02	AF03	AF04	AF05	AF06	AF07	AF08	AF09	AF10	AF11	AF12	AF13	AF14	AF15
	SYS_AF	TIM1/TIM2	TIM3/ TIM4/ TIM5	TIM8/ TIM9/ TIM11	I2C1/I2C2/ I2C3	SPI1/I2S1S PI2/ I2S2/SPI3/ I2S3	SPI2/I2S2/ SPI3/ I2S3/SPI4/ I2S4/SPI5/ I2S5	SP03/I2S3/ USART1/ USART2	USART6	I2C2/ I2C3	OTG1_FS		SDIO			
PA2	-	TIM2_CH1/ TIM2_ETR	TIM5_CH1	-	-	-	-	USART2_ CTS	-	-	-	-	-	-	-	EVENT OUT
PA1	-	TIM2_CH2	TIM5_CH2	-	-	SPI4_MOSI/ I2S4_SD	-	USART2_ RTS	-	-	-	-	-	-	-	EVENT OUT
PA2	-	TIM2_CH3	TIM5_CH3	TIM9_CH1	-	I2S2_CKIN	-	USART2_ TX	-	-	-	-	-	-	-	EVENT OUT

Figure 10.4: PA2 alternate function

🔍 **Quick tip**: Need to see a high-resolution version of this image? Open this book in the next-gen Packt Reader or view it in the PDF/ePub copy.

🔖 **The next-gen Packt Reader** and a **free PDF/ePub copy** of this book are included with your purchase. Unlock them by scanning the QR code below or visiting `https://www.packtpub.com/unlock/9781835460818`.

We now have all the information we need to develop the UART2 transmitter driver.

Create a copy of your previous project and rename it UART. Next, create a new file named uart.c in the Src folder and another file named uart.h in the Inc folder. Populate your uart.c file with the following code:

```c
#include <stdint.h>
#include "uart.h"

#define GPIOAEN            (1U<<0)
#define UART2EN            (1U<<17)

#define DBG_UART_BAUDRATE        115200
#define SYS_FREQ             16000000
#define APB1_CLK             SYS_FREQ
#define CR1_TE             (1U<<3)
#define CR1_UE             (1U<<13)
```

```c
#define SR_TXE                    (1U<<7)


static void uart_set_baudrate(uint32_t periph_clk,uint32_t baudrate);
static void uart_write(int ch);

int __io_putchar(int ch)
{
    uart_write(ch);
    return ch;
}

void uart_init(void)
{
    /*Enable clock access to GPIOA*/
    RCC->AHB1ENR |= GPIOAEN;

    /*Set the mode of PA2 to alternate function mode*/
    GPIOA->MODER &=~(1U<<4);
    GPIOA->MODER |=(1U<<5);

    /*Set alternate function type to AF7(UART2_TX)*/
    GPIOA->AFR[0] |=(1U<<8);
    GPIOA->AFR[0] |=(1U<<9);
    GPIOA->AFR[0] |=(1U<<10);
    GPIOA->AFR[0] &=~(1U<<11);

    /*Enable clock access to UART2*/
     RCC->APB1ENR |=    UART2EN;

    /*Configure uart baudrate*/

      uart_set_baudrate(APB1_CLK,DBG_UART_BAUDRATE);

    /*Configure transfer direction*/
     USART2->CR1 = CR1_TE;

    /*Enable UART Module*/
     USART2->CR1 |= CR1_UE;
}
```

```
static void uart_write(int ch)
{
    /*Make sure transmit data register is empty*/
    while(!(USART2->SR & SR_TXE)){}

    /*Write to transmit data register*/
    USART2->DR =(ch & 0xFF);
}
static uint16_t compute_uart_bd(uint32_t periph_clk,uint32_t baudrate)
{
    return((periph_clk + (baudrate/2U))/baudrate);
}

static void uart_set_baudrate(uint32_t periph_clk,uint32_t baudrate)
{
    USART2->BRR = compute_uart_bd(periph_clk,baudrate);
}
```

Let's break it down.

First, we have the necessary includes and macros.

```
#include <stdint.h>
#include "uart.h"

#define GPIOAEN (1U<<0)
#define UART2EN (1U<<17)

#define DBG_UART_BAUDRATE 115200
#define SYS_FREQ 16000000
#define APB1_CLK SYS_FREQ
#define CR1_TE (1U<<3)
#define CR1_UE (1U<<13)
#define SR_TXE (1U<<7)
```

Here are the uses of the macros:

- GPIOAEN: This macro enables the clock for GPIOA by setting bit 0 in the AHB1ENR register.

- UART2EN: This macro enables the clock for UART2 by setting bit 17 in the APB1ENR register.

- DBG_UART_BAUDRATE: This macro defines the baud rate for UART communication, set to 115200 bps.

- SYS_FREQ: This macro defines the system frequency, set to 16 MHz, and the default frequency of the STM32F411 microcontroller on the NUCLEO development board.

- APB1_CLK: This macro sets the APB1 peripheral clock frequency to the system frequency (16 MHz).

- CR1_TE: This macro enables the transmitter by setting bit 3 in the USART_CR1 register.

- CR1_UE: This macro enables the UART module by setting bit 13 in the USART_CR1 register.

- SR_TXE: This macro represents the TXE bit in the USART_SR register.

Next, we have the helper functions for computing and setting the baud rate:

```
static uint16_t compute_uart_bd(uint32_t periph_clk, uint32_t
baudrate)
{
    return ((periph_clk + (baudrate / 2U)) / baudrate);
}
```

This helper function calculates the baud rate divisor. It uses the peripheral clock and desired baud rate to compute the value to be set in the **Baud Rate Register** (**BRR**):

```
static void uart_set_baudrate(uint32_t periph_clk, uint32_t baudrate)
{
    USART2->BRR = compute_uart_bd(periph_clk, baudrate);
}
```

This function sets the baud rate for UART2 by writing the computed divisor to the BRR. Let's turn our focus to the initialization function:

```
RCC->AHB1ENR |= GPIOAEN;
```

This line enables the clock for GPIOA by setting the appropriate bit in the AHB1 peripheral clock enable register:

```
GPIOA->MODER &= ~(1U << 4);
GPIOA->MODER |= (1U << 5);
```

These lines configure pin PA2 to operate in alternate function mode, which is necessary for UART functionality:

```
GPIOA->AFR[0] |= (1U << 8);
GPIOA->AFR[0] |= (1U << 9);
GPIOA->AFR[0] |= (1U << 10);
GPIOA->AFR[0] &= ~(1U << 11);
```

These lines configure PA2 as an alternate function (AF7), which corresponds to UART2_TX:

```
RCC->APB1ENR |= UART2EN;
```

This line enables the clock for UART2 by setting the appropriate bit in the APB1 peripheral clock enable register:

```
uart_set_baudrate(APB1_CLK, DBG_UART_BAUDRATE);
```

This function call sets the baud rate for UART2 using the uart_set_baudrate() function:

```
USART2->CR1 = CR1_TE;
```

This configures UART2 for transmission by setting the transmitter enable bit in the control register:

```
USART2->CR1 |= CR1_UE;
```

This enables the UART2 module by setting the UART enable bit in the control register.

Next, we have the function for writing to UART:

```
static void uart_write(int ch)
{
    /* Make sure transmit data register is empty */
    while (!(USART2->SR & SR_TXE)) {}

    /* Write to transmit data register */
    USART2->DR = (ch & 0xFF);
}
```

Let's break it down:

```
while (!(USART2->SR & SR_TXE)) {}
```

This loop ensures that the transmit data register is empty before we write new data:

```
USART2->DR = (ch & 0xFF);
```

This line writes the character to the data register for transmission.

Finally, we have a useful function that allows us to redirect printf output to our UART transmitter:

```
int __io_putchar(int ch)
{
    uart_write(ch);
    return ch;
}
```

It calls `uart_write()` to send the character and then returns the character.

After sending the character, `__io_putchar` returns the same character, `ch`.

Returning the character is a standard practice, allowing the function to comply with the typical `putchar` function signature, which returns the character written as an `int` variable.

Our next task is to populate the `uart.h` file. Here's the code:

```
#ifndef __UART_H__
#define __UART_H__
#include "stm32f4xx.h"
void uart_init(void);
#endif
```

Here, we are simply exposing the uart initialization function implemented in `uart.c`, making it callable from other files. We are now ready to test our driver in `main.c`. Update your `main.c` file, like so:

```
#include <stdio.h>
#include "uart.h"
int main(void)
{
    /*Initialize debug UART*/
    uart_init();
    while(1)
    {
        printf("Hello from STM32...\r\n");
    }
}
```

This main function simply initializes the UART2 peripheral and then continuously prints the sentence `Hello from STM32....`

Let's test the project. To do so, we'll need to install a program on our computer that can display the data that's received through the computer's serial port. In this setup, our development board acts as the transmitter, while the computer is the receiver.

1. **Install a serial terminal program**:

   - Choose a serial terminal program that's appropriate for your operating system. Options include *Realterm*, *Tera Term*, *Hercules*, and *Cool Term*.

   - If you're using Windows, I recommend Realterm. You can download it from SourceForge: `https://sourceforge.net/projects/realterm/`.

   - Follow the installation wizard to complete the setup.

2. **Prepare to identify your development board's serial port**:

   I.   Disconnect your development board from your computer.

   II.   Open Realterm and navigate to the **Port** tab.

   III. Click on the **Port** drop-down menu; you'll see a list of available ports. Since your development board is currently disconnected, its port won't appear in the list. **Take note** of the listed ports.

3. **Identify the development board's port**:

   I.   Close Realterm and connect your development board to the computer.

   II.   Reopen Realterm and go back to the **Port** drop-down menu. You should now see a new port in the list, which corresponds to your development board.

   III. Select this newly added port.

4. **Set the baud rate**:

   Click the **Baud** drop-down menu and select **115200**. This is the baud rate we configured in our driver.

5. **Build and run the project**:

   Return to your IDE, build the project, and run the firmware on your microcontroller.

6. **Test the setup**:

   - Go back to Realterm and click the **Open** button to start the communication.

   - You should see a message stating `Hello from STM32...` continuously being printed in the Terminal window.

Figure 10.5 shows the settings described for Realterm:

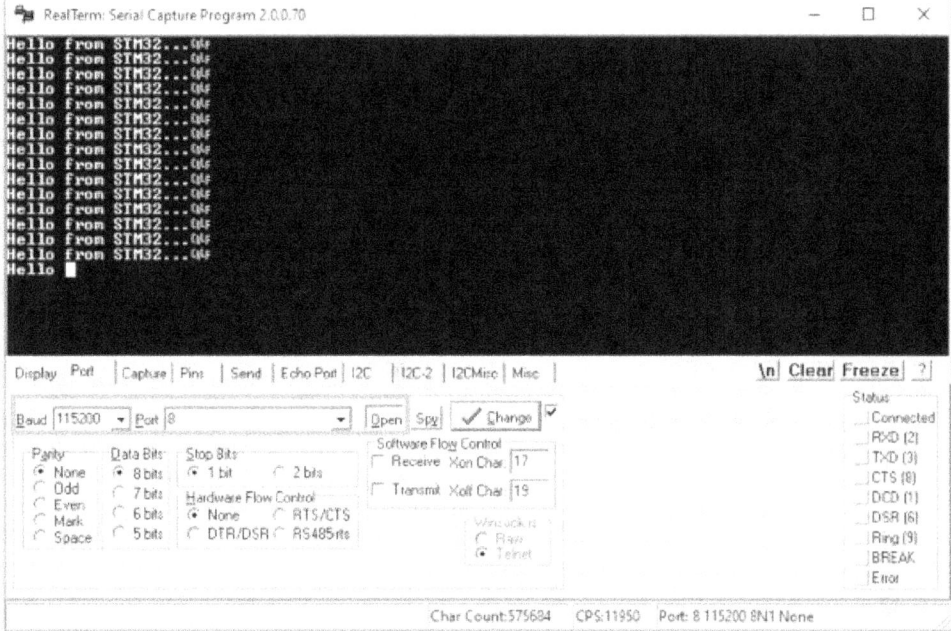

Figure 10.5: Realterm settings

Summary

In this chapter, we learned about the UART protocol, a fundamental communication method that's widely used in embedded systems. We began by discussing the importance of communication protocols in embedded systems, emphasizing how UART, alongside SPI and I2C, facilitates seamless communication between microcontrollers and peripheral devices.

Next, we provided a detailed overview of the UART protocol while covering its operational principles, including how data is transmitted asynchronously using start and stop bits, and the role of parity in error checking. We also discussed how the baud rate, a critical aspect of UART communication, is configured to ensure synchronized data transfer between devices.

Then, we delved into the specifics of the STM32 UART peripheral, examining key registers such as the Status Register (USART_SR), Data Register (USART_DR), Baud Rate Register (USART_BRR), and Control Register 1 (USART_CR1). Understanding these registers is essential for configuring UART for effective communication in STM32 microcontrollers.

Finally, we applied our theoretical understanding by developing a bare-metal UART driver for the STM32F4 microcontroller. This involved initializing the UART peripheral, setting the baud rate, and implementing functions for transmitting data. We also demonstrated how to redirect `printf` output to the UART, enabling easy debugging and data logging through a serial terminal.

In the next chapter, we will learn about the **analog-to-digital converter (ADC)**.

Unlock this book's exclusive benefits now

This book comes with additional benefits designed to elevate your learning experience.

Note: Have your purchase invoice ready before you begin.

https://www.packtpub.com/
unlock/9781835460818

11

Analog-to-Digital Converter (ADC)

In this chapter, we will learn about the **analog-to-digital converter** (ADC), an important peripheral in embedded systems that enables the microcontroller to interface with the analog world. We will start by providing an overview of the analog-to-digital conversion process, the importance of the ADC, and its key specifications.

Following this, we will extract and analyze the relevant registers from the STM32F411 reference manual that are necessary for ADC operations. Finally, we will develop a bare-metal ADC driver to demonstrate the practical application of the theoretical concepts we've discussed.

In this chapter, we're going to cover the following main topics:

- Overview of analog-to-digital conversion
- The STM32F4 ADC peripheral
- The key ADC registers and flags
- Developing the ADC driver

By the end of this chapter, you will have a comprehensive understanding of the STM32 ADC peripheral and the skills necessary to develop efficient drivers for it, enabling you to effectively integrate analog-to-digital conversion capabilities into your embedded systems projects.

Technical requirements

All the code examples for this chapter can be found on GitHub at `https://github.com/PacktPublishing/Bare-Metal-Embedded-C-Programming`.

Overview of analog-to-digital conversion

ADC is a critical process in embedded systems that allows our microcontrollers to interpret and process real-world analog signals. In this section, we will walk through this process, explaining each step involved in converting an analog signal into digital values.

What is analog-to-digital conversion?

Analog-to-digital conversion is the process of converting a continuous analog signal into a discrete digital representation. Analog signals, which can have any value within a certain range, are transformed into digital signals, which have specific, quantized levels. This conversion is essential because microcontrollers and digital systems can only process digital data.

The conversion process typically involves several key steps: sampling, quantization, and encoding. Let's break down these steps, starting with sampling.

Sampling

Sampling involves measuring the amplitude of an analog signal at regular intervals, called **sampling intervals**. The result is a series of discrete values that approximate the original analog signal. *Figure 11.1* depicts this process:

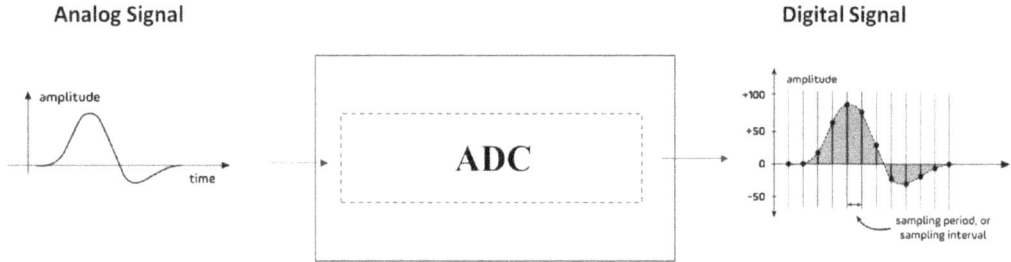

Figure 11.1: The sampling process

The rate at which the analog signal is sampled is known as the **sampling rate** or **sampling frequency**. This is typically measured in samples per second (Hz). According to the **Nyquist Theorem**, the sampling rate must be at least twice the highest frequency present in the analog signal to accurately reconstruct the original signal.

The next step in the process is quantization.

Quantization

Quantization is the process of mapping the sampled analog values to the nearest discrete levels available in the digital domain. Each discrete level corresponds to a unique digital code.

The number of discrete levels available for quantization is determined by the resolution of the analog-to-digital converter. For example, an *8-bit ADC has 256 levels (28)*, while a *12-bit ADC has 4096 levels (212)*.

The quantization process inherently introduces an error, known as **quantization error** or **quantization noise**, because the exact analog value is approximated to the nearest digital level. We can minimize this error by increasing the resolution of the ADC. For example, if an analog signal ranges *from 0 to 3.3V* and an *8-bit ADC* is used, the quantization step size is approximately *12.9 mV (3.3V / 256)*. An analog input of *1.5V* might be quantized to the closest digital level, which could be slightly higher or lower than *1.5V*.

The final step in the process is encoding.

Encoding

Encoding is the final step and is where the quantized levels are converted into a binary code that can be processed by the digital system. Each quantized level is represented by a unique binary value.

The number of bits used in the ADC determines the binary code length. For example, a *10-bit* ADC will produce a *10-bit* binary number for each sampled value. Continuing with our previous example, if the quantized level for *1.5V* is determined to be level *116*, the binary representation would be *01110100* for an *8-bit* ADC.

Figure 11.2 shows the encoding process of a *6-bit ADC*. The columns in the table show the 6-bit binary representation of the quantization levels. For a 6-bit ADC, the digital output ranges from `000000` for the lowest quantization level to `111111` for the highest. Each binary value corresponds to a specific quantization level:

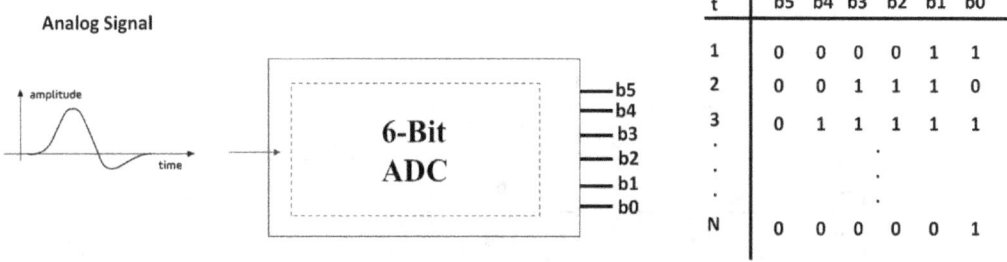

t	b5	b4	b3	b2	b1	b0
1	0	0	0	0	1	1
2	0	0	1	1	1	0
3	0	1	1	1	1	1
.					.	
.			.			
.	.					
N	0	0	0	0	0	1

Figure 11.2: The encoding process

In summary, the analog-to-digital conversion process begins with an analog input signal, which can vary continuously over time. This signal could be a voltage from a temperature sensor, an audio signal, or any other analog signal. The analog-to-digital converter typically includes a sample-and-hold circuit that captures and holds the analog signal at each sampling interval. This ensures that the signal remains constant during the conversion process. The core of the ADC performs quantization and encoding. It compares the held analog value to a set of reference voltages to determine the closest matching digital level. The resulting digital code is output from the ADC and can be read by our microcontroller or digital system for further processing.

In the next section, we will explain some of the key terms that were used in this section, including resolution and **reference voltage** (**VREF**).

Key specifications of the ADC – resolution, step size, and VREF

To effectively use ADCs, it's important to understand their key specifications, which define their performance for various applications. Let's start with resolution.

Resolution

The resolution of an ADC determines the number of distinct output levels it can produce, corresponding to the number of intervals the input voltage range is divided into. It is typically expressed in bits. Higher resolution allows us to have a more precise representation of the analog input signal, reducing quantization error and improving the accuracy of measurements.

For an *N-bit* ADC, the number of discrete output levels is *2N*. For example, an *8-bit* ADC has *256(28)* levels, while a *12-bit* ADC has *4,096* levels.

Table 11.1 shows common ADC resolutions and their corresponding number of discrete levels. The following table highlights how the number of discrete levels increases exponentially with the resolution, providing finer granularity in the digital representation of the analog input signal:

ADC Resolution (bits)	Number of Discrete Levels (2^N)
8	256
10	1,024
12	4,096
14	16,384
16	65,536
18	262,144
20	1,048,576
24	16,777,216

Table 11.1: Common ADC resolutions

The next key specification is the **VREF**.

VREF

The VREF is the maximum voltage that our ADC can convert. The analog input voltage is compared to this VREF to produce a digital value. The stability and accuracy of the VREF directly impact the accuracy of the ADC as any fluctuations in the VREF can cause corresponding errors in the digital output.

We can choose to derive the VREF from the microcontroller or provide an external one for more precise applications. Internal references are convenient but might have higher variability, while external references can offer better stability and accuracy. The choice of VREF depends on our application's accuracy requirements and the nature of the analog signal being measured.

For example, if the VREF is 5V, the ADC can accurately convert any analog input signal within the range of 0V to 5V.

The last specification we'll examine is the step size.

Step size

The step size is the smallest change in analog input that can be distinguished by the ADC. It is determined by the VRED and the resolution and it determines the granularity of the ADC's output. A smaller step size indicates finer resolution, allowing the ADC to detect smaller changes in the input signal.

The step size is calculated by *dividing the VREF by 2 raised to the power of the ADC's resolution* (number of bits):

$Step\ size= \frac{VREF}{2^N}$

Here, *VREF* is the reference voltage and *N* is the resolution bits.

For example, for a 10-bit ADC with VREF = 3.3V:

$Step\ size= \frac{3.3V}{2^{10}} = \frac{3.3V}{1024} = 3.22mV$

This means that each increment in the quantized digital output corresponds to a *3.22mV* change in the analog input. *Table 11.2* lists common ADC resolutions, the corresponding number of steps, and the step size using a VREF of 3.3V:

ADC Resolution (Bits)	Number of Steps (2^N)	Step Size (mV)
8	256	12.9
10	1,024	3.22
12	4,096	0.805
14	16,384	0.201
16	65,536	0.0504

ADC Resolution (Bits)	Number of Steps (2^N)	Step Size (mV)
18	262,144	0.0126
20	1,048,576	0.0032
24	16,777,216	0.000197

Table 11.2: ADC resolutions and step sizes at 3.3V VREF

This table provides a clear view of how the step size decreases with increasing resolution, allowing for finer granularity in the digital representation of the analog input.

This concludes our overview of the analog-to-digital conversion process. In the next section, we will examine the ADC peripheral of our STM32 microcontroller.

The STM32F4 ADC peripheral

Our STM32F411 microcontroller features a 12-bit ADC capable of measuring signals from up to *19 multiplexed channels*. The ADC can operate in various modes, such as single, continuous, scan, or discontinuous, with the results stored in *a 16-bit data register*. Additionally, the ADC has an analog watchdog feature that allows the system to detect when the input voltage exceeds predefined thresholds.

Before we explain the various ADC modes, let's understand what we mean by ADC channels.

The ADC channels

An **ADC channel** is a dedicated pathway through which an analog signal is fed into the ADC so that it can be converted into a digital value. Each ADC channel corresponds to a specific GPIO pin configured to operate in analog mode.

Sensors, which produce analog signals representing physical phenomena (such as temperature, light, or pressure), are interfaced with our microcontroller through these GPIO pins. By configuring a GPIO pin as an analog input, the microcontroller can receive the sensor's analog output signal on the corresponding ADC channel. The ADC then converts this continuous analog signal into a discrete digital representation that the microcontroller can process, analyze, and use for further decision-making tasks in our embedded systems applications.

You might be wondering, does having 19 channels mean we have 19 separate ADC modules? This is where multiplexing comes into play.

Multiplexing ADC channels

Multiplexing allows the ADC to switch between different input signals, sampling each one sequentially. This is achieved using an analog multiplexer (MUX) within the ADC peripheral. As we learned earlier, each of the ADC channels is connected to a specific GPIO pin configured for analog input. The analog MUX selects which analog input signal (from the GPIO pins or internal sources) is connected to the ADC's sampling circuitry at any given time. This selection is controlled by the ADC's configuration registers.

Figure 11.3 shows the ADC channels' connection to the analog multiplexor within the ADC peripheral block:

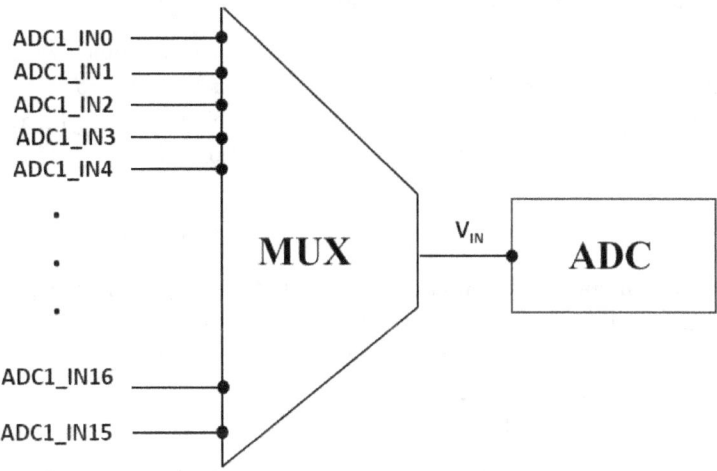

Figure 11.3: ADC channel multiplexing

Now, let's examine the available ADC modes.

The ADC modes

The ADC in our STM32F411 microcontroller can operate in several modes, each tailored to specific application requirements. The primary modes of operation include single conversion mode, continuous conversion mode, scan mode, discontinuous mode, and injected conversion mode.

Let's break them down:

- **Single conversion mode**: In this mode, the ADC performs a single conversion of the selected channel and then stops. This mode is useful for applications where periodic or event-driven sampling of an analog signal is required. To select this mode, we have to set the **CONT** bit to 0 in the ADC_CR2 register.

  Example use case: Reading the value from a temperature sensor at specific intervals.

- **Continuous conversion mode**: This mode allows the ADC to repeatedly convert the input signal. After each conversion, the next conversion starts automatically. This mode is enabled by setting the **CONT** bit to 1 in the ADC_CR2 register.

  Example Use Case: Continuously monitoring a potentiometer to track its position in real time.

- **Scan mode**: This mode is used to convert a sequence of analog channels. We use this mode when we need to sample multiple signals in a defined order. This mode is enabled by setting the **SCAN** bit in the ADC_CR1 register and then configuring the sequence of channels in the ADC_SQRx registers. If the **CONT** bit is set in ADC_CR2, the sequence restarts after the last channel is converted.

  Example use case: Sampling multiple sensor inputs in a data acquisition system.

- **Discontinuous mode**: This mode allows us to convert a subset of channels within a sequence. It reduces the number of channels converted in each sequence, which can be useful for power saving or reducing processing load. This mode can be enabled by setting the **DISCEN** bit in the ADC_CR1 register and then defining the number of channels to convert in each group by setting the **DISCNUM** bits in the ADC_CR1 register.

  Example use case: Reducing the sampling rate for a subset of channels in a multi-channel system.

- **Injected conversion mode**: This mode is designed for higher-priority conversions that can interrupt regular conversions and is useful for applications that require precise timing for specific signal measurements. We can use this mode by configuring the injected channels and their sequence in the ADC_JSQR register, and then starting the conversion by setting the **JSWSTART** bit in the ADC_CR2 register or via an external trigger.

  Example use case: Prioritizing critical cell voltage measurements in a **battery management system** (**BMS**) during rapid charging or discharging.

Before exploring the common ADC registers, let's understand the two types of channels available in the STM32F411 microcontroller.

Understanding regular channels versus injected channels in STM32F411 ADC

In the STM32F411 microcontroller, the ADC offers a versatile approach to handling multiple analog inputs through two main types of channels: regular channels and injected channels. Regular channels are configured for routine, sequential conversions, ideal for periodic data acquisition from sensors where timing is not extremely critical. These channels follow a predefined sequence set by the ADC_SQRx registers and can be triggered by software or external events.

In contrast, injected channels, such as those configured in injected conversion mode, are designed for *high-priority, time-sensitive tasks*, interrupting the regular sequence to perform immediate conversions when specific conditions are met. This makes injected channels perfect for capturing critical measurements with precise timing, such as motor current sensing in control applications. Additionally, the ADC includes an *Analog Watchdog feature*, which can monitor both regular and injected channels for values that exceed predefined thresholds, generating interrupts to handle out-of-range conditions. This dual-channel capability, combined with the Analog Watchdog, provides a robust framework for diverse applications, from routine environmental monitoring to critical real-time data processing and safety monitoring.

In the next section, we will examine the key registers of the ADC peripheral and some of the flags associated with the ADC operations.

The key ADC registers and flags

In this section, we will explore the characteristics and functions of some of the crucial registers within the ADC peripheral.

Let's start with ADC Control Register 1 (`ADC_CR1`).

ADC Control Register 1 (ADC_CR1)

This is one of the main control registers that's used to configure the ADC's operational settings. It provides various configuration options, such as resolution, scan mode, discontinuous mode, and interrupt enable.

The following are the key bits in this register:

- **RES[1:0]** (**resolution bits**): These bits set the resolution of the ADC (12-bit, 10-bit, 8-bit, or 6-bit)
- **SCAN** (**scan mode**): Setting this bit enables scan mode, allowing the ADC to convert multiple channels in sequence
- **DISCEN** (**discontinuous mode**): When set, this bit enables discontinuous mode on regular channels
- **AWDEN** (**Analog Watchdog enable**): This bit enables the Analog Watchdog on all regular channels
- **EOCIE** (**end of conversion interrupt enable**): When set, this bit allows an interrupt to be generated when the EOC flag is set

You can find detailed information about this register on page 229 of the STM32F411 reference manual (RM0383).

Next, we have ADC Control Register 2 (`ADC_CR2`).

ADC Control Register 2 (ADC_CR2)

This is another crucial control register that handles different aspects of ADC operation, including the start of conversion, data alignment, and external triggers.

Here are the key bits in this register:

- **ADON** (**ADC on**): This bit turns the ADC on or off
- **CONT** (**continuous conversion**): Setting this bit enables continuous conversion mode
- **SWSTART** (**start conversion of regular channels**): Setting this bit starts the conversion of regular channels
- **ALIGN** (**data alignment**): This bit sets the alignment of the converted data (right or left)
- **EXTEN[1:0]**: This is an external trigger that enables polarity selection for regular channels

Further information about this register can be found on page 231 of the reference manual.

Let's move on to the ADC Regular Sequence Register (ADC_SQRx).

ADC Regular Sequence Register (ADC_SQRx)

The ADC_SQRx registers define the sequence in which the ADC converts the channels. There are multiple SQR registers to handle the sequence for up to 16 regular channels.

Here are the key bits in this register:

- **L[3:0]**: Regular channel sequence length. These bits set the total number of conversions in the regular sequence.
- **SQ1-SQ16**: Regular channel sequence. These bits specify the order of the channels to be converted.

You can read more about this register on page 235 of the reference. The next crucial register is the ADC Data Register (ADC_DR)

ADC Data Register (ADC_DR)

The ADC_DR register holds the result of the conversion. This is where the digital representation of the analog input is stored after the conversion is complete. The register is read-only and the data is stored in the lower 16 bits of the register.

The final register we will examine is the ADC Status Register (ADC_SR).

ADC Status Register (ADC_SR)

This register holds various status flags that indicate the state of the ADC. These flags are essential for monitoring the ADC's operation and handling interrupts. We'll examine these flags in the next section.

The key ADC flags

ADC flags are status indicators that inform the system about the state of the ADC operations. These flags are essential for monitoring the ADC's progress, handling interrupts, and managing errors.

The key ADC flags in the STM32F411 are as follows:

- **End of conversion (EOC) flag**: The **EOC** flag indicates that the ADC has completed a conversion, and the result is available in the data register. It is located in the ADC_SR register at bit position 1 (EOC) and is set by hardware when a regular conversion finishes.

  If the **EOCIE** bit in the ADC_CR1 register is set, the EOC flag can trigger an interrupt. In this case, an interrupt service routine can be triggered to process the converted data.

- **End of injected conversion (JEOC) flag**: This flag indicates that an injected conversion sequence has been completed and the result is available in the injected data register. It is located in the ADC_SR register at bit position 2 (**JEOC**) and is set by hardware at the end of the conversion of all injected channels in the group. Similar to **EOC**, this flag can generate an interrupt if its corresponding interrupt enable bit (**JEOCIE**) is set in the ADC_CR1 register.

- **Analog Watchdog (AWD) flag**: This flag indicates that the converted value has exceeded the predefined high or low thresholds set for the analog watchdog. It is located in the ADC_SR register at bit position 0 (**AWD**). It can also generate an interrupt if its corresponding interrupt enable bit (**AWDIE**) is set in the ADC_CR1 register.

- **Overrun (OVR) flag**: This flag indicates that a new conversion result has overwritten the previous data before it was read. It is located in the ADC_SR register at bit position 5 (**OVR**). It can also generate an interrupt if its corresponding interrupt enable bit (**OVRIE**) is set in the ADC_CR1 register.

- **Start conversion (STRT) flag**: The **STRT** flag indicates that an ADC conversion has started. We can use this flag to verify that the ADC has initiated a conversion process.

Understanding and effectively using ADC flags is crucial for managing ADC operations in our STM32 microcontroller. Flags such as EOC, JEOC, AWD, OVR, and STRT provide essential information about the status of conversions, data integrity, and threshold monitoring. By leveraging these flags, we can enhance the reliability and functionality of our ADC implementations, ensuring accurate and timely data acquisition and processing in our embedded systems projects.

In the next section, we will apply the information we've learned to develop an ADC driver for reading analog sensor values.

Developing the ADC driver

In this section, we will apply everything we have learned about the ADC peripheral to develop a driver for reading sensor values from a sensor connected to one of the ADC channels.

Identifying the GPIO pins for the ADC

Let's begin by identifying the GPIO pins connected to the ADC channels. To do this, refer to the table on *page 39* of the *STM32F411RE datasheet*. This table lists all the GPIO pins of the microcontroller, along with their descriptions and additional functionalities. As shown in *Figure 11.4*, part of this table reveals that **PA1** has an additional function labeled as ADC1_IN1. This indicates that **PA1** is connected to ADC1, **channel 1**:

Table 8. STM32F411xC/xE pin definitions (continued)

Pin number					Pin name (function after reset)[1]	Pin type	I/O structure	Notes	Alternate functions	Additional functions
UFQFPN48	LQFP64	WLCSP49	LQFP100	UFBGA100						
10	14	F6	23	L2	PA0-WKUP	I/O	TC	(5)	TIM2_CH1/TIM2_ET, TIM5_CH1, USART2_CTS, EVENTOUT	ADC1_0, WKUP1
11	15	G7	24	M2	PA1	I/O	FT	-	TIM2_CH2, TIM5_CH2, SPI4_MOSI/I2S4_SD, USART2_RTS, EVENTOUT	ADC1_1

Figure 11.4: Pin definitions

Let's configure **PA1** so that it functions as an ADC pin.

First, create a copy of your previous project in your IDE, following the steps outlined in earlier chapters. Rename this copied project to ADC. Next, create a new file named adc.c in the Src folder and another file named adc.h in the Inc folder.

Populate your adc.c file with the following code:

```
#include "adc.h"

#define GPIOAEN          (1U<<0)
#define ADC1EN           (1U<<8)
#define ADC_CH1          (1U<<0)
#define ADC_SEQ_LEN_1    0x00
```

```c
#define CR2_ADCON       (1U<<0)
#define CR2_CONT            (1U<<1)
#define CR2_SWSTART     (1U<<30)
#define SR_EOC          (1U<<1)

void pa1_adc_init(void)
{
    /****Configure the ADC GPIO Pin**/
    /*Enable clock access to GPIOA*/
    RCC->AHB1ENR |= GPIOAEN;

    /*Set PA1 mode to analog mode*/
    GPIOA->MODER |=(1U<<2);
    GPIOA->MODER |=(1U<<3);

    /****Configure the ADC Module**/
    /*Enable clock access to the ADC module*/
    RCC->APB2ENR |=ADC1EN;

    /*Set conversion sequence start*/
    ADC1->SQR3 = ADC_CH1;

    /*Set conversion sequence length*/
    ADC1->SQR1 = ADC_SEQ_LEN_1;

    /*Enable ADC module*/
    ADC1->CR2 |=CR2_ADCON;

}


void start_conversion(void)
{
    /*Enable continuous conversion*/
    ADC1->CR2 |=CR2_CONT;

    /*Start ADC conversion*/
    ADC1->CR2 |=CR2_SWSTART;
}


uint32_t adc_read(void)
{
```

```
/*Wait for conversion to be complete*/
while(!(ADC1->SR & SR_EOC)){}

/*Read converted value*/
return (ADC1->DR);
}
```

Let's break down the source code, starting with the macro definitions:

```
#define GPIOAEN        (1U<<0)
#define ADC1EN         (1U<<8)
#define ADC_CH1         (1U<<0)
#define ADC_SEQ_LEN_1   0x00
#define CR2_ADCON      (1U<<0)
#define CR2_CONT      (1U<<1)
#define CR2_SWSTART     (1U<<30)
#define SR_EOC          (1U<<1)
```

Let's break down the macros:

- GPIOAEN: This macro enables the clock for GPIOA by setting bit 0 in the AHB1ENR register
- ADC1EN: This enables the clock for ADC1 by setting bit 8 in the APB2ENR register
- ADC_CH1: This selects channel 1 for the ADC conversion in the SQR3 register
- ADC_SEQ_LEN_1: This sets the conversion sequence length to 1 in the SQR1 register
- CR2_ADCON: This enables the ADC module by setting bit 0 in the CR2 register
- CR2_CONT: This enables continuous conversion mode by setting bit 1 in the CR2 register
- CR2_SWSTART: This starts the ADC conversion by setting bit 30 in the CR2 register
- SR_EOC: This macro waits for the end of conversion by reading bit 1 in the **status register** (**SR**)

Next, we must analyze the configuration sequence of the GPIO pin that's used for ADC functionality:

```
/* Enable clock access to GPIOA */
RCC->AHB1ENR |= GPIOAEN;
```

This line enables the clock for GPIOA by setting the appropriate bit in the AHB1ENR register using the **GPIOAEN** macro:

```
/* Set PA1 mode to analog mode */
    GPIOA->MODER |= (1U<<2);
    GPIOA->MODER |= (1U<<3);
```

These lines configure **PA1** as an analog input by setting bits 2 and 3 in the GPIOA_MODER register.

Let's move on to the part of the code that configures the ADC parameters:

```
/* Enable clock access to the ADC module */
RCC->APB2ENR |= ADC1EN;
```

This line enables the clock for ADC1 by setting the appropriate bit in the APB2ENR register using the **ADC1EN** macro.

```
/* Set conversion sequence start */
ADC1->SQR3 = ADC_CH1;
```

This line sets channel 1 as the start of the conversion sequence in the ADC_SQR3 register using the **ADC_CH1** macro:

```
/* Set conversion sequence length */
ADC1->SQR1 = ADC_SEQ_LEN_1;
```

This line sets the sequence length to 1 in the ADC_SQR1 register using the **ADC_SEQ_LEN_1** macro, meaning only one channel will be converted:

```
/* Enable ADC module */
ADC1->CR2 |= CR2_ADCON;
```

This line enables the ADC module by setting the **ADCON** bit in the ADC_CR2 register.

Next, we can start the conversion:

```
/* Enable continuous conversion */
ADC1->CR2 |= CR2_CONT;
```

This line enables continuous conversion mode by setting the **CONT** bit in the ADC_CR2 register using the **CR2_CONT** macro:

```
/* Start ADC conversion */
ADC1->CR2 |= CR2_SWSTART;
```

This line starts the ADC conversion by setting the **SWSTART** bit in the ADC_CR2 register using the **CR2_SWSTART** macro.

Next, we must wait for the results to be ready:

```
/* Wait for conversion to be complete */
while (!(ADC1->SR & SR_EOC)) {}
```

This line waits until the conversion is complete by checking the **EOC** flag in the ADC_SR register:

```
/* Read converted value */
return (ADC1->DR);
```

This line reads the converted digital value from the ADC_DR register.

In summary, our code performs the following actions:

1. **Initializes the ADC GPIO pin**:

   - Enables the clock for GPIOA

   - Sets PA1 to analog mode

2. **Configures the ADC module**:

   - Enables the clock for ADC1

   - Sets channel 1 as the start of the conversion sequence

   - Sets the conversion sequence's length to 1

   - Enables the ADC module

3. **Starts the ADC conversion process**:

   - Enables continuous conversion mode

   - Starts the ADC conversion process

4. **Reads the ADC value**:

   - Waits for the conversion to complete

   - Reads the converted value from the ADC data register

Our next task is to populate the adc.h file. Here's the code:

```
#ifndef ADC_H__
#define ADC_H__
```

```
#include <stdint.h>
#include "stm32f4xx.h"

void pa1_adc_init(void);
void start_conversion(void);
uint32_t adc_read(void);

#endif
```

Here, we are simply exposing the functions implemented in adc.c, making them callable from other files.

Let's move on to the main.c file. Update your main.c file, like so:

```
#include <stdio.h>
#include "adc.h"
#include "uart.h"

int sensor_value;

int main(void)
{

    /*Initialize debug UART*/
    uart_init();

    /*Initialize ADC*/
    pa1_adc_init();

    /*Start conversion*/
    start_conversion();

    while(1)
    {
        sensor_value = adc_read();

        printf("Sensor Value: %d\r\n", sensor_value);


    }
}
```

Let's break it down:

- **Including header files**:

```
#include <stdio.h>
#include "adc.h"
#include "uart.h"
```

Let's take a closer look:

- `#include <stdio.h>`: This includes the standard input/output library, which provides the `printf()` function for printing the sensor values

- `#include "adc.h"`: This includes the header file for the ADC functions, ensuring that the `pa1_adc_init`, `start_conversion`, and `adc_read` functions from our `adc.c` file are available

- `#include "uart.h"`: This includes the header file for the UART functions we developed in the previous chapter, ensuring that the `uart_init` function is available

- **Global variable declaration**:

```
int sensor_value;
```

This declares a global variable to store the ADC value that's read from the sensor.

- **Main function**:

```
/* Initialize debug UART */
uart_init();
```

This line initializes the UART peripheral, allowing us to print the sensor value:

```
/* Initialize ADC */
pa1_adc_init();
```

This line initializes the ADC:

```
/* Start conversion */
 start_conversion();
```

This line starts the ADC conversion process.

- **Infinite loop**:

```
sensor_value = adc_read();
```

This line reads the latest converted value from the ADC and stores it in the `sensor_value` variable:

```
printf("Sensor Value: %d\r\n", sensor_value);
```

This line prints the sensor value to the terminal or console using the UART. The `\r\n` part at the end of the string ensures that the printed value starts on a new line each time.

We are now ready to test the project.

Testing the project

To test your project, you must connect your sensor or a potentiometer to the development board. Follow these steps:

1. **Connect a sensor**:

   - **Signal pin**: Connect the signal pin of your sensor to **PA1**.

   - **GND pin**: Connect the GND pin of the sensor to one of the GND pins on the development board.

   - **VCC pin**: Connect the VCC pin to either the 3.3V or 5V pin on the development board. Ensure you verify the required voltage from your sensor's documentation as different sensors may need either 3.3V or 5V.

2. **Use a potentiometer**:

   - If a sensor is not available, you can use a potentiometer instead. A potentiometer is an adjustable resistor that's used to vary the voltage. It has three terminals: two fixed and one variable (wiper).

   - **Middle terminal**: Connect the middle terminal (wiper) of the potentiometer to PA1.

   - **Left terminal**: Connect the left terminal to 3.3V.

   - **Right terminal**: Connect the right terminal to GND.

See *Figure 11.4* for the connection diagram:

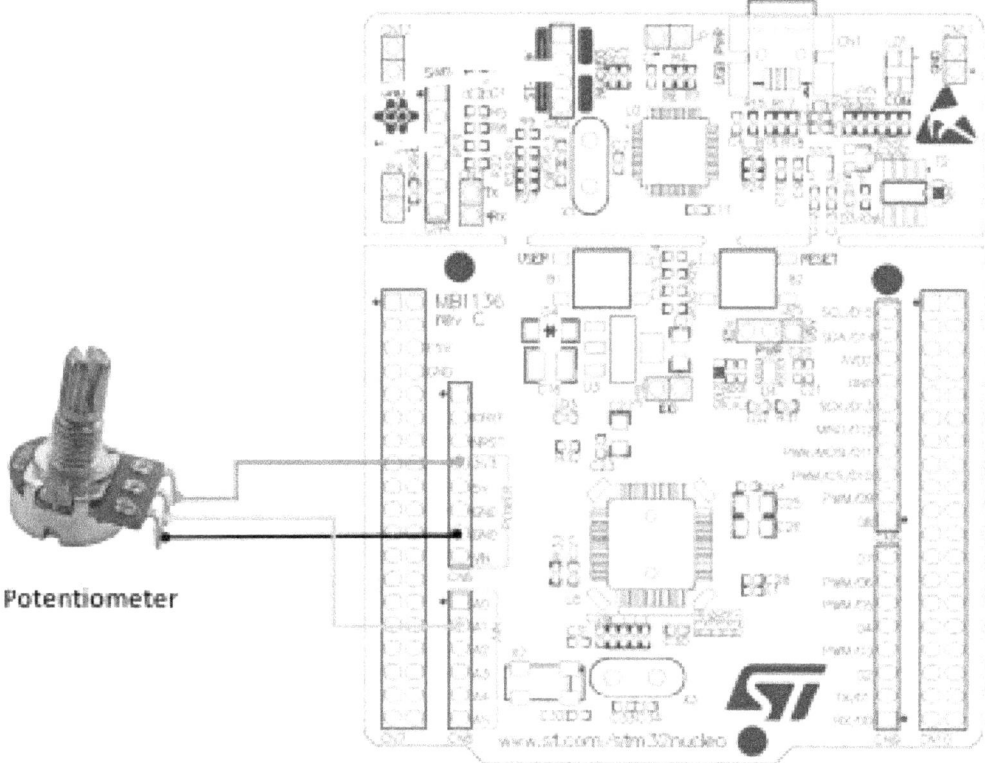

Figure 11.5: Potentiometer connection

🔍 **Quick tip**: Need to see a high-resolution version of this image? Open this book in the next-gen Packt Reader or view it in the PDF/ePub copy.

🔖 **The next-gen Packt Reader** and a **free PDF/ePub copy** of this book are included with your purchase. Unlock them by scanning the QR code below or visiting https://www.packtpub.com/unlock/9781835460818.

As you turn the knob of the potentiometer, the resistance between the middle terminal and the fixed terminals (3.3V and GND) will change, which, in turn, changes the voltage output at the middle terminal. This varying voltage will be measured by the ADC.

1. **Run the project**:

    - Build and run the project on the development board.

    - Open **RealTerm** or another serial terminal program and select the appropriate port and baud rate.

    - You should see the sensor values being printed in real time on the terminal. As you turn the potentiometer knob, the displayed value should change, reflecting the varying output voltage.

Summary

In this chapter, we explored the ADC, a vital peripheral in embedded systems that enables microcontrollers to interface with the analog world. We started with an overview of the analog-to-digital conversion process, highlighting its importance and discussing key specifications such as resolution, step size, and VREF.

Then, we delved into the STM32F411 microcontroller's ADC peripheral, examining its capabilities and the relevant registers required for ADC operations. This included an overview of key ADC registers, such as ADC_CR1, ADC_CR2, ADC_SQRx, ADC_SR, and ADC_DR, as well as important ADC flags, such as EOC, JEOC, AWD, OVR, and STRT.

This chapter also explained the different ADC modes, including single conversion mode, continuous conversion mode, scan mode, discontinuous mode, and injected conversion mode. Each mode was explained with practical use cases to illustrate their applications.

Next, we examined how multiplexing allows the ADC to switch between multiple input signals, enabling the microcontroller to handle multiple analog inputs efficiently.

Finally, we applied the theoretical concepts by developing a bare-metal ADC driver. This involved configuring a GPIO pin for ADC input, configuring the ADC module, starting conversions, and reading the ADC values.

In the next chapter, we will focus on the **Serial Peripheral Interface** (**SPI**), another commonly used communication protocol known for its speed and efficiency in embedded systems.

Unlock this book's exclusive benefits now

This book comes with additional benefits designed to elevate your learning experience.

Note: Have your purchase invoice ready before you begin.

https://www.packtpub.com/
unlock/9781835460818

12

Serial Peripheral Interface (SPI)

In this chapter, we will learn about the **Serial Peripheral Interface** (**SPI**) protocol, another important communication protocol widely used in embedded systems.

We will start by delving into the basics of the SPI protocol, understanding its master-slave architecture, data transfer modes, and typical use cases. Next, we will examine the key registers of the SPI peripheral in STM32 microcontrollers, providing detailed insights into their configuration and usage. Finally, we will apply this knowledge to develop a bare-metal SPI driver, demonstrating practical implementation and testing.

In this chapter, we will cover the following main topics:

- Overview of the SPI protocol
- The STM32F4 SPI peripherals
- Developing the SPI driver

By the end of this chapter, you will have a good understanding of the SPI protocol and be equipped to develop bare-metal drivers for SPI.

Technical requirements

All code examples for this chapter can be found on GitHub at the following link:

https://github.com/PacktPublishing/Bare-Metal-Embedded-C-Programming

Overview of the SPI protocol

Let's dive into what SPI is, its key features, how it works, and some of the nuances that make it so powerful.

What is SPI?

SPI is a *synchronous* serial communication protocol developed by Motorola. Unlike **Universal Asynchronous Receiver-Transmitter (UART)**, which is asynchronous, SPI *relies on a clock signal* to synchronize data transfer between devices. It's designed for short-distance communication (usually no more than 30 cm), primarily between a microcontroller and peripheral devices such as sensors, SD cards, and display modules. Let's see its key features.

Key features of SPI

SPI stands out due to its efficiency. Here are some of its key features:

- **Full-duplex communication**: SPI supports simultaneous data transmission and reception
- **High speed**: SPI can operate at much higher speeds compared to protocols such as **Inter-Integrated Circuit (I2C)** and UART
- **Master-slave architecture**: One master device controls communication, while one or more slave devices respond
- **Flexible data length**: Can handle various data lengths, commonly 8 bits, but not limited to that

To be able to connect two SPI devices, we must understand the SPI interface.

The SPI interface

SPI uses four primary lines for communication, each with several alternative names you might encounter:

- **Master In Slave Out (MISO)**: Also known as **Serial Data Out (SDO)** or **Data Out (DOUT)**, this line carries data from the slave device to the master
- **Master Out Slave In (MOSI)**: Also known as **Serial Data In (SDI)** or **Data In (DIN)**, this line carries data from the master device to the slave
- **Serial Clock (SCK)**: Also referred to as **SCLK** (or simply **CLK**), this is the clock signal generated by the master to synchronize data transfer
- **Slave Select (SS)**: Also known as **Chip Select (CS)** or **Not Slave Select (NSS)**, this line is used by the master to select which slave device to communicate with

When multiple slaves are used, each slave typically has its own SS line, allowing the master to control communication with each slave individually. *Figure 12.1* illustrates the SPI connection between a single master and a single slave:

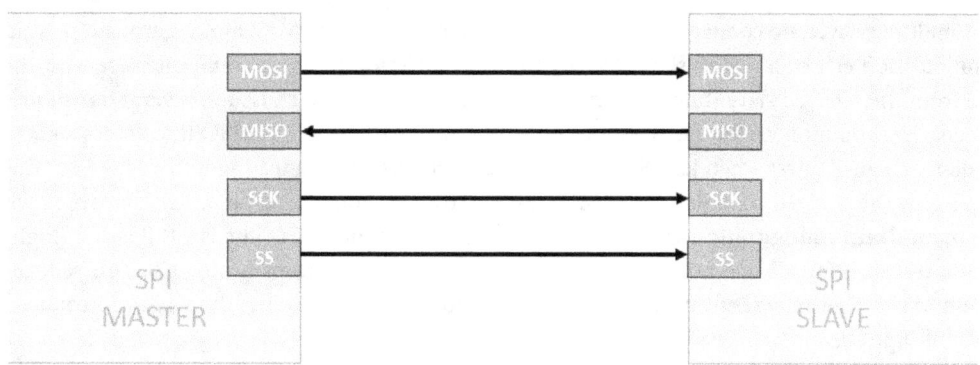

Figure 12.1: The SPI interface

Figure 12.2 depicts the SPI setup with a single master controlling multiple slaves:

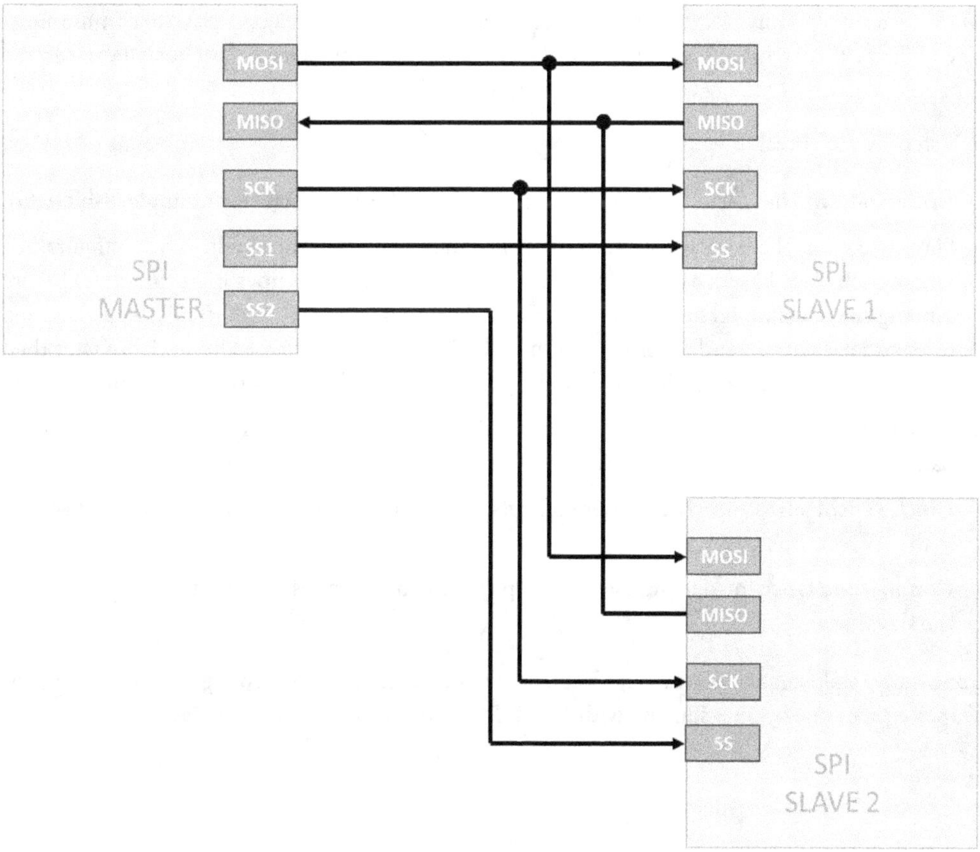

Figure 12.2: The SPI interface – multiple slaves

When multiple slaves are connected to a single SPI bus, managing the MISO line is crucial to avoid communication errors. Since all slaves share this line, non-selected slaves could interfere with the signal from the selected slave if not properly controlled. To prevent such issues, several techniques are used. One common method is **tri-state buffering**, where each slave's MISO line enters a high-impedance (high-Z) state when its CS line is inactive, effectively disconnecting it from the bus. This ensures only the selected slave drives the MISO line, preventing bus contention. Another approach is the **open-drain configuration** with a pull-up resistor, where the MISO line is left floating (high-Z) when transmitting a 1 and pulled low by the selected slave when transmitting a 0. This reduces contention risks but may result in slower communication speeds due to the time delay introduced by the pull-up resistor.

Let's see how the SPI protocol works.

How SPI works

SPI works on a simple principle: the master generates a clock signal and selects a slave to communicate with by *pulling the corresponding SS line low*. Data is then exchanged simultaneously between the master and the slave over the MOSI and MISO lines.

Here's a step-by-step breakdown:

1. **Initialization**: The master sets the clock frequency and data format (for example, 8-bit data).

2. **Slave selection**: The master pulls the SS line of the target slave low. In a multi-slave configuration, where each slave has its own CS line, the master first sets all CS lines high (inactive) before sending any initialization messages. This ensures that uninitialized slaves don't mistakenly respond to commands not intended for them. Once all CS lines are confirmed high, the master then activates the CS line of the desired slave by pulling it low to begin controlled communication.

3. **Data transmission**: The master sends data to the slave on the MOSI line, while the slave sends data to the master on the MISO line.

4. **Clock synchronization**: The master controls the clock, ensuring data is sampled and shifted at the correct times.

5. **Completion**: Once the data transfer is complete, the master pulls the SS line high, deselecting the slave.

To successfully implement an SPI driver, it's essential to understand key SPI configuration parameters. Let's explore them one by one, starting with **Clock Phase (CPHA)** and **Clock Polarity (CPOL)**.

CPHA and CPOL

In SPI communication, the settings for CPHA and CPOL determine the timing and characteristics of the clock signal used to synchronize data transfer between the master and slave devices. These settings are crucial for ensuring that data is correctly sampled and interpreted by both the master and the slave. Here's a detailed look at how CPHA and CPOL affect SPI communication.

CPOL

CPOL determines the idle state of the clock signal (SCK). It controls whether the clock signal is high or low when no data is being transferred:

- **CPOL = 0**: The clock signal is low (0) when idle. This means that the clock line remains low between data transmissions.

- **CPOL = 1**: The clock signal is high (1) when idle. This means that the clock line remains high between data transmissions.

CPHA

CPHA determines when data is sampled and when it is shifted out. It controls the edge of the clock signal on which data is read and written:

- **CPHA = 0**: Data is sampled on the leading edge (first edge) of the clock pulse and shifted out on the trailing edge (second edge)

- **CPHA = 1**: Data is shifted out on the leading edge (first edge) of the clock pulse and sampled on the trailing edge (second edge)

The combination of CPOL and CPHA results in four different SPI modes, each affecting the timing of data sampling and shifting.

Selecting the appropriate SPI mode is crucial for ensuring proper communication between the master and slave devices. Both devices must be configured to use the same CPOL and CPHA settings to correctly interpret the data being exchanged. The choice of mode depends on the specific requirements of the devices and the timing constraints of the application. Let's move on to SPI data modes.

Data modes

SPI is flexible with the data length it can handle. While **8-bit** data transfers are common, SPI can be configured to handle different data lengths, such as **16-bit** or **32-bit** transfers, depending on the application. The master and slave devices need to agree on the data length to ensure accurate communication. The last configuration parameter is the SPI speed.

SPI speed

One of SPI's significant advantages is its speed. SPI can operate at very high frequencies, typically up to several tens of **MHz**, depending on the hardware capabilities of the master and slave devices. The actual speed used in an application depends on several factors:

- **Device capabilities**: The maximum speed supported by both the master and the slave

- **Signal integrity**: Higher speeds can lead to signal integrity issues such as crosstalk and reflections, especially over longer distances

- **Power consumption**: Higher speeds consume more power, which might be a consideration in battery-powered applications

This concludes our overview of the SPI protocol. In the next section, we will analyze the SPI peripheral in the STM32F4 microcontroller.

The STM32F4 SPI peripherals

As with other peripherals, STM32 microcontrollers often include several SPI peripherals; the number varies depending on the specific model. The STM32F411 microcontroller has five SPI peripherals, namely the following:

- SPI1

- SPI2

- SPI3

- SPI4

- SPI5

Key features

Here are some of the key features:

- **Full-duplex and half-duplex communication**: Supports simultaneous two-way communication (full-duplex) or one-way communication (half-duplex)

- **Master/slave configuration**: Each SPI peripheral can be configured as either a master or a slave device

- **Flexible data size**: Supports data sizes ranging from 4 to 16 bits

- **High-speed communication**: Capable of operating at speeds up to 42 MHz in master mode and up to 21 MHz in slave mode

- **Direct Memory Access (DMA) support**: DMA support for efficient data transfer without CPU intervention

- **Negative SS (NSS) pin management**: Hardware management of the NSS pin for multi-slave configurations

- **Cyclic Redundancy Check (CRC) calculation**: Built-in hardware CRC calculation for data integrity verification

- **Bidirectional mode**: Supports bidirectional data mode, allowing a single data line to be used for both sending and receiving data

Let's examine the key registers of this peripheral.

Key SPI registers

To get SPI up and running on the STM32F411 microcontroller, we need to configure several registers that control various aspects of the SPI peripheral. Let's break down the main registers we'll be working with, starting with the **Control Register 1** register.

SPI Control Register 1 (SPI_CR1)

The SPI_CR1 register is central to configuring the SPI peripheral. It includes settings that define the **SPI mode**, **data format**, **clock settings**, and more. Key bits in this register include the following:

- **CPHA**: This bit determines on which clock edge the data is sampled. Setting CPHA to 0 means data is sampled on the first edge (leading edge), while setting it to 1 means data is sampled on the second edge (trailing edge).

- **CPOL**: This bit sets the idle state of the clock line. Setting CPOL to 0 means the clock is low when idle, and setting it to 1 means the clock is high when idle.

- **Master Selection (MSTR)**: This bit configures the SPI peripheral as either a master or a slave. Setting MSTR to 1 makes the SPI peripheral a master, while 0 sets it as a slave.

- **Baud Rate Control (BR[2:0])**: These bits configure the baud rate for SPI communication.

- **SPI Enable (SPE)**: This bit enables the SPI peripheral. We need to set SPE to 1 to activate SPI communication.

- **Frame Format (LSBFIRST)**: This bit determines the bit order in data transmission. Setting LSBFIRST to 0 transmits the **most significant bit** (MSB) first, while 1 transmits the **least significant bit** (LSB) first.

- **SS Internal (SSI)**: This bit is used in master mode to internally control the SS line.

- **Software Slave Management (SSM)**: Setting this bit to 1 enables software management of the SS line, allowing the master to control it manually.

Next, we have the SPI Status Register.

SPI Status Register (SPI_SR)

The `SPI_SR` register provides real-time status updates on the SPI peripheral, informing us about various operational states and flags. Key bits in this register include the following:

- **Receive Buffer Not Empty (RXNE)**: This flag indicates that the receive buffer contains unread data
- **Transmit Buffer Empty (TXE)**: This flag signals that the transmit buffer is empty and ready for new data
- **CRC Error Flag (CRCERR)**: This flag is set when a CRC error is detected, indicating possible data corruption
- **Mode Fault (MODF)**: This flag signals a mode fault, often due to incorrect master/slave configuration
- **Overrun Flag (OVR)**: This flag indicates an overrun condition, where the receive buffer wasn't read in time
- **Busy Flag (BSY)**: This flag indicates that the SPI peripheral is currently engaged in a transmission or reception

The last key register is the Data Register.

SPI Data Register (SPI_DR)

The `SPI_DR` register is the conduit for data transmission and reception. It's where we write data to be sent out and read data that's been received:

- **Transmitting data**: When we write to the `SPI_DR` register, data is sent out over the `MOSI` line
- **Receiving data**: When we read from the `SPI_DR`, you get the data that was received on the `MISO` line

With these registers in mind, we're now ready to develop the SPI driver. Let's jump into that in the next section.

Developing the SPI driver

Create a copy of your previous project in your IDE and rename this copied project to `SPI`. Next, create a new file named `spi.c` in the `Src` folder and another file named `spi.h` in the `Inc` folder. Populate your `spi.c` file with the following code:

```
#include "spi.h"
#define SPI1EN              (1U<<12)
#define GPIOAEN             (1U<<0)
#define SR_TXE              (1U<<1)
```

```c
#define SR_RXNE              (1U<<0)
#define SR_BSY               (1U<<7)
void spi_gpio_init(void)
{
    /*Enable clock access to GPIOA*/
    RCC->AHB1ENR  |= GPIOAEN;

    /*Set PA5,PA6,PA7 mode to alternate function*/

    /*PA5*/
    GPIOA->MODER &=~(1U<<10);
    GPIOA->MODER  |=(1U<<11);

    /*PA6*/
    GPIOA->MODER &=~(1U<<12);
    GPIOA->MODER  |=(1U<<13);

    /*PA7*/
    GPIOA->MODER &=~(1U<<14);
    GPIOA->MODER  |=(1U<<15);


    /*Set PA9 as output pin*/
    GPIOA->MODER  |=(1U<<18);
    GPIOA->MODER &=~(1U<<19);

    /*Set PA5,PA6,PA7 alternate function type to SPI1*/
    /*PA5*/
    GPIOA->AFR[0]  |=(1U<<20);
    GPIOA->AFR[0]  &=  ~(1U<<21);
    GPIOA->AFR[0]  |=(1U<<22);
    GPIOA->AFR[0]  &=  ~(1U<<23);

    /*PA6*/
    GPIOA->AFR[0]  |=(1U<<24);
    GPIOA->AFR[0]  &=  ~(1U<<25);
    GPIOA->AFR[0]  |=(1U<<26);
    GPIOA->AFR[0]  &=  ~(1U<<27);

    /*PA7*/
    GPIOA->AFR[0]  |=(1U<<28);
    GPIOA->AFR[0]  &=  ~(1U<<29);
```

```
GPIOA->AFR[0]  |=(1U<<30);
GPIOA->AFR[0]  &= ~(1U<<31);



}
```

> ♡ **Quick tip**: Enhance your coding experience with the **AI Code Explainer** and **Quick Copy** features. Open this book in the next-gen Packt Reader. Click the **Copy** button (**1**) to quickly copy code into your coding environment, or click the **Explain** button (**2**) to get the AI assistant to explain a block of code to you.
>
> ```
>                                              Copy      Explain
> function calculate(a, b) {
>   return {sum: a + b};                        1           2
> };
> ```
>
> 📖 **The next-gen Packt Reader** is included for free with the purchase of this book. Unlock it by scanning the QR code below or visiting `https://www.packtpub.com/unlock/9781835460818`.
>
>

Next, we have the function for configuring the SPI parameters:

```
void spi1_config(void)
{
    /*Enable clock access to SPI1 module*/
    RCC->APB2ENR |= SPI1EN;

    /*Set clock to fPCLK/4*/
    SPI1->CR1 |=(1U<<3);
    SPI1->CR1 &=~(1U<<4);
    SPI1->CR1 &=~(1U<<5);
```

```c
    /*Set CPOL to 1 and CPHA to 1*/
    SPI1->CR1 |=(1U<<0);
    SPI1->CR1 |=(1U<<1);

    /*Enable full duplex*/
    SPI1->CR1 &=~(1U<<10);

    /*Set MSB first*/
    SPI1->CR1 &= ~(1U<<7);

    /*Set mode to MASTER*/
    SPI1->CR1 |= (1U<<2);

    /*Set 8 bit data mode*/
    SPI1->CR1 &= ~(1U<<11);

    /*Select software slave management by
     * setting SSM=1 and SSI=1*/
    SPI1->CR1 |= (1<<8);
    SPI1->CR1 |= (1<<9);

    /*Enable SPI module*/
    SPI1->CR1 |= (1<<6);

}
void spi1_transmit(uint8_t *data,uint32_t size)
{
    uint32_t i=0;
    uint8_t temp;
    while(i<size)
    {
        /*Wait until TXE is set*/
        while(!(SPI1->SR & (SR_TXE))){}

        /*Write the data to the data register*/
        SPI1->DR = data[i];
        i++;
    }
    /*Wait until TXE is set*/
    while(!(SPI1->SR & (SR_TXE))){}
```

```
    /*Wait for BUSY flag to reset*/
    while((SPI1->SR & (SR_BSY))){}

    /*Clear OVR flag*/
    temp = SPI1->DR;
    temp = SPI1->SR;
}
```

Here is the function for receiving data:

```
void spi1_receive(uint8_t *data,uint32_t size)
{
    while(size)
    {
        /*Send dummy data*/
        SPI1->DR =0;
        /*Wait for RXNE flag to be set*/
        while(!(SPI1->SR & (SR_RXNE))){}
        /*Read data from data register*/
        *data++ = (SPI1->DR);
        size--;
    }
}
```

Finally, we have the functions for controlling the CS pin:

```
void cs_enable(void)
{
    GPIOA->ODR &=~(1U<<9);
}
```

And then, the function for deselecting the slave:

```
/*Pull high to disable*/
void cs_disable(void)
{
    GPIOA->ODR |=(1U<<9);
}
```

Let's walk through each part of the SPI initialization and communication code. We'll start by looking at the defined macros and then dive into each function.

Defined macros

Let's break down the meaning of the macros and their functions:

```
#define SPI1EN        (1U<<12)
#define GPIOAEN       (1U<<0)

#define SR_TXE        (1U<<1)
#define SR_RXNE       (1U<<0)
#define SR_BSY        (1U<<7)
```

Over here, we see the following:

- SPI1EN: This is defined as (1U<<12), which sets bit 12. It's used to enable the clock for the SPI1 peripheral.

- GPIOAEN: This is defined as (1U<<0), which sets bit 0. This enables the clock for GPIOA.

- SR_TXE: This is defined as (1U<<1). This indicates that the transmit buffer is empty.

- SR_RXNE: This is defined as (1U<<0). This indicates that the receive buffer is not empty.

- SR_BSY: This is defined as (1U<<7). This indicates that the SPI interface is busy with a transfer.

Let's break down the initialization function.

GPIO initialization for SPI

Let's analyze the configuration of the SPI1 GPIO pins:

```
RCC->AHB1ENR |= GPIOAEN;
```

This line enables the clock for GPIOA by setting the appropriate bit in the AHB1 peripheral clock enable register:

```
/*PA5*/
GPIOA->MODER &=~(1U<<10);
GPIOA->MODER |=(1U<<11);

/*PA6*/
GPIOA->MODER &=~(1U<<12);
GPIOA->MODER |=(1U<<13);

/*PA7*/
GPIOA->MODER &=~(1U<<14);
GPIOA->MODER |=(1U<<15);
```

These lines configure PA5, PA6, and PA7 pins to alternate function modes, necessary for SPI:

```
GPIOA->MODER |= (1U<<18);
GPIOA->MODER &= ~(1U<<19);
```

This configures PA9 as a general-purpose output pin, which will be used for SS:

```
/*PA5*/
GPIOA->AFR[0] |=(1U<<20);
GPIOA->AFR[0] &= ~(1U<<21);
GPIOA->AFR[0] |=(1U<<22);
GPIOA->AFR[0] &= ~(1U<<23);

/*PA6*/
GPIOA->AFR[0] |=(1U<<24);
GPIOA->AFR[0] &= ~(1U<<25);
GPIOA->AFR[0] |=(1U<<26);
GPIOA->AFR[0] &= ~(1U<<27);

/*PA7*/
GPIOA->AFR[0] |=(1U<<28);
GPIOA->AFR[0] &= ~(1U<<29);
GPIOA->AFR[0] |=(1U<<30);
GPIOA->AFR[0] &= ~(1U<<31);
```

These lines set the alternate function registers to configure PA5, PA6, and PA7 for SPI1.

SPI1 configuration

Next, we have the code for configuring the SPI parameters:

```
RCC->APB2ENR |= SPI1EN;
```

This line enables the clock for SPI1 by setting the appropriate bit in the APB2 peripheral clock enable register:

```
SPI1->CR1 |=(1U<<3);
SPI1->CR1 &=~(1U<<4);
SPI1->CR1 &=~(1U<<5);
```

These lines configure the SPI clock prescaler to set the baud rate by dividing the APB2 peripheral clock by 4, as SPI1 is connected to the APB2 bus. The baud rate is determined by the **Baud Rate (BR)** bits in the SPI Control Register (SPI1->CR1). In this case, setting the BR bits (bit 5 to bit 3) to 001 results in the peripheral clock being divided by 4, which dictates the speed at which data is transferred over the SPI bus:

```
SPI1->CR1 |=(1U<<0);
SPI1->CR1 |=(1U<<1);
```

These lines set the clock polarity and phase to ensure correct data sampling:

```
SPI1->CR1 &=~(1U<<10);
```

This line ensures that full-duplex mode is enabled for simultaneous transmit and receive:

```
SPI1->CR1 &= ~(1U<<7);
```

This line configures SPI to transmit the MSB first:

```
SPI1->CR1 |= (1U<<2);
```

This line sets SPI1 to master mode, making it the controller of the SPI bus:

```
SPI1->CR1 &= ~(1U<<11);
```

This line configures the SPI data frame size to 8 bits:

```
SPI1->CR1 |= (1<<8);
SPI1->CR1 |= (1<<9);
```

These lines enable **software management** of the SS line.

```
SPI1->CR1 |= (1<<6);
```

This line enables the SPI peripheral for operation:

Let's move on to the spi1_transmit() function.

Transmitting data with SPI

This snippet deals with transmitting the data:

```
while (!(SPI1->SR & (SR_TXE))) {}
```

This loop waits until the transmit buffer is empty before sending the next byte:

```
SPI1->DR = data[i];
```

This line sends the current byte of data:

```
while ((SPI1->SR & (SR_BSY))) {}
```

This ensures the SPI bus is not busy before continuing:

```
temp = SPI1->DR;
temp = SPI1->SR;
```

These two lines play a crucial role in managing the SPI communication process. After the master transmits data through the SPI Data Register, the same register captures the data received from the slave. To ensure incoming data is properly processed, we read the Data Register, even if we don't need the value. This read operation automatically clears the OVR flag. It's also advisable to read the Status Register as part of this process.

Next, we have the `spi1_receive()` function.

SPI data reception

This deals with receiving the data:

```
SPI1->DR = 0;
```

This line sends dummy data to generate clock pulses:

```
while (!(SPI1->SR & (SR_RXNE))) {}
```

This line waits until data is received:

```
*data++ = (SPI1->DR);
```

This line reads the received data:

The last functions are the `cs_enable()` and `cs_disable()` functions.

CS management

This line pulls the SS line low to enable the slave device:

```
GPIOA->ODR &= ~(1U << 9);
```

This line pulls the SS line high to disable the slave device:

```
GPIOA->ODR |= (1U << 9);
```

Our next task is to populate the spi.h file.

The header file

Here is the code:

```
#ifndef SPI_H_
#define SPI_H_
#include "stm32f4xx.h"
#include <stdint.h>

void spi_gpio_init(void);
void spi1_config(void);
void spi1_transmit(uint8_t *data,uint32_t size);
void spi1_receive(uint8_t *data,uint32_t size);
void cs_enable(void);
void cs_disable(void);

#endif
```

Over here, we are simply exposing the functions to make them accessible in other files.

To effectively test the SPI driver, we need a suitable slave device. In the next section, we'll dive into the **ADXL345 accelerometer**, which we'll use as our slave device to test the SPI driver.

Getting to know the ADXL345 accelerometer

ADXL345 is a gem in the world of digital accelerometers, and it's perfect for testing our SPI module. Let's dive into what makes this device so special and how it fits into our embedded system projects.

What is ADXL345?

ADXL345 is a small, thin, ultralow power, **3-axis accelerometer** that can measure **static acceleration of gravity** in tilt-sensing applications, as well as **dynamic acceleration** resulting from motion or shock. It's versatile, highly accurate, and can handle a variety of tasks with ease. This accelerometer offers high-resolution (up to 13-bit) measurements with a selectable measurement range of **±2 g**, **±4 g**, **±8 g**, or **±16 g**:

Figure 12.3: The ADXL345

Let's analyze its key features.

Key features of the ADXL345

Following is a list of the ADXL345's features:

- **Ultralow power**: The device consumes as little as 23 µA in measurement mode and just 0.1 µA in standby mode, making it ideal for battery-powered applications.

- **User-selectable resolution**: We can choose a resolution from 10 to 13 bits, providing a scale factor of 4 mg/LSB across all g ranges.

- **Flexible interface**: The ADXL345 supports both SPI (3- and 4-wire) and I2C digital interfaces, giving us flexibility in how you integrate it into your system.

- **Special sensing functions**: It includes single tap, double tap, and free-fall detection, along with activity/inactivity monitoring. These functions can be individually mapped to two interrupt output pins, making it highly responsive to physical events.

- **Wide supply voltage range**: It operates from 2.0 V to 3.6 V, accommodating various power configurations.

- **Robust performance**: The ADXL345 can withstand a shock of up to 10,000 g, ensuring durability in rugged applications.

Let's see some of its common applications.

Applications

Given its robust feature set, the ADXL345 is well suited for a range of applications:

- **Industrial equipment**: For machinery monitoring and fault detection
- **Aerospace equipment**: In systems where reliability and precision are paramount
- **Consumer electronics**: Examples are smartphones, gaming devices, and wearable technology
- **Health and sports**: For tracking motion and activity in health monitoring devices

Let's take a closer look at its sensing function.

Sensing function

At its core, the ADXL345 measures acceleration along three axes: x, y, and z. The data is available in a 16-bit two's complement format and can be accessed via either the SPI or I2C interface.

The following are its sensing functions:

- **Activity and inactivity monitoring**: The accelerometer can detect movement or the absence thereof, making it great for sleep monitoring and fitness applications
- **Tap detection**: It can recognize single and double taps in any direction, which is useful for gesture-based controls
- **Free-fall detection**: The device can detect if it's in free fall, which can be used in safety systems to trigger an alert or a response

Figure 12.4 shows the x, y, and z axes:

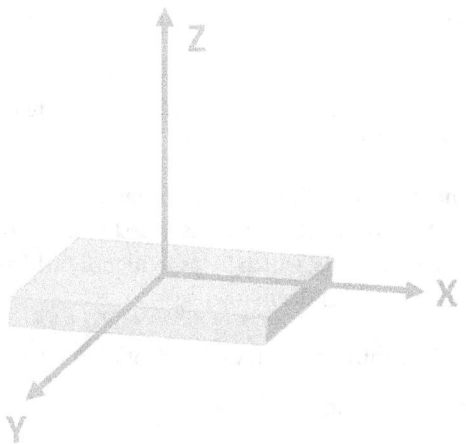

Figure 12.4: The x, y, and z axes

The ADXL345 also offers various low-power modes to help manage power consumption intelligently. These modes allow the device to enter sleep or standby states based on our defined thresholds and activity levels.

It also includes a 32-level **FIFO buffer**, which helps in storing data temporarily to reduce the load on the host processor. This buffer is especially useful in applications requiring high data throughput or when the processor is busy with other tasks. Lastly, its pinout is straightforward:

- **VDD I/O**: Digital interface supply voltage
- **GND**: Ground
- **CS**: CS for SPI communication
- **INT1 and INT2**: Interrupt output pins
- **SDA/SDI/SDIO**: Serial data line for I2C or SPI input
- **SCL/SCLK**: Serial clock line for I2C or SPI

Before we dive into developing the driver for this slave device, let's first explore some key concepts of acceleration measurement.

Understanding key concepts – static acceleration of gravity, tilt-sensing, and dynamic acceleration

When working with accelerometers such as the ADXL345, it's important to grasp some fundamental concepts that underpin their operation and applications. Let's break down what static acceleration of gravity, tilt-sensing, and dynamic acceleration mean.

Static acceleration of gravity

Static acceleration of gravity refers to the constant acceleration due to gravity that acts on an object at rest. This acceleration is always present and has a magnitude of approximately **9.8 meters per second squared (m/s²)** on the surface of the Earth.

In the context of an accelerometer such as the ADXL345, static acceleration is used to determine the orientation of the device. When the accelerometer is at rest and positioned flat, it measures the static acceleration of gravity along the *z* axis, which helps to identify which direction is "down." This capability is crucial for applications such as the following:

- **Orientation detection**: Determining the device's orientation relative to the Earth's surface
- **Tilt-sensing**: Measuring the tilt angle of the device by observing how gravity's force changes across different axes

The next important concept is tilt-sensing.

Tilt-sensing

Tilt-sensing is the process of measuring the angle at which an object is tilted with respect to the force of gravity. This is achieved by analyzing the static acceleration readings from the accelerometer.

Imagine holding a tablet. When you tilt it forward, backward, or sideways, the accelerometer inside detects changes in the static acceleration along its x, y, and z axes. By comparing these changes, the device can calculate the tilt angle. Here's how it works:

- **X-axis tilt**: If the device is tilted along the x axis, the static acceleration detected on the x axis will increase or decrease depending on the direction of the tilt.

- **Y-axis tilt**: Similarly, tilting along the y axis will cause variations in the static acceleration readings on the y axis.

- **Z-axis stability**: The z axis usually detects the full force of gravity when the device is lying flat. Changes in tilt cause redistributions of this force among the x and y axes.

Tilt-sensing is widely used in applications such as the following:

- **Screen orientation**: Automatically adjusting the display from portrait to landscape mode

- **Gaming controllers**: Detecting movements and tilts to enhance gameplay

- **Industrial equipment**: Monitoring the tilt of machinery or vehicles for stability and safety

The final key concept is dynamic acceleration.

Dynamic acceleration

Dynamic acceleration refers to the acceleration that results from motion or external forces acting on the device. Unlike static acceleration, which is constant, dynamic acceleration varies based on how the device is moving.

For instance, if you shake or move the accelerometer, it measures these changes as dynamic acceleration. This type of acceleration is crucial for the following:

- **Motion detection**: Identifying when the device is moved, which can be used in fitness trackers to count steps

- **Shock or impact sensing**: Detecting sudden impacts or vibrations, useful in crash detection systems or drop tests

- **Vibration monitoring**: Measuring vibrations in industrial machinery to predict failures or maintenance needs

Before wrapping up this section, let's clarify one more concept we introduced earlier: "g."

When dealing with accelerometers such as the ADXL345, you often come across terms such as ±2 g, ±4 g, ±8 g, or ±16 g. These terms are crucial for understanding the measurement capabilities and limits of the device. Let's break down what g means and how these ranges affect the performance and application of an accelerometer.

What is g?

The term *g* refers to the acceleration due to gravity at the Earth's surface, which is approximately 9.8 **meters per second squared (m/s^2)**. It is used as a unit of measurement for acceleration. When we say an accelerometer can measure ±2 g, it means it can detect accelerations up to twice the force of gravity in either direction along an axis.

With this clarified, we are now ready to develop the driver for the ADXL345 device.

Developing the ADXL345 driver

Create a new file named adxl345.c in the Src folder and another file named adxl345.h in the Inc folder.

The header file

Populate the adxl345.h file with this:

```
#ifndef ADXL345_H_
#define ADXL345_H_
#include "spi.h"
#include <stdint.h>

#define ADXL345_REG_DEVID                (0x00)
#define ADXL345_REG_DATA_FORMAT          (0x31)
#define ADXL345_REG_POWER_CTL            (0x2D)
#define ADXL345_REG_DATA_START           (0x32)

#define ADXL345_RANGE_4G                 (0x01)
#define ADXL345_RESET                    (0x00)
#define ADXL345_MEASURE_BIT              (0x08)

#define ADXL345_MULTI_BYTE_ENABLE        (0x40)
#define ADXL345_READ_OPERATION           (0x80)
void adxl_init (void);
void adxl_read(uint8_t address, uint8_t * rxdata);
#endif
```

The adxl345.h file begins by including our SPI driver with #include "spi.h" and proceeds to define the necessary macros. Let's break down the macros:

- ADXL345_REG_DEVID (0x00): This macro defines the register address for the device ID of the ADXL345

- ADXL345_REG_DATA_FORMAT (0x31): This macro defines the register address for setting the data format of the ADXL345

- ADXL345_REG_POWER_CTL (0x2D): This macro defines the register address for the power control settings of the ADXL345

- ADXL345_REG_DATA_START (0x32): This macro defines the starting register address for reading acceleration data from the ADXL345

- ADXL345_RANGE_4G (0x01): This macro defines the value to set the measurement range of the ADXL345 to ±4g

- ADXL345_RESET (0x00): This macro defines the reset value for certain registers

- ADXL345_MEASURE_BIT (0x08): This macro defines the bit value to enable measurement mode in the power control register

- ADXL345_MULTI_BYTE_ENABLE (0x40): This macro defines the bit to enable multi-byte operations

- ADXL345_READ_OPERATION (0x80): This macro defines the bit to specify a read operation

Next, we populate the adxl345.c file:

```
#include "adxl345.h"


void adxl_read(uint8_t address, uint8_t * rxdata)
{

        /*Set read operation*/
        address |= ADXL345_READ_OPERATION;

        /*Enable multi-byte*/
        address |= ADXL345_MULTI_BYTE_ENABLE;

        /*Pull cs line low to enable slave*/
        cs_enable();

        /*Send address*/
        spi1_transmit(&address,1);
```

```
        /*Read 6 bytes */
        spi1_receive(rxdata,6);

        /*Pull cs line high to disable slave*/
        cs_disable();

}


void adxl_write (uint8_t address, uint8_t value)
{
  uint8_t data[2];

  /*Enable multi-byte, place address into buffer*/
  data[0] = address|ADXL345_MULTI_BYTE_ENABLE;

  /*Place data into buffer*/
  data[1] = value;

  /*Pull cs line low to enable slave*/
  cs_enable();

  /*Transmit data and address*/
  spi1_transmit(data, 2);

  /*Pull cs line high to disable slave*/
  cs_disable();


}


void adxl_init (void)
{
    /*Enable SPI gpio*/
    spi_gpio_init();

    /*Config SPI*/
    spi1_config();

    /*Set data format range to +-4g*/
    adxl_write (ADXL345_REG_DATA_FORMAT, ADXL345_RANGE_4G);
```

```
    /*Reset all bits*/
    adxl_write (ADXL345_REG_POWER_CTL, ADXL345_RESET);

    /*Configure power control measure bit*/
    adxl_write (ADXL345_REG_POWER_CTL, ADXL345_MEASURE_BIT);
}
```

Let's analyze the functions line by line, starting with the `adxl_read()` function.

Function – adxl_read()

Let's break down the read function:

- `address |= ADXL345_READ_OPERATION;`: This line sets the MSB of the address to indicate a read operation

- `address |= ADXL345_MULTI_BYTE_ENABLE;`: This sets the multi-byte bit to enable multi-byte operations

- `cs_enable();`: This function pulls the CS line low, enabling communication with the ADXL345

- `spi1_transmit(&address, 1);`: This transmits the address (with read and multi-byte bits set) to the ADXL345

- `spi1_receive(rxdata, 6);`: This line reads 6 bytes of data from the ADXL345 and stores it in the buffer pointed to by `rxdata`

- `cs_disable();`: This function pulls the CS line high, ending communication with the ADXL345

Next, we have the `adxl_write()` function.

Function – adxl_write

Let's go through each line of this function:

- `data[0] = address | ADXL345_MULTI_BYTE_ENABLE;`: This sets the multi-byte bit and stores the modified address in the buffer

- `data[1] = value;`: This stores the data to be written in the buffer

- `cs_enable();`: This function pulls the CS line low, enabling communication with the ADXL345

- `spi1_transmit(data, 2);`: This transmits the address and data to the ADXL345 in one transaction

- `cs_disable();`: This function pulls the CS line high, ending communication with the ADXL345

Finally, we have the `adxl_init()` function.

Function – adxl_init

Let's analyze the initialization function:

- `spi_gpio_init();`: This function initializes the GPIO pins needed for SPI communication

- `spi1_config();`: This function configures the SPI settings (clock speed, mode, etc.)

- `adxl_write(ADXL345_REG_DATA_FORMAT, ADXL345_RANGE_4G);`: This line writes to the data format register to set the measurement range of the ADXL345 to ±4g

- `adxl_write(ADXL345_REG_POWER_CTL, ADXL345_RESET);`: This line writes to the power control register to reset all bits

- `adxl_write(ADXL345_REG_POWER_CTL, ADXL345_MEASURE_BIT);`: This line writes to the power control register to set the measure bit, enabling measurement mode

We are now ready to test the driver inside the `main.c` file. Update your `main.c` file as shown next:

```c
#include <stdio.h>
#include <stdint.h>
#include "stm32f4xx.h"
#include "uart.h"
#include "adxl345.h"

//Variables for storing accelerometer data
int16_t accel_x, accel_y, accel_z;
double accel_x_g, accel_y_g, accel_z_g;

uint8_t data_buffer[6];

int main(void)
{
    uart_init();

    // Initialize the ADXL345 accelerometer
    adxl_init();

    while (1)
    {
        // Read accelerometer data starting from the data start
        // register
        adxl_read(ADXL345_REG_DATA_START, data_buffer);

        // Combine high and low bytes to form the accelerometer data
        accel_x = (int16_t)((data_buffer[1] << 8) | data_buffer[0]);
```

```
        accel_y = (int16_t)((data_buffer[3] << 8) | data_buffer[2]);
        accel_z = (int16_t)((data_buffer[5] << 8) | data_buffer[4]);

        // Convert raw data to g values
        accel_x_g = accel_x * 0.0078;
        accel_y_g = accel_y * 0.0078;
        accel_z_g = accel_z * 0.0078;


        //Print values for debugging purposes
        printf("accel_x : %d accel_y : %d  accel_z : %d\n\
        r",accel_x,accel_y,accel_z);



    }

    return 0;
}
```

Let's break down the main() function:

- accel_x, accel_y, accel_z: These are variables to store the raw accelerometer data for each axis.

- accel_x_g, accel_y_g, accel_z_g: These are variables to store the converted accelerometer data in g units.

- data_buffer[6]: This is a buffer to hold the raw data bytes read from the ADXL345.

- adxl_init(): This initializes the ADXL345 accelerometer.

- adxl_read(ADXL345_REG_DATA_START, data_buffer);: This line reads data from the ADXL345 starting at the specified register (ADXL345_REG_DATA_START). The data is stored in data_buffer.

Finally, we have the lines for constructing the final 16-bit values:

```
accel_x = (int16_t)((data_buffer[1] << 8) | data_buffer[0]);
accel_y = (int16_t)((data_buffer[3] << 8) | data_buffer[2]);
accel_z = (int16_t)((data_buffer[5] << 8) | data_buffer[4]);
```

The data read from the ADXL345 is in 2 bytes (high and low) for each axis. These lines combine the bytes to form 16-bit values for each axis:

```
accel_x_g = accel_x * 0.0078;
accel_y_g = accel_y * 0.0078;
accel_z_g = accel_z * 0.0078;
```

These lines convert the raw accelerometer values to g values:

```
printf("accel_x : %d accel_y : %d  accel_z : %d\n\
r",accel_x,accel_y,accel_z);
```

This line outputs the raw accelerometer data for debugging purposes:

Now, let's test the project. To test the project, compile the code and run it on your microcontroller. Open RealTerm or another serial terminal application and configure it with the appropriate port and baud rate to view the debug messages. Press the black pushbutton on the development board to reset the microcontroller. You should see the *x*, *y*, and *z* accelerometer values continuously being printed. Try moving the accelerometer to observe the values change significantly.

Summary

In this chapter, we explored the SPI protocol, a widely used communication protocol in embedded systems for efficient data transfer between microcontrollers and peripherals. We began by understanding the basic principles of SPI, including its master-slave architecture, data transfer modes, and typical use cases, emphasizing its advantages such as full-duplex communication and high-speed operation.

Next, we examined the SPI peripheral in STM32F4 microcontrollers, focusing on critical registers such as SPI Control Register 1 (`SPI_CR1`), SPI Status Register (`SPI_SR`), and SPI Data Register (`SPI_DR`). We detailed how to configure these registers to set up the SPI peripheral for communication, covering important aspects such as **clock polarity (CPOL)** and **clock phase (CPHA)**, data frame size, and master/slave configuration.

We then applied this theoretical knowledge by developing a bare-metal SPI driver. The development process included initializing the SPI peripheral, implementing data transmission and reception functions, and handling CS management. We also integrated the SPI driver with an ADXL345 accelerometer, using SPI to communicate with the sensor and retrieve acceleration data. Finally, we tested the driver by reading and displaying the accelerometer data in real time.

In the next chapter, we will explore the final of the three most common communication protocols in embedded systems: I2C.

13

Inter-Integrated Circuit (I2C)

In this chapter, we will learn about the **Inter-Integrated Circuit (I2C)** communication protocol. We will begin by exploring the fundamental principles of the I2C protocol, covering its modes of operation, addressing methods, and the communication process. Then, we will examine the key registers of the I2C peripheral in STM32 microcontrollers and apply this knowledge to develop a bare-metal I2C driver.

In this chapter, we will cover the following main topics:

- An overview of the I2C protocol
- The STM32 I2C peripheral
- Developing the I2C Driver

By the end of this chapter, you will have a solid grasp of the I2C protocol and be equipped with the skills to develop bare-metal drivers for I2C.

Technical requirements

All the code examples for this chapter can be found on GitHub at `https://github.com/PacktPublishing/Bare-Metal-Embedded-C-Programming`.

An overview of the I2C protocol

I2C is another commonly used protocol. Let's explore what it is, its key features, how it works, and its data format.

What is I2C?

I2C is a *multi-master*, multi-slave, packet-switched, single-ended, serial communication bus invented by Philips Semiconductor (now NXP Semiconductors). It's designed for short-distance communication within a single device or between multiple devices on the same board. I2C is known for its simplicity and ease of use, making it a popular choice for communication between microcontrollers and other ICs. Let's see its key features.

The key features of I2C

I2C has a number of unique features, which makes it ideal for various applications in embedded systems:

- **A two-wire interface**: I2C uses only two wires, **Serial Data** (**SDA**) and **Serial Clock** (**SCL**), which simplifies the wiring and reduces the number of pins required on the microcontroller.

- **Multi-master and multi-slave**: Multiple master devices can initiate communication on the bus, and multiple slave devices can respond. This flexibility allows for complex communication setups. The I2C protocol supports up to 128 devices with 7-bit addressing, although the practical limit is 119 due to reserved addresses. With 10-bit addressing, the protocol theoretically allows for 1,024 devices, but again, reserved addresses reduce the practical maximum slightly. The 10-bit mode, while less common, enables a higher number of devices on the same bus.

- **Addressable devices**: Each device on the I2C bus has a unique address, enabling the master to communicate with specific slaves.

- **Synchronous communication**: The SCL line provides the clock signal, ensuring that data is transferred synchronously between devices.

- **Speed variants**: I2C supports various speed modes, including standard mode (100 kHz), fast mode (400 kHz), fast mode plus (1 MHz), and high-speed mode (3.4 MHz), catering to different speed requirements.

- **Simple and low-cost**: The protocol's simplicity and minimal hardware requirements make it cost-effective and easy to implement.

Let's look at the I2C interface.

The I2C interface

The I2C interface consists of two main lines:

- **Serial Data (SDA)**: This line carries the data being transferred between devices. It's a bidirectional line, meaning that both the master and slave can send and receive data.

- **Serial Clock (SCL)**: This line carries the clock signal generated by the master device. It synchronizes the data transfer between the master and the slave.

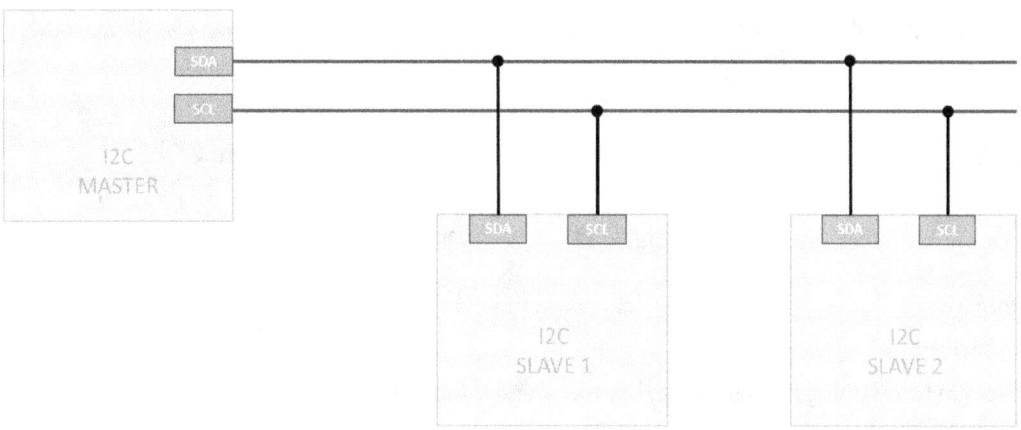

Figure 13.1: The I2C interface – multiple slaves

These two lines are connected to all devices on the bus, with pull-up resistors to ensure that the lines are pulled to a high state when idle. Let's see how it works.

How I2C works

Understanding how I2C works involves looking at the **roles** of master and slave devices, the **addressing scheme**, and the **communication process**.

The following are the roles and addressing scheme:

- **Master device**: The master device initiates communication and generates a clock signal. It controls the flow of data and can address multiple slaves.

- **Slave device**: The slave device responds to the master's commands and performs the requested operations. Each slave has a unique **7-bit** or **10-bit address** that the master uses to identify it.

The communication process is as follows:

1. **A start condition**: Communication begins with the master generating a start condition. This involves pulling the SDA line low while the SCL line is high.

2. **Address frame**: The master sends the address of the target slave device, followed by a read/write bit indicating the operation type (0 for write and 1 for read).

3. **Acknowledge (ACK) bit**: The addressed slave responds with an ACK bit by pulling the SDA line low during the next clock pulse.

4. **Data Frames**: Data is transferred in 8-bit frames. Each byte is followed by an ACK bit from the receiver.

5. **A stop condition**: The master ends the communication by generating a stop condition, which involves pulling the SDA line high while the SCL line is high.

Start	Address Frame	R/W	ACK	Data Frame	ACK	Data Frame	ACK	Stop
1 bit	7 to 10 bits	1 bit	1 bit	8 bits	1 bit	8 bits	1 bit	1 bit

Figure 13.2: The I2C packet

Before we proceed to the next section, let's take a moment to touch on the I2C data transfer, using an example.

Let's begin by revisiting the role of the data frame and the start condition:

- **Data frames**: Data is transferred in **8-bit bytes**. After each byte, the receiver sends an ACK bit to confirm successful reception.

- **Repeated start condition**: If the master needs to communicate with another slave or continue communication without releasing the bus, it can generate a **repeated start condition** instead of a stop condition.

Let's see the data transfer:

- **Write operation**: The master sends a **start condition**, the **address frame** with the **write bit**, and the **data frames**. Each data byte is followed by an ACK bit from the slave.

- **Read operation**: The master sends a start condition, the address frame with the **read bit**, and then reads the data frames from the slave. Each data byte is acknowledged by the master with an ACK bit, except for the last byte, which is followed by a **NACK** to indicate the end of the read operation.

For a better understanding, let's analyze *Figures 13.3* to *13.6*, starting with the start and stop conditions.

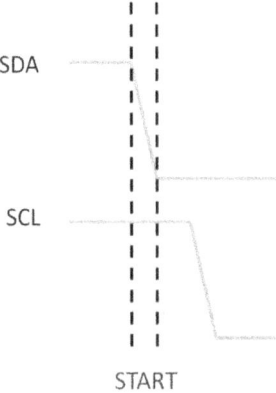

Figure 13.3: The start condition

The following is the stop condition:

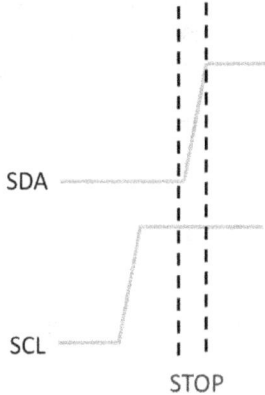

Figure 13.4: The stop condition

The start condition occurs when the master device pulls the SDA line low while the SCL line remains high. This sequence signals all devices on the I2C bus that a communication session is about to begin, allowing the master to claim the bus for its intended operations. Without a valid start condition, the I²C communication cannot commence.

Conversely, the stop condition signals the end of communication. The master device releases the SDA line to a high state while the SCL line is high, indicating that the communication session is complete and the bus is now free for other devices. The proper use of stop conditions is essential for ensuring that no devices remain active on the bus, which could lead to conflicts or communication errors.

Next, let's see how the I2C protocol distinguishes between zeros and ones.

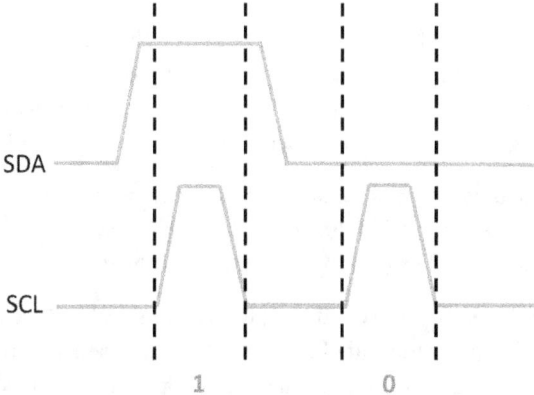

Figure 13.5: The data transmission process

Figure 13.5 illustrates the data transmission process. Data is sent bit by bit, synchronized with the clock pulses on the SCL line. As shown, each bit of data is placed on the SDA line while the SCL line is low. When the SCL line transitions to high, the state of the SDA line is read by the receiving device. This particular figure shows the transmission of **1**, followed by a **0**.

Finally, let's examine the complete packet and how it interacts with the SDA and SCL lines.

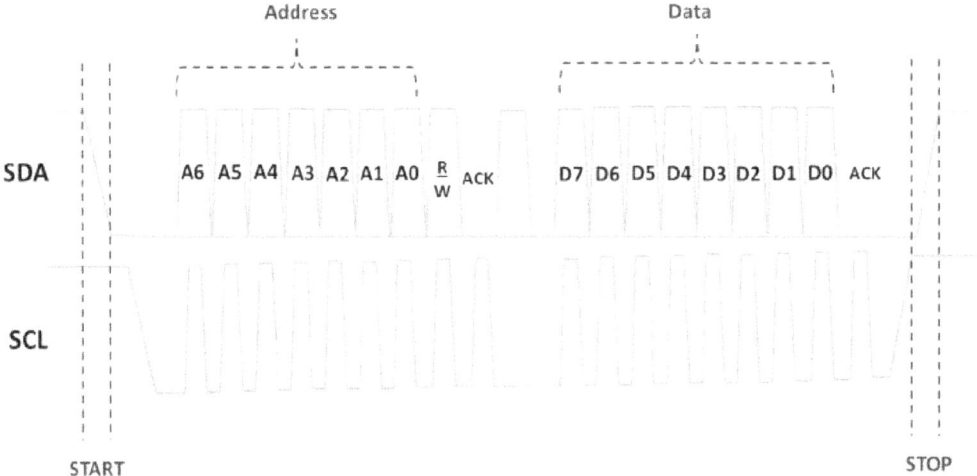

Figure 13.6: The complete packet

Figure 13.6 provides a comprehensive view of a complete I²C communication packet, showcasing the relationship between the SDA and SCL lines throughout the transaction. The communication begins with a start condition, where the SDA line is pulled low while the SCL line remains high, signaling the initiation of a new communication sequence.

Following the start condition, the address frame is transmitted. This frame contains the 7-bit address of the target device, followed by the **read/write (R/W)** bit that indicates whether the master intends to read from or write to the slave device. The address frame is then acknowledged by the slave device with an ACK bit, confirming that it is ready to proceed with the communication.

After the address frame, the data frame is transmitted. The data is sent in 8-bit bytes, with each bit being placed on the SDA line while the SCL line clocks each bit in sync. After each byte of data, the receiving device responds with another ACK bit, ensuring that the data was received correctly.

The communication concludes with a stop condition, where the SDA line is released to go high while the SCL line is also high. This signals the end of the communication session, freeing the bus for other potential communications. This complete cycle, from start to stop, forms the backbone of data exchange in the I²C protocol, ensuring structured and reliable communication between devices.

This concludes our overview of the I2C protocol. In the next section, we shall analyze the I2C peripheral in the STM32F4 microcontroller.

The STM32F4 I2C peripherals

Depending on the specific model of STM32F4 you are working with, you can typically find up to three I2C peripherals labeled I2C1, I2C2, and I2C3. These peripherals enable the microcontroller to communicate with I2C-compatible devices using the standard two-wire interface.

The I2C peripherals in STM32F4 microcontrollers come packed with features that enhance their versatility and performance:

- **Multi-master and multi-slave capabilities**: Each I2C peripheral can operate as both master and slave, supporting multiple master configurations where more than one master device can control the bus

- **Standard, fast, and fast mode plus**: The peripherals support multiple speed modes, including **standard mode** (100 kHz), **fast mode** (400 kHz), and **fast mode plus** (1 MHz), allowing for flexibility in communication speed

- **10-bit addressing**: In addition to standard **7-bit addressing**, the I2C peripherals also support **10-bit addressing**, enabling communication with a broader range of devices

- **Dual addressing mode**: Each I2C peripheral can be configured to respond to two different addresses, useful for complex multi-device setups

- **DMA support**: **Direct Memory Access** (**DMA**) support is available, enabling efficient data transfer without CPU intervention

Let's examine the key registers of this peripheral.

The key I2C registers

Configuring the I2C peripheral on an STM32 microcontroller involves several key registers that control various aspects of its operation.

Each register has specific bits that need to be set correctly to ensure proper functionality. Let's break down the main registers we'll be working with, starting with **Control Register 1**.

I2C Control Register 1 (I2C_CR1)

I2C_CR1 is one of the primary control registers used to configure the I2C peripheral's basic operational settings. It provides options to **enable the peripheral**, manage the start and stop conditions, and control the acknowledge feature.

The key bits in this register include the following:

- **Peripheral enable (PE)**: This bit enables or disables the I2C peripheral. Setting this bit to 1 turns on the I2C peripheral, while clearing it turns it off.

- **Start generation (START)**: Setting this bit generates a START condition, initiating communication.

- **Stop generation (STOP)**: Setting this bit generates a STOP condition, terminating communication.

- **Acknowledge enable (ACK)**: When set, this bit enables the ACK after each byte received.

- **Acknowledge/PEC position (POS)**: This bit controls the position of the ACK bit.

- **Software reset (SWRST)**: Setting this bit resets the I2C peripheral.

You can find detailed information about this register on *page 492* of the STM32F4 reference manual (RM0383). Next, let's look at I2C **Control Register 2**.

I2C Control Register 2 (I2C_CR2)

I2C_CR2 is another crucial control register that handles different aspects of I2C operation, including **clock frequency**, **interrupt** enable, and **DMA** control. Key bits in this register include the following:

- **FREQ[5:0] (peripheral clock frequency)**: These bits set the I2C peripheral clock frequency in MHz

- **DMAEN (DMA requests enable)**: When set, this bit enables the DMA requests for the I2C peripheral

You can find detailed information about this register on *page 494* of the STM32F4 reference manual (RM0383). Next, let's look at the I2C **Clock Control Register**.

I2C Clock Control Register (I2C_CCR)

I2C_CCR configures the clock control settings for standard, fast, and fast mode plus operations. Key bits in this register include the following:

- **CCR[11:0] (clock control)**: These bits set the clock control value, determining the I2C clock speed

- **DUTY (fast mode duty cycle)**: This bit selects the duty cycle for fast mode

- **F/S (I2C master mode selection)**: This bit selects between standard mode (0) and fast mode (1)

You can find detailed information about this register on *page 502* of the STM32F4 reference manual (RM0383). The next register is the I2C Rise Time Register.

I2C TRISE register (I2C_TRISE)

I2C_TRISE configures the maximum rise time for the I2C signals, ensuring compliance with I2C specifications. This register has only one field – **TRISE[5:0] (maximum rise time)**. These bits set the maximum rise time for the SDA and SCL signals in nanoseconds.

The final register is the I2C **Data Register**.

I2C Data Register (I2C_DR)

I2C_DR is the data register used for both **transmitting** and **receiving** data. Data written to this register is transmitted, and data received from the bus is stored in this register. This register has only one field – **DR[7:0] (8-bit data register)**: This register holds the 8-bit data to be transmitted or the data received from the bus.

With these registers in mind, we're now ready to develop the I2C driver. Let's do that in the next section.

Developing the I2C driver

Let's develop the I2C driver. Create a copy of your previous project in your IDE and rename this copied project I2C. Next, create a new file named i2c.c in the Src folder and another file named i2c.h in the Inc folder.

The initialization function

Let's populate the i2c.c file, starting with the macros and initialization function:

```
#include "stm32f4xx.h"




#define     GPIOBEN                 (1U<<1)
#define     I2C1EN                  (1U<<21)

#define     I2C_100KHZ              80
#define     SD_MODE_MAX_RISE_TIME   17
#define     CR1_PE                  (1U<<0)

#define     SR2_BUSY                (1U<<1)
#define     CR1_START               (1U<<8)
#define     SR1_SB                  (1U<<0)
#define     SR1_ADDR                (1U<<1)
#define     SR1_TXE                 (1U<<7)
#define     CR1_ACK                 (1U<<10)
#define     CR1_STOP                (1U<<9)
```

```
#define    SR1_RXNE            (1U<<6)
#define    SR1_BTF             (1U<<2)

/*
 * PB8 ---- SCL
 * PB9 ----- SDA
 * */

void i2c1_init(void)
{
    /*Enable clock access to GPIOB*/
     RCC->AHB1ENR |=GPIOBEN;

    /*Set PB8 and PB9 mode to alternate function*/
    GPIOB->MODER &=~(1U<<16);
    GPIOB->MODER |=(1U<<17);

    GPIOB->MODER &=~(1U<<18);
    GPIOB->MODER |=(1U<<19);

    /*Set PB8 and PB9 output type to  open drain*/
    GPIOB->OTYPER |=(1U<<8);
    GPIOB->OTYPER |=(1U<<9);

    /*Enable Pull-up for PB8 and PB9*/
    GPIOB->PUPDR |=(1U<<16);
    GPIOB->PUPDR &=~(1U<<17);

    GPIOB->PUPDR |=(1U<<18);
    GPIOB->PUPDR &=~(1U<<19);

    /*Set PB8 and PB9 alternate function type to I2C (AF4)*/
    GPIOB->AFR[1] &=~(1U<<0);
    GPIOB->AFR[1] &=~(1U<<1);
    GPIOB->AFR[1] |=(1U<<2);
    GPIOB->AFR[1] &=~(1U<<3);

    GPIOB->AFR[1] &=~(1U<<4);
    GPIOB->AFR[1] &=~(1U<<5);
    GPIOB->AFR[1] |=(1U<<6);
    GPIOB->AFR[1] &=~(1U<<7);
```

```
    /*Enable clock access to I2C1*/
     RCC->APB1ENR |= I2C1EN;

    /*Enter reset mode  */
    I2C1->CR1 |= (1U<<15);

    /*Come out of reset mode   */
    I2C1->CR1 &=~(1U<<15);

    /*Set Peripheral clock frequency*/
    I2C1->CR2 = (1U<<4);    //16 Mhz

    /*Set I2C to standard mode, 100kHz clock */
    I2C1->CCR = I2C_100KHZ;

    /*Set rise time */
    I2C1->TRISE = SD_MODE_MAX_RISE_TIME;

    /*Enable I2C1 module */
    I2C1->CR1 |= CR1_PE;



}
```

Let's break down what we have so far:

```
RCC->AHB1ENR |= GPIOBEN;
```

This line enables the clock for GPIOB by setting the corresponding bit.

```
GPIOB->MODER &=~(1U<<16);
GPIOB->MODER |=(1U<<17);
GPIOB->MODER &=~(1U<<18);
GPIOB->MODER |=(1U<<19);
```

These lines configure PB8 and PB9 pins to an alternate function mode for I2C.

```
GPIOB->OTYPER |=(1U<<8);
GPIOB->OTYPER |=(1U<<9);
```

These lines configure the pins as **open-drain**, which is required for I2C communication.

```
GPIOB->PUPDR |=(1U<<16);
GPIOB->PUPDR &=~(1U<<17);
```

```
GPIOB->PUPDR |=(1U<<18);
GPIOB->PUPDR &=~(1U<<19);
```

These lines enable **pull-up resistors** for the I2C pins.

```
GPIOB->AFR[1] &=~(1U<<0);
GPIOB->AFR[1] &=~(1U<<1);
GPIOB->AFR[1] |=(1U<<2);
GPIOB->AFR[1] &=~(1U<<3);

GPIOB->AFR[1] &=~(1U<<4);
GPIOB->AFR[1] &=~(1U<<5);
GPIOB->AFR[1] |=(1U<<6);
GPIOB->AFR[1] &=~(1U<<7);
```

These lines configure the **alternate function** for the pins to I2C1.

```
RCC->APB1ENR |= I2C1EN;
```

This line enables the clock for the I2C1 peripheral.

```
I2C1->CR1 |= (1U<<15);
I2C1->CR1 &=~(1U<<15);
```

These lines **reset** the I2C1 peripheral.

```
I2C1->CR2 = (1U<<4);
```

This configures the I2C1 clock.

```
I2C1->CCR = I2C_100KHZ;
```

This line sets the clock control register for 100 kHz standard mode, using the macro we defined.

```
I2C1->TRISE = SD_MODE_MAX_RISE_TIME;
```

This line sets the rise time for the I2C signals using the macro we defined. The TRISE register specifies the maximum time the signal is allowed to take to transition from a low to a high state on the I2C bus. Setting this value correctly is important to ensure that the I2C communication adheres to the timing requirements of the I2C standard, which helps maintain reliable and stable communication.

```
I2C1->CR1 |= CR1_PE;
```

This line enables the I2C1 peripheral by setting the PE bit. Next, we will add and analyze the function to read a byte from an I2C slave device.

The read function

Let's analyze the read function:

```c
void i2c1_byte_read(char saddr, char maddr, char* data) {

    volatile int tmp;

    /* Wait until bus not busy */
    while (I2C1->SR2 & (SR2_BUSY)){}

    /* Generate start */
    I2C1->CR1 |= CR1_START;

    /* Wait until start flag is set */
    while (!(I2C1->SR1 & (SR1_SB))){}

    /* Transmit slave address + Write */
    I2C1->DR = saddr << 1;

    /* Wait until addr flag is set */
    while (!(I2C1->SR1 & (SR1_ADDR))){}

    /* Clear addr flag */
    tmp = I2C1->SR2;

    /* Send memory address */
    I2C1->DR = maddr;

    /*Wait until transmitter empty */
    while (!(I2C1->SR1 & SR1_TXE)){}

    /*Generate restart */
    I2C1->CR1 |= CR1_START;

    /* Wait until start flag is set */
    while (!(I2C1->SR1 & SR1_SB)){}

    /* Transmit slave address + Read */
    I2C1->DR = saddr << 1 | 1;

    /* Wait until addr flag is set */
    while (!(I2C1->SR1 & (SR1_ADDR))){}
```

```
    /* Disable Acknowledge */
    I2C1->CR1 &= ~CR1_ACK;

    /* Clear addr flag */
    tmp = I2C1->SR2;

    /* Generate stop after data received */
    I2C1->CR1 |= CR1_STOP;

    /* Wait until RXNE flag is set */
    while (!(I2C1->SR1 & SR1_RXNE)){}

    /* Read data from DR */
      *data++ = I2C1->DR;
}
```

Let's break down what we have so far:

- `while (I2C1->SR2 & (SR2_BUSY)){}`: This line waits for the I2C bus to be free by checking the state of the BUSY bit in I2C Status Register 2.

- `I2C1->CR1 |= CR1_START;`: This line initiates a start condition on the I2C bus.

- `while (!(I2C1->SR1 & (SR1_SB))){}`: This line waits until the start condition is acknowledged by checking the SB bit in I2C Status Register 1.

- `I2C1->DR = saddr << 1;`: This line sends the slave address with the write bit. The 7-bit address of the device is left-shifted by 1 bit to make room for the R/W bit in the **least significant bit (LSB)** position. By only shifting saddr left by 1, we prepare the address for a subsequent write operation to the slave device.

- `while (!(I2C1->SR1 & (SR1_ADDR))){}`: This line waits until the address is acknowledged by checking the ADDR bit in I2C Status Register 1.

- `tmp = I2C1->SR2;`: This line clears the address flag by simply reading I2C Status Register 2.

- `2C1->DR = maddr;`: Here, we send the memory address to read from the slave device.

- `while (!(I2C1->SR1 & SR1_TXE)){}`: This line waits until the data register is empty by reading the transmit buffer empty (TXE) bit in I2C Status Register 1.

- `I2C1->CR1 |= CR1_START;`: This line initiates a **restart condition** on the I2C bus.

- `while (!(I2C1->SR1 & SR1_SB)){}`: Here, we wait until the restart condition is acknowledged by checking the SB bit in I2C Status Register 1.

- `I2C1->DR = saddr << 1 | 1;`: This line prepares the I2C data register for a read operation by setting up the 7-bit I2C address of the slave device and appending the R/W bit. Specifically, `saddr << 1` shifts the 7-bit address left by one bit to make room for the LSB, which is then set to 1 using the bitwise `OR` operator (`| 1`). This final value, with the LSB set to 1, indicates a read operation when loaded into the I2C1 **data register** (**DR**). Hence, this line configures the I2C peripheral to initiate communication with the slave device, requesting to read data from it.

- `while (!(I2C1->SR1 & (SR1_ADDR))){}`: This line waits until the address is acknowledged.

- `I2C1->CR1 &= ~CR1_ACK;`: This line disables the acknowledge bit to prepare for a stop condition.

- `tmp = I2C1->SR2;`: This line clears the address flag by reading I2C Status Register 2.

- `I2C1->CR1 |= CR1_STOP;`: This initiates a stop condition on the I2C bus.

- `while (!(I2C1->SR1 & SR1_RXNE)){}`: This line waits until the receive buffer is not empty by reading the **Receive Buffer Register Not Empty** (RXNE) flag in I2C Status Register 1. This flag indicates that new data has been received and is available in the data register.

- `*data++ = I2C1->DR;`: This line is responsible for storing the received byte of data from the I2C DR in the memory location pointed to by the `data` pointer.

Next, we have a function to read multiple bytes from the slave device:

```
void i2c1_burst_read(char saddr, char maddr, int n, char* data) {

    volatile int tmp;

    /* Wait until bus not busy */
    while (I2C1->SR2 & (SR2_BUSY)){}

    /* Generate start */
    I2C1->CR1 |= CR1_START;

    /* Wait until start flag is set */
    while (!(I2C1->SR1 & SR1_SB)){}

    /* Transmit slave address + Write */
    I2C1->DR = saddr << 1;

    /* Wait until addr flag is set */
    while (!(I2C1->SR1 & SR1_ADDR)){}

    /* Clear addr flag */
```

```c
 tmp = I2C1->SR2;

/* Wait until transmitter empty */
while (!(I2C1->SR1 & SR1_TXE)){}

/*Send memory address */
I2C1->DR = maddr;

/*Wait until transmitter empty */
while (!(I2C1->SR1 & SR1_TXE)){}

/*Generate restart */
I2C1->CR1 |= CR1_START;

/* Wait until start flag is set */
while (!(I2C1->SR1 & SR1_SB)){}

/* Transmit slave address + Read */
I2C1->DR = saddr << 1 | 1;

/* Wait until addr flag is set */
while (!(I2C1->SR1 & (SR1_ADDR))){}

/* Clear addr flag */
tmp = I2C1->SR2;

/* Enable Acknowledge */
  I2C1->CR1 |=  CR1_ACK;

while(n > 0U)
{
    /*if one byte*/
    if(n == 1U)
    {
        /* Disable Acknowledge */
        I2C1->CR1 &= ~CR1_ACK;

        /* Generate Stop */
        I2C1->CR1 |= CR1_STOP;

        /* Wait for RXNE flag set */
        while (!(I2C1->SR1 & SR1_RXNE)){}
```

```
        /* Read data from DR */
        *data++ = I2C1->DR;
        break;
    }
    else
    {
        /* Wait until RXNE flag is set */
        while (!(I2C1->SR1 & SR1_RXNE)){}

        /* Read data from DR */
        (*data++) = I2C1->DR;

        n--;
    }
}


}
```

This function reads multiple bytes of data from a specified memory address in the I2C slave device. Here's a breakdown of what it does:

1. **Waits for bus availability**: The function starts by ensuring that the I2C bus is not busy, waiting until it is free to initiate communication.

2. **Generates a start condition**: It generates a start condition to begin communication with the slave device.

3. **Transmits a slave address for write**: The function sends the slave device address with a write bit, indicating that it will initially write data to specify the memory address.

4. **Waits for the address flag and clears it**: It waits for the address flag to be set and then clears it by reading the SR2 register.

5. **Transmits the memory address**: The memory address from which to start reading is sent to the slave device.

6. **Generates a restart condition**: A repeated start condition is generated to switch the communication mode from write to read.

7. **Transmits a slave address for read**: The function sends the slave address with a read bit, indicating that it will read data from the slave device.

8. **Waits for the address flag and clears it**: Again, it waits for the address flag to be set and clears it by reading the SR2 register.

9. **Enables acknowledge**: The acknowledge bit is set.

10. **Reads a data loop**: The function enters a loop to read the specified number of bytes:

- **A single-byte read**: If only one byte is left to read, the acknowledge bit is cleared, a stop condition is generated, and the data is read into the buffer

- **A multiple-byte read**: If more than one byte is to be read, it waits for the RXNE flag to indicate that data is ready, reads the data into the buffer, and decrements the byte counter

Finally, we add the function to write data to the slave device.

The write function

Let's break down the function to write multiple bytes to the slave device:

```
void i2c1_burst_write(char saddr, char maddr, int n, char* data) {

    volatile int tmp;

    /* Wait until bus not busy */
    while (I2C1->SR2 & (SR2_BUSY)){}

    /* Generate start */
    I2C1->CR1 |= CR1_START;

    /* Wait until start flag is set */
    while (!(I2C1->SR1 & (SR1_SB))){}

    /* Transmit slave address */
    I2C1->DR = saddr << 1;

    /* Wait until addr flag is set */
    while (!(I2C1->SR1 & (SR1_ADDR))){}

    /* Clear addr flag */
    tmp = I2C1->SR2;

    /* Wait until data register empty */
    while (!(I2C1->SR1 & (SR1_TXE))){}

    /* Send memory address */
    I2C1->DR = maddr;

    for (int i = 0; i < n; i++) {
```

```
    /* Wait until data register empty */
       while (!(I2C1->SR1 & (SR1_TXE))){}

    /* Transmit memory address */
    I2C1->DR = *data++;
    }

    /* Wait until transfer finished */
    while (!(I2C1->SR1 & (SR1_BTF))){}

    /* Generate stop */
    I2C1->CR1 |= CR1_STOP;
}
```

This function writes **multiple bytes** of data to a specific memory address in the I2C slave device. The function begins by waiting for the I2C bus to be free, ensuring that there is no ongoing communication. It then generates a start condition to initiate communication with the slave device. The slave address is transmitted with a write bit, and then the function waits for the address flag to be set and cleared by reading the SR2 register. After ensuring the data register is empty, it sends the memory address where the data writing should begin. The function enters a loop to transmit each byte of data, waiting for the data register to empty before each byte is sent. Once all bytes have been transmitted, it waits for the byte transfer to finish and then generates a stop condition to end the communication.

Our next task is to populate the i2c.h file.

The header file

Here is the code for the header file:

```
#ifndef I2C_H_
#define I2C_H_


void i2c1_init(void);
void i2c1_byte_read(char saddr, char maddr, char* data);
void i2c1_burst_read(char saddr, char maddr, int n, char* data);
void i2c1_burst_write(char saddr, char maddr, int n, char* data);

#endif
```

Let's update our driver for the adxl345 device to use the I2C driver we developed.

The ADXL345 I2C driver

Update the current adxl345.c in the Src folder:

```c
#include "adxl345.h"

// Variable to store single byte of data
char data;

// Buffer to store multiple bytes of data from the ADXL345
uint8_t data_buffer[6];

void adxl_read_address (uint8_t reg)
{
    i2c1_byte_read( ADXL345_DEVICE_ADDR, reg, &data);
}

void adxl_write (uint8_t reg, char value)
{
    char data[1];
    data[0] = value;
    i2c1_burst_write( ADXL345_DEVICE_ADDR, reg,1, data) ;
}

void adxl_read_values (uint8_t reg)
{
    // Read 6 bytes into wthe data buffer
    i2c1_burst_read(ADXL345_DEVICE_ADDR, reg, 6,(char *)data_buffer);
}


void adxl_init (void)
{
    /*Enable I2C*/
    i2c1_init();

    /*Read the DEVID, this should return 0xE5*/
    adxl_read_address(ADXL345_REG_DEVID);

    /*Set data format range to +-4g*/
    adxl_write (ADXL345_REG_DATA_FORMAT, ADXL345_RANGE_4G);
    /*Reset all bits*/
    adxl_write (ADXL345_REG_POWER_CTL, ADXL345_RESET);
```

```
        /*Configure power control measure bit*/
        adxl_write (ADXL345_REG_POWER_CTL, ADXL345_MEASURE_BIT);
}
```

Here is a breakdown of the code:

- The adxl_read_address function reads a single byte of data from a specified register in the ADXL345 accelerometer. It uses the i2c1_byte_read function to communicate over the I2C bus, fetching the data from the register identified by the reg parameter and storing it in the data variable.

- The adxl_write function writes a single byte of data to a specific register in the ADXL345. It prepares a single-element array, containing the value to be written, and then uses i2c1_burst_write to send this data to the register specified by the reg parameter, over the I2C interface.

- The adxl_read_values function reads a block of data from the ADXL345 – specifically, **6 bytes**, which is the size required to capture the accelerometer's *X*, *Y*, and *Z* axis data. It uses i2c1_burst_read to pull this data, starting from the register specified by the reg parameter, and stores it in data_buffer for further processing.

- The adxl_init function initializes the ADXL345 accelerometer. It first enables I2C communication by calling i2c1_init, and then it checks the device's identity by reading the DEVID register. Following this, it configures the data format to a range of ±4g, resets the power control register, and finally, sets the power control register to start measuring acceleration.

Next, we update the current adxl345.h in the Inc folder:

```
#ifndef ADXL345_H_
#define ADXL345_H_


#include "i2c.h"
#include <stdint.h>

#define    ADXL345_REG_DEVID                (0x00)
#define    ADXL345_DEVICE_ADDR           (0x53)
#define    ADXL345_REG_DATA_FORMAT          (0x31)
#define    ADXL345_REG_POWER_CTL            (0x2D)
#define    ADXL345_REG_DATA_START           (0x32)
#define    ADXL345_REG_DATA_FORMAT          (0x31)


#define        ADXL345_RANGE_4G               (0x01)
```

```
#define          ADXL345_RESET                        (0x00)
#define      ADXL345_MEASURE_BIT                       (0x08)
void adxl_init(void);
void adxl_read_values(uint8_t reg);


#endif
```

We are now ready to test the driver inside the `main.c` file.

The main function

Update your `main.c` file, as shown here:

```c
#include <stdio.h>
#include <stdint.h>
#include "stm32f4xx.h"
#include "uart.h"
#include "adxl345.h"


//Variables for storing accelerometer data
int16_t accel_x, accel_y, accel_z;
double accel_x_g, accel_y_g, accel_z_g;

extern uint8_t data_buffer[6];

int main(void)
{

    uart_init();

    // Initialize the ADXL345 accelerometer
    adxl_init();

    while (1)
    {
        // Read accelerometer data starting from the data start
        // register
        adxl_read_values(ADXL345_REG_DATA_START);

        // Combine high and low bytes to form the accelerometer data
        accel_x = (int16_t)((data_buffer[1] << 8) | data_buffer[0]);
        accel_y = (int16_t)((data_buffer[3] << 8) | data_buffer[2]);
```

```
        accel_z = (int16_t)((data_buffer[5] << 8) | data_buffer[4]);

        // Convert raw data to g values
        accel_x_g = accel_x * 0.0078;
        accel_y_g = accel_y * 0.0078;
        accel_z_g = accel_z * 0.0078;


        printf("accel_x : %d accel_y : %d  accel_z : %d\n\
        r",accel_x,accel_y,accel_z);


    }

    return 0;
}
```

It's time to test the project. To do so, compile the code and run it on your microcontroller. Open RealTerm or another serial terminal application, and then configure it with the appropriate port and baud rate to view the debug messages. Press the black push button on the development board to reset the microcontroller. You should see the *X*, *Y*, and *Z* accelerometer values continuously being printed. Try moving the accelerometer to see how the values change significantly.

Summary

In this chapter, we learned about the I2C communication protocol. We began by discussing the fundamental principles of the I2C protocol, including its modes of operation, addressing schemes, and the step-by-step communication process.

We then delved into the specifics of the STM32 I2C peripheral, highlighting key registers such as Control Register 1 (**I2C_CR1**), Control Register 2 (**I2C_CR2**), the Clock Control Register (**I2C_CCR**), and the Data Register (**I2C_DR**).

Finally, we applied this theoretical knowledge to develop a bare-metal I2C driver. This driver allows us to initialize the I2C peripheral, perform both single-byte and burst data transfers, and handle communication with an external device such as the ADXL345 accelerometer.

In the next chapter, we will learn about interrupts, a critical feature in modern microcontrollers that enables responsive and efficient handling of real-time events.

Unlock this book's exclusive benefits now

This book comes with additional benefits designed to elevate your learning experience.

Note: Have your purchase invoice ready before you begin.

```
https://www.packtpub.com/
unlock/9781835460818
```

14

External Interrupts and Events (EXTI)

In this chapter, we will learn about interrupts and their critical role in embedded systems development. Interrupts are pivotal for creating responsive and efficient firmware, allowing microcontrollers to handle real-time events effectively. By understanding interrupts, you can develop systems that can react promptly to external stimuli, making your embedded applications more robust and versatile.

We will begin by exploring the fundamental role of interrupts in firmware, contrasting them with exceptions to highlight their unique purposes and handling mechanisms. Following this, we will dive into the specifics of the **Interrupt Service Routine (ISR)**, the **Interrupt Vector Table (IVT)**, and the **Nested Vectored Interrupt Controller (NVIC)**, which collectively form the backbone of interrupt handling in Arm Cortex-M microcontrollers.

Next, we will focus on the STM32 **External Interrupt (EXTI)** controller, an essential peripheral for managing external interrupts in STM32 microcontrollers. We will examine the key features and registers of the EXTI controller, learning how to configure and utilize it for various applications.

Finally, we will apply this knowledge by developing an EXTI driver, providing you with practical experience in implementing interrupt-driven firmware. This hands-on approach will solidify your understanding and enable you to create responsive, interrupt-based systems.

In this chapter, we will cover the following main topics:

- Interrupts and their role in firmware
- The STM32 EXTI controller
- Developing the EXTI driver

By the end of this chapter, you will have a comprehensive understanding of interrupts and how to develop bare-metal EXTI drivers for STM32 microcontrollers, empowering you to create responsive and efficient embedded systems.

Technical requirements

All the code examples for this chapter can be found on GitHub at the following link:

`https://github.com/PacktPublishing/Bare-Metal-Embedded-C-Programming`

Interrupts and their role in firmware

Interrupts are one of the most critical mechanisms in embedded systems, allowing microcontrollers to react to real-time events efficiently. To fully appreciate their role in firmware, it's essential to understand what interrupts are, how they work, and the various scenarios where they prove indispensable. So, let's dive in and explore the fascinating world of interrupts, their operation, and their practical applications.

What are interrupts?

Imagine you're deeply engrossed in reading a book but then the doorbell rings. You momentarily stop reading, attend to the visitor, and then return to your book. Interrupts in microcontrollers work similarly. They are signals that temporarily halt the current execution of a program to allow a special routine, known as an ISR, to run. Once the ISR completes, the microcontroller resumes its previous task right where it left off.

Interrupts can be triggered by hardware events, such as a timer overflow, a key press, or data reception on a communication interface. They can also be generated by software, providing a flexible way to manage both external and internal events. Now, let's see how they work.

How do interrupts work?

At the heart of interrupt handling is the concept of context switching. When an interrupt occurs, the microcontroller saves its current state—essentially, a snapshot of all the important information, such as the program counter and CPU registers. This allows the microcontroller to pause its current task, execute the ISR, and then restore the saved state to continue where it left off. The process typically follows these steps:

1. **Interrupt request**: An event triggers an **Interrupt Request (IRQ)**.
2. **Acknowledge and prioritize**: The interrupt controller acknowledges the request and prioritizes it based on predefined levels.
3. **Context save**: The CPU saves its current execution context.
4. **Vector fetch**: The CPU fetches the address of the ISR from the IVT.
5. **ISR execution**: The ISR runs to handle the interrupt.

6. **Context restore**: After the ISR completes, the CPU restores the saved context.

7. **Resume execution**: The CPU resumes the interrupted task.

You may wonder why interrupts are important. Let's find out.

Importance of interrupts in firmware

Interrupts are important for creating efficient and responsive embedded systems. Here are some key reasons why they are so important:

- **Real-time response**: Interrupts allow a microcontroller to react almost instantaneously to critical events. For instance, in a motor control system, an interrupt can immediately handle a sensor signal indicating that the motor has reached its desired position.

- **Resource optimization**: Instead of constantly polling for events (which wastes CPU cycles and power), interrupts enable the CPU to remain in a low-power state or focus on other tasks until an event occurs. This optimization is crucial for battery-powered devices such as wearables or remote sensors.

- **Prioritization and preemption**: Interrupts can be prioritized, allowing more critical tasks to preempt less critical ones. This ensures that high-priority tasks, such as emergency stop signals in industrial machinery, are addressed immediately.

When discussing interrupts, we often encounter another key term: exceptions. Although they share similarities, they serve different purposes in embedded systems. Let's explore the differences between interrupts and exceptions.

Interrupts versus exceptions

Interrupts are signals from hardware or software indicating an event that needs immediate attention. Examples include timer overflows, GPIO pin changes, and peripheral data reception.

As we learned earlier, interrupts enable embedded systems to handle real-time events and are essential for responsive and efficient system behavior. When an interrupt occurs, the CPU stops executing the main program and jumps to a predefined address to execute the ISR.

Exceptions are events that disrupt the normal execution flow, often due to errors such as divide-by-zero operations or accessing invalid memory addresses. While similar to interrupts, exceptions typically handle error conditions and system-level events.

Differences between interrupts and exceptions

Here are some of the differences between the two:

- **Source**: Interrupts usually originate from external hardware devices or other peripherals within the microcontroller, while exceptions are typically the result of internal CPU operations

- **Purpose**: Interrupts manage real-time events, whereas exceptions handle error conditions and system anomalies

- **Handling**: Both use ISRs, but exceptions often involve more complex error handling and recovery mechanisms

To properly understand how interrupts are handled, we need to examine the three key components involved: the NVIC, the ISR, and the IVT.

The NVIC, ISR, and IVT

The NVIC in Arm Cortex-M microcontrollers, such as the STM32 series, plays a pivotal role in managing interrupts. Let's explore what the NVIC is, how it works, and why it's so important for embedded development.

The NVIC

The NVIC is a hardware module integrated into Arm Cortex-M microcontrollers that manages the prioritization and handling of interrupts. It enables the microcontroller to respond to interrupts quickly and efficiently, while also allowing for nested interrupts, where higher-priority interrupts can preempt lower-priority ones. This capability is essential for real-time applications where timely responses to events are critical.

Its key features include the following:

- **Interrupt prioritization**: The NVIC supports multiple priority levels, allowing us to assign different priorities to different interrupts. This ensures that more critical tasks are handled first.

- **Nested interrupts**: The NVIC allows higher-priority interrupts to interrupt lower-priority ones. This feature is crucial for maintaining system responsiveness in real-time applications.

- **Dynamic priority adjustment**: We can dynamically adjust the priority of interrupts during runtime, providing flexibility to adapt to changing conditions.

    The diagram in *Figure 14.1* illustrates the NVIC and its connections to various components within the microcontroller.

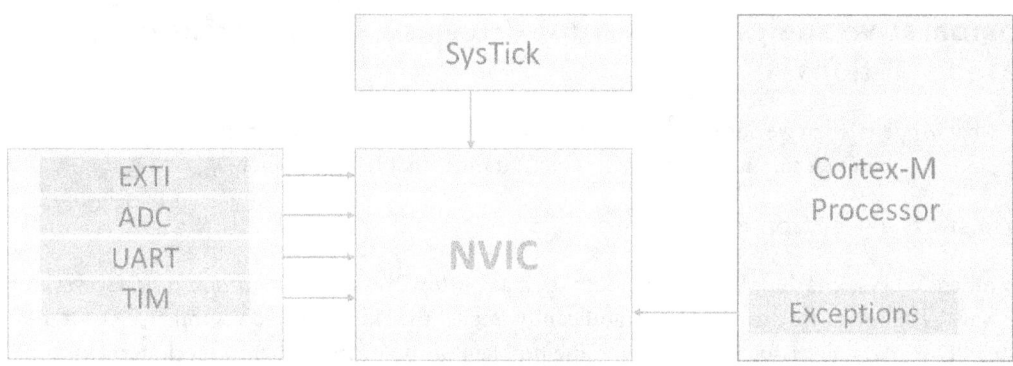

Figure 14.1: The NVIC

Next, we have the ISR.

The ISR

The ISR is a crucial piece of the puzzle when it comes to handling interrupts in embedded systems. The ISR is a specialized function that the CPU executes in response to an interrupt. Every interrupt has its own ISR. When an interrupt occurs, the CPU temporarily halts its current task, saves its state, and jumps to the ISR's predefined address (function) to execute the necessary code.

The final critical component is the IVT.

The IVT

The IVT is like the roadmap for the CPU when an interrupt occurs. It's a data structure that holds the addresses of all the ISRs. Each interrupt source is assigned a specific entry in this table, mapping it to its corresponding ISR. When an interrupt is triggered, the CPU consults the IVT to find the address of the ISR associated with that interrupt. This lookup ensures that the CPU can quickly and efficiently jump to the right piece of code to handle the event.

Its key features include the following:

- **Fixed location**: The IVT is typically located at a fixed address in memory. For Arm Cortex-M microcontrollers, it starts at address 0x00000000.

- **ISR addresses**: Each entry in the IVT contains the address of an ISR. When an interrupt occurs, the CPU uses the IVT to quickly locate and jump to the appropriate ISR.

- **Vector numbers**: Each interrupt source is assigned a unique vector number, which corresponds to an entry in the IVT. For example, vector number 0 might be for a reset interrupt, vector number 1 for a **non-maskable interrupt** (**NMI**), and so on.

- **Configurable**: In many systems, the IVT can be configured during system initialization to point to the ISRs defined in your firmware.

Comparative analysis—interrupt-driven solutions versus polling-based solutions

But what happens when we don't use interrupts? Let's explore some real-world case studies to illustrate the difference between solutions using interrupts and those that rely on polling.

Case study 1—button debouncing

In embedded systems, handling user inputs such as button presses is a common task. However, the way these inputs are managed can significantly impact the efficiency and responsiveness of the system. One particular issue is dealing with the problem of "bouncing," where a mechanical button generates multiple rapid signals due to its physical characteristics. This can lead to incorrect readings and erratic behavior if not properly managed. In this case study, we will explore two approaches to handling button debouncing: one without interrupts and the other using interrupts.

We'll start with the approach that doesn't use interrupts.

Imagine you have a simple user interface with a button that, when pressed, toggles an LED. Without interrupts, the most straightforward approach is to continuously check (or "poll") the button's state in the main loop. This involves repeatedly reading the button's input pin to see whether it has changed from high to low, indicating a press. The problem here is that mechanical buttons can generate spurious signals due to physical bouncing, leading to multiple detections of a single press. To handle this, you need to add a delay after detecting a press, effectively ignoring further signals for a short period.

There are a couple of drawbacks:

- **Inefficiency**: The CPU is constantly busy checking the button state, wasting valuable processing time that could be used for other tasks

- **Delayed response**: Adding a delay to handle debouncing means the system might miss other important tasks while waiting for the button to stabilize

Let's look at the approach with interrupts.

Using interrupts, you configure the button pin to generate an interrupt on a falling edge (when the button is pressed). The ISR is triggered immediately when the button is pressed, handling the debouncing logic. The main loop remains free to perform other tasks without the overhead of constantly polling the button.

There are a couple of benefits:

- **Efficiency**: The CPU can focus on other tasks and only respond to the button press when necessary

- **Immediate response**: The ISR responds instantly to the button press, making the system more responsive

Case study 2—sensor data acquisition

Another common task in embedded systems is acquiring data from sensors, especially in applications such as weather stations, where multiple sensors continuously monitor environmental conditions. The method used to handle sensor data acquisition can greatly affect the system's complexity and efficiency. Let's compare two approaches: one without using interrupts and the other leveraging interrupts to optimize the process.

We'll start with the approach that doesn't use interrupts.

Consider a weather station that reads data from various sensors such as temperature, humidity, and pressure at regular intervals. Without interrupts, the main loop would include code to periodically read data from each sensor. This could be done using timers to create delays between readings, ensuring data is acquired at the right intervals.

There are a couple of drawbacks:

- **Complexity**: Managing multiple sensors with precise timing using polling can lead to a complicated main loop
- **Inefficiency**: The main loop might spend a lot of time waiting for timers to expire—again, wasting CPU resources

Let's look at the approach with interrupts.

Using interrupts, each sensor can trigger an interrupt when new data is available. The ISR for each sensor reads the data and stores it in a buffer for the main loop to process later. This approach decouples data acquisition from the main loop, allowing it to focus on data processing and other tasks.

There are a couple of benefits:

- **Simplified code**: The main loop is cleaner and easier to manage, as it doesn't need to handle timing and sensor polling directly.
- **Resource efficiency**: The CPU spends less time waiting and more time processing, leading to more efficient use of resources.

Case study 3—communication protocols

Now, let's see how interrupts can improve communication. Effective communication between a microcontroller and external devices is crucial for many embedded systems, whether you're dealing with sensors, displays, or other peripherals. The approach you take to manage data transmission and reception can have a significant impact on your system's performance, particularly in terms of CPU load and latency. Let's analyze two methods for handling communication over **Universal Asynchronous Receiver/Transmitter** (**UART**): one without using interrupts and the other leveraging interrupts to optimize the process.

We'll start with the approach that doesn't use interrupts.

Let's look at a scenario where a microcontroller communicates with another device over UART. Without interrupts, the firmware would continuously check the UART status register to see whether new data has arrived or whether the transmitter is ready to send data. This polling approach ensures no data is missed, but it can be very CPU-intensive.

There are a couple of drawbacks:

- **Resource efficiency**: Continuous polling keeps the CPU busy, leaving less processing power for other tasks
- **Latency**: The time between data arrival and processing depends on how frequently the UART status is checked

Let's look at the approach with interrupts.

Enabling UART interrupts allows the microcontroller to handle data reception and transmission events automatically. When new data arrives, an interrupt is triggered, and the ISR reads the data and processes it. Similarly, when the transmitter is ready, another interrupt can handle sending data.

There are a couple of benefits:

- **Low CPU load**: The CPU can perform other tasks and only deal with UART events when necessary
- **Real-time handling**: Data is processed immediately upon arrival, reducing latency and improving communication efficiency

Interrupts provide a powerful and efficient way to handle real-time events in embedded systems. Compared to polling methods, interrupts offer significant benefits in terms of responsiveness, efficiency, and code simplicity. By understanding and leveraging interrupts, you can develop more robust and efficient firmware, capable of handling a wide range of real-time applications. Whether it's managing user inputs, acquiring sensor data, handling communication protocols, or maintaining accurate timekeeping, interrupts are an indispensable tool.

In the next section, we shall explore the STM32 EXTI controller peripheral.

The STM32 EXTI controller

The EXTI module in STM32 microcontrollers is designed to manage external interrupt lines. These lines can be triggered by signals on GPIO pins, enabling your microcontroller to react to changes in the external environment quickly and efficiently.

The EXTI controller is equipped with a range of features that enhance its flexibility and utility in embedded systems. Let's explore these features in detail and understand their practical applications.

Key features of the EXTI

Here are the key features of the EXTI module that make it a versatile and powerful tool for managing external and internal events in STM32 microcontrollers:

- Provides up to 23 independent interrupt/event lines, with up to 16 from GPIO pins, and the rest from internal signals

- Each line can be independently configured as either an interrupt or an event

- Edge detection options for each line: rising, falling, or both edges

- Dedicated status flags for each line to indicate pending interrupts/events

- Ability to generate software interrupts/events

An event registers that something happened by setting a corresponding status flag bit, but does not trigger an interrupt or execute any code (ISR).

For example, events can be used to wake the system from sleep mode without executing an ISR.

To use the EXTI for generating interrupts, we need to configure and enable the interrupt line properly. This involves programming the trigger registers to detect the desired edge (rising, falling, or both) and enabling the IRQ by setting the appropriate bit in the interrupt mask register. When the specified edge is detected on the external interrupt line, an IRQ is generated, and the corresponding pending bit is set. This pending bit must be cleared by writing a *1* to it in the pending register.

To generate events, we simply configure the event line by setting the appropriate trigger registers and enabling the corresponding bit in the event mask register.

We can also generate interrupts/events by software by writing a *1* to the software interrupt/event register (EXTI_SWIER).

Configuring the lines as interrupt sources involves three steps:

1. **Configure mask bits**: Set the mask bits for the 23 interrupt lines using the **Interrupt Mask Register (EXTI_IMR)**.

2. **Configure trigger selection bits**: Use the **Rising Trigger Selection Register (EXTI_RTSR)** and **Falling Trigger Selection Register (EXTI_FTSR)** to set the desired trigger conditions for the interrupt lines.

3. **Enable the NVIC IRQ channel**: Over here, we simply configure the **enable** and **mask** bits that control the NVIC IRQ channel mapped to the EXTI.

Before we develop the EXTI driver, let's first understand how the EXTI lines are mapped to the GPIO pins.

External interrupt/event line mapping

The EXTI controller can connect up to 81 GPIOs (in the STM32F411xC/E series) to the 16 external interrupt/event lines. The GPIO pins are mapped to the EXTI lines through the SYSCFG_EXTICR registers.

GPIO pins and EXTI lines

Each GPIO pin on the STM32 microcontroller can be connected to an EXTI line, allowing it to generate external interrupts. This flexibility means you can enable interrupts for any GPIO pin, but there's a catch: multiple pins share the same EXTI line. This sharing is based on the pin number, not the port.

Here's a breakdown of how the pins are connected:

- Pin 0 of every port is connected to EXTI0_IRQ
- Pin 1 of every port is connected to EXTI1_IRQ
- Pin 2 of every port is connected to EXTI2_IRQ
- Pin 3 of every port is connected to EXTI3_IRQ
- And so on...

This mapping means that all pins with the same number across different ports share the same EXTI line. For example, PA0, PB0, PC0, PD0, PE0, and PH0 are all connected to EXTI0.

> **Important note**
>
> **Shared EXTI lines**: Since multiple pins share the same EXTI line, you cannot enable interrupts on two pins with the same number across different ports simultaneously. For instance, if you enable an interrupt on PB0, you cannot also enable an interrupt on PA0 because both pins share EXTI0.
>
> **Configuration in SYSCFG_EXTICR**: The SYSCFG **external interrupt configuration registers (EXTICRs)** are used to select which port's pin will be connected to a particular EXTI line. This selection ensures that only one pin from one port can be the source for a given EXTI line.

Developing the EXTI driver

The STM32 EXTI module relies on several key registers to configure its operation. These registers allow you to set up trigger conditions, enable interrupts, and manage pending interrupt requests. Understanding these registers is crucial for effectively using the EXTI module in your embedded projects.

EXTI_IMR

We use the EXTI_IMR to enable or disable interrupts on each EXTI line.

The bits in this register are named **MRx**. They are mask bits for each EXTI line (x = 0 to 22). Setting a bit to 1 unmasks the interrupt line, allowing it to generate an interrupt request. Conversely, setting it to 0 masks the line, preventing it from generating interrupts.

EXTI_RTSR

The EXTI_RTSR configures the rising edge trigger for each EXTI line. When a rising edge is detected on a line configured in this register, it can generate an interrupt or an event.

The bits in this register are named **TRx**. They are trigger selection bits for each EXTI line (x = 0 to 22). Setting a bit to 1 configures the line to trigger on a rising edge.

EXTI_FTSR

The EXTI_FTSR is used to configure the falling edge trigger for each EXTI line. When a falling edge is detected on a line set in this register, it can generate an interrupt or an event.

The bits in this register are named **TRx**. They are trigger selection bits for each EXTI line (x = 0 to 22). Setting a bit to 1 configures the line to trigger on a falling edge.

Pending Register (EXTI_PR)

The EXTI_PR indicates which EXTI lines have pending interrupt requests. This register is also used to clear pending interrupts by writing a 1 to the appropriate bit.

The bits in this register are named **PRx**. They are pending bits for each EXTI line (x = 0 to 22). A bit set to 1 indicates a pending interrupt request. Writing 1 to the bit clears the pending request.

Let's configure PC13 as an EXTI pin.

The EXTI driver

Create a copy of your previous project in your IDE and rename this copied project EXTI. Next, create a new file named gpio_exti.c in the Src folder and another file named gpio_exti.h in the Inc folder.

Populate gpio_exti.c with the following code:

```
#include "gpio_exti.h"


#define GPIOCEN        (1U<<2)
```

```
#define SYSCFGEN          (1U<<14)

void pc13_exti_init(void)
{
    /*Disable global interrupts*/
    __disable_irq();

    /*Enable clock access for GPIOC*/
    RCC->AHB1ENR |=GPIOCEN;

    /*Set PC13 as input*/
    GPIOC->MODER &=~(1U<<26);
    GPIOC->MODER &=~(1U<<27);


    /*Enable clock access to SYSCFG*/
    RCC->APB2ENR |=SYSCFGEN;

    /*Select PORTC for EXTI13*/
    SYSCFG->EXTICR[3] |=(1U<<5);

    /*Unmask EXTI13*/
    EXTI->IMR |=(1U<<13);

    /*Select falling edge trigger*/
    EXTI->FTSR |=(1U<<13);

    /*Enable EXTI13 line in NVIC*/
    NVIC_EnableIRQ(EXTI15_10_IRQn);

    /*Enable global interrupts*/
    __enable_irq();

}
```

Let's break down each step within the pc13_exti_init function:

```
__disable_irq();
```

This line disables global interrupts to ensure that the configuration process is not interrupted, which is crucial for maintaining consistency and avoiding race conditions.

```
RCC->AHB1ENR |= GPIOCEN;
```

This line enables the clock for GPIOC by setting the appropriate bit in the AHB1 peripheral clock enable register (AHB1ENR).

```
GPIOC->MODER &= ~(1U<<26);
GPIOC->MODER &= ~(1U<<27);
```

These lines configure pin PC13 as an input by clearing the appropriate bits in the GPIO mode register (MODER).

```
RCC->APB2ENR |= SYSCFGEN;
```

This line enables the clock for SYSCFG by setting the appropriate bit in the APB2 peripheral clock enable register (APB2ENR). SYSCFG is required for configuring the EXTI lines to map to the appropriate GPIO pins.

```
SYSCFG->EXTICR[3] |= (1U<<5);
```

This line configures the SYSCFG external interrupt configuration register to map EXTI line 13 to PORTC. The EXTICR[3] register controls EXTI lines 12 to 15, and setting the correct bits ensures that EXTI13 is connected to PC13.

```
EXTI->IMR |= (1U<<13);
```

This line unmasks EXTI line 13 by setting the appropriate bit in the interrupt mask register (IMR). Unmasking the line allows it to generate interrupt requests.

```
EXTI->FTSR |= (1U<<13);
```

This line sets EXTI line 13 to trigger on a falling edge by setting the appropriate bit in the FTSR. This configuration is essential for detecting when the signal transitions from high to low.

```
NVIC_EnableIRQ(EXTI15_10_IRQn);
```

This line enables the EXTI15_10 interrupt line in the NVIC. EXTI lines 10 to 15 share an IRQ in the NVIC, and enabling it allows the microcontroller to handle interrupts from these lines.

```
__enable_irq();
```

This line re-enables global interrupts after the configuration is complete, allowing the microcontroller to respond to interrupts.

Our next task is to populate the gpio_exti.h file.

Here is the code:

```
#ifndef  GPIO_EXTI_H__
#define GPIO_EXTI_H__

#include <stdint.h>
#include "stm32f4xx.h"
#define   LINE13         (1U<<13)
void pc13_exti_init(void);
#endif
```

Now, let's test the driver in the main.c file:

```
#include <stdio.h>
#include "adc.h"
#include "uart.h"
#include "gpio.h"
#include "gpio_exti.h"

uint8_t g_btn_press;

int main(void)
{

    /*Initialize debug UART*/
    uart_init();

    /*Initialize LED*/
    led_init();

    /*Initialize EXTI*/
    pc13_exti_init();


    while(1)
    {


    }
}
static void exti_callback(void)
{
    printf("BTN Pressed...\n\r");
    led_toggle();
```

```
}
void EXTI15_10_IRQHandler(void) {
    if((EXTI->PR & LINE13)!=0)
    {
        /*Clear PR flag*/
        EXTI->PR |=LINE13;

        //Do something...
        exti_callback();
    }

}
```

Let's break down the code:

- The main function initializes the system components and then enters an infinite loop.
- The `exti_callback` function is called when the external interrupt occurs.
- `EXTI15_10_IRQHandler` handles interrupts for EXTI lines 10 to 15, including line 13 (PC13)

```
if((EXTI->PR & LINE13) != 0)
```

The preceding line checks whether the pending bit for EXTI line 13 is set, indicating an interrupt has occurred.

```
EXTI->PR |= LINE13;
```

The preceding line clears the pending bit by writing a 1 to it, acknowledging the interrupt, and allowing it to be processed again.

```
exti_callback();
```

This calls the `exti_callback` function to handle the interrupt, which, in our case, prints a message and toggles the LED.

To test on the microcontroller, simply build the project and run it. To generate the EXTI interrupt, press the blue push button.

Open RealTerm and configure the appropriate port and baud rate to view the printed message that confirms the EXTI interrupts occur when the blue push button is pressed.

Summary

In this chapter, we learned about the important role of interrupts in embedded systems development. Interrupts are essential for creating responsive and efficient firmware, allowing microcontrollers to handle real-time events effectively. By mastering the concepts of interrupts, you can develop systems that react promptly to external stimuli, enhancing the robustness and versatility of your embedded applications.

We started by exploring the fundamental role of interrupts in firmware, comparing them with exceptions to highlight their unique purposes and handling mechanisms. We then examined the specifics of the ISR, the IVT, and the NVIC, which together form the backbone of interrupt handling in Arm Cortex-M microcontrollers.

Next, we focused on the STM32 EXTI controller, a vital peripheral for managing external interrupts in STM32 microcontrollers. We discussed the key features and registers of the EXTI, and how to configure and utilize it for various applications.

Finally, we applied this knowledge by developing an EXTI driver, providing practical experience in implementing interrupt-driven firmware.

In the next chapter, we will learn about the **Realtime Clock (RTC)** peripheral.

Unlock this book's exclusive benefits now

This book comes with additional benefits designed to elevate your learning experience.

Note: Have your purchase invoice ready before you begin.

```
https://www.packtpub.com/
    unlock/9781835460818
```

The Real-Time Clock (RTC)

In this chapter, we will explore the **Real-Time Clock** (**RTC**) peripheral, an essential component for timekeeping in embedded systems. This peripheral is crucial for applications that require accurate time and date maintenance, making it fundamental for a wide range of embedded applications.

We will start by introducing RTCs and understanding how they function. Following this, we will delve into the STM32 RTC module, examining its features and capabilities. Next, we will analyze the relevant registers from the STM32 reference manual, providing a detailed understanding of the configuration and operation of the RTC. Finally, we will apply this knowledge to develop an RTC driver, enabling precise timekeeping in your embedded projects.

In this chapter, we will cover the following main topics:

- Understanding RTCs
- The STM32 RTC module
- Some key RTC registers
- Developing the RTC driver

By the end of this chapter, you will have a solid understanding of how RTCs work and will be equipped with the skills to develop bare-metal RTC drivers, allowing you to implement accurate timekeeping in your embedded systems projects.

Technical requirements

All the code examples for this chapter can be found on GitHub at https://github.com/PacktPublishing/Bare-Metal-Embedded-C-Programming.

Understanding RTCs

In this section, we will enter the world of RTCs, understanding what they are and how they work before exploring common use cases through a few interesting case studies. Let's get started!

RTCs are specialized hardware devices found in many microcontrollers and embedded systems. Their primary function is to keep track of the current time and date, even when the main power supply is turned off. Imagine them as the little timekeepers of the digital world, ensuring that the clock never stops ticking, no matter what.

RTCs are crucial in applications where timekeeping is essential. This includes everything from simple alarm clocks to complex data logging systems, where accurate timestamps are necessary. An RTC continues to operate on a small battery when the main system is powered down, maintaining accurate time and date information. Let's see how they work.

How do RTCs work?

At the heart of an RTC is a crystal oscillator, which provides a stable clock signal. This oscillator typically runs at **32.768 kHz**, a frequency chosen because it is easily divisible by powers of two, making it convenient for binary counting.

Here's a simplified breakdown of how an RTC works:

- **Crystal oscillator**: The RTC contains a crystal oscillator that generates a precise clock signal.
- **Counter**: This clock signal drives a counter. The counter increments at a rate determined by the oscillator's frequency.
- **Time and date registers**: The counter's value is used to update time and date registers, which hold the current time (hours, minutes, seconds) and date (day, month, year).
- **Battery backup**: To ensure continuous operation, RTCs often have a battery backup. This keeps the oscillator running and the counter active even when the main power is off.

Let's consider some common use cases for RTCs.

Common use cases for RTCs

RTCs are incredibly versatile and are used in a wide variety of applications. Let's explore some of the common use cases through a few case studies.

Case study 1 – data logging

One of the most common applications of RTCs is in data logging. Imagine that you're designing a weather station that collects temperature, humidity, and pressure data. Accurate timestamps are crucial for analyzing trends and patterns over time. Here's how an RTC plays a vital role in this scenario:

- **Initialization**: The RTC is initialized and set to the current time and date
- **Data collection**: Every time a sensor reading is taken, the RTC provides a timestamp

- **Storage**: The sensor data, along with the timestamp, is stored in memory
- **Analysis**: When the data is retrieved for analysis, the timestamps ensure that each reading can be accurately placed on a timeline

In this case, the RTC ensures that every piece of data is accurately timestamped, making it possible to track changes and trends with precision.

Case study 2 – alarm clocks

RTCs are also fundamental in designing alarm clocks. Be it a simple bedside alarm clock or a complex scheduling system, the RTC provides the accurate timekeeping needed to trigger events at the right moment. Let's look at a typical alarm clock scenario:

- **Timekeeping**: The RTC keeps track of the current time, continuously updating the time registers.
- **Alarm setting**: The user sets an alarm for a specific time. This information is stored in the RTC alarm registers.
- **Alarm trigger**: When the RTC time matches the alarm time, an interrupt is triggered, activating the alarm mechanism (such as sounding a buzzer or turning on a light).

In this case, the RTC ensures that the alarm goes off at the precise time set by the user, making it an essential component for reliable time-based alerts. At this point, your next question might be, "*What's so special about RTCs?*"

Why are RTCs important?

You might be wondering why we can't just use the system clock or general-purpose timers for timekeeping. The answer lies in the RTC's ability to keep accurate time, even when the main system is powered down. Here are some key reasons why RTCs are indispensable:

- **Accuracy**: RTCs use crystal oscillators, which provide highly accurate timekeeping
- **Low power consumption**: RTCs are designed to operate on very low power, often running for years on a small battery
- **Battery backup**: RTCs continue to keep time even when the main power is off, thanks to their battery backup
- **Independence from the main system**: RTCs operate independently of the main microcontroller, ensuring continuous timekeeping

By understanding how RTCs work and their common use cases, we can appreciate their importance and effectively incorporate them into our embedded projects. Whether you're building a simple alarm clock or a complex data logging system, the RTC is an important component that ensures your system always knows the right time. In the next section, we will explore the RTC peripheral in our STM32F411 microcontroller.

The STM32 RTC module

In this section, we will explore the RTC module in the STM32F4 microcontroller family. Let's start by looking at its features.

The main features of the STM32F4 RTC module

The STM32F4 RTC module is like the Swiss Army knife of timekeeping, offering a rich set of features designed to meet the needs of numerous applications. Here are some of the standout features:

- **Calendar with sub-seconds**: The RTC module doesn't just keep track of hours, minutes, and seconds; it also maintains sub-second accuracy. This is particularly useful for applications that require precise time measurements.

- **Alarm functionality**: Imagine that you have two alarm clocks within your microcontroller. The STM32F4 RTC module provides two programmable alarms, **Alarm A** and **Alarm B**, which can trigger events at specific times. This is perfect for tasks that need to be performed at regular intervals or at a specific time of day.

- **Low power consumption**: One of the biggest advantages of the RTC module is its low power usage. This makes it ideal for battery-operated devices, where conserving power is paramount.

- **Backup domain**: The RTC can operate independently of the main power supply thanks to a backup battery. This means that even if your device loses power, the RTC keeps running, maintaining accurate time.

- **Daylight saving time**: With the RTC module, you can program adjustments for daylight saving time automatically. No more manual resets twice a year!

- **Automatic wakeup**: The RTC can generate periodic wakeup signals, bringing your system out of low-power modes at preset intervals. This feature is invaluable for applications that need to perform regular checks or updates.

- **Tamper detection**: Security is a critical aspect of many applications, and the RTC module has you covered with tamper detection. It can log tamper events, providing an added layer of security for your system.

- **Digital calibration**: Accuracy is king when it comes to timekeeping. The RTC module includes a digital calibration feature to compensate for deviations in the crystal oscillator frequency, ensuring your timekeeping remains spot-on.

- **Synchronization with external clocks**: To enhance precision, the RTC can synchronize with an external clock source. This is great for applications that need to maintain very high accuracy over long periods.

Now, let's analyze some of the key components of the STM32F4 RTC module.

The key components of the STM32F4 RTC module

Let's take a closer look at the key components of the RTC module in the STM32F4 microcontroller family. We'll break down each part to understand how they work together to provide accurate timekeeping and versatile functionality, starting with the clock sources.

Clock sources

The driver of the RTC module is its clock sources. *Figure 15.1* presents a detailed block diagram of the RTC module, highlighting the RTC clock sources. This diagram, sourced from the reference manual, provides a clear visual representation of the various components and their interactions within the RTC module:

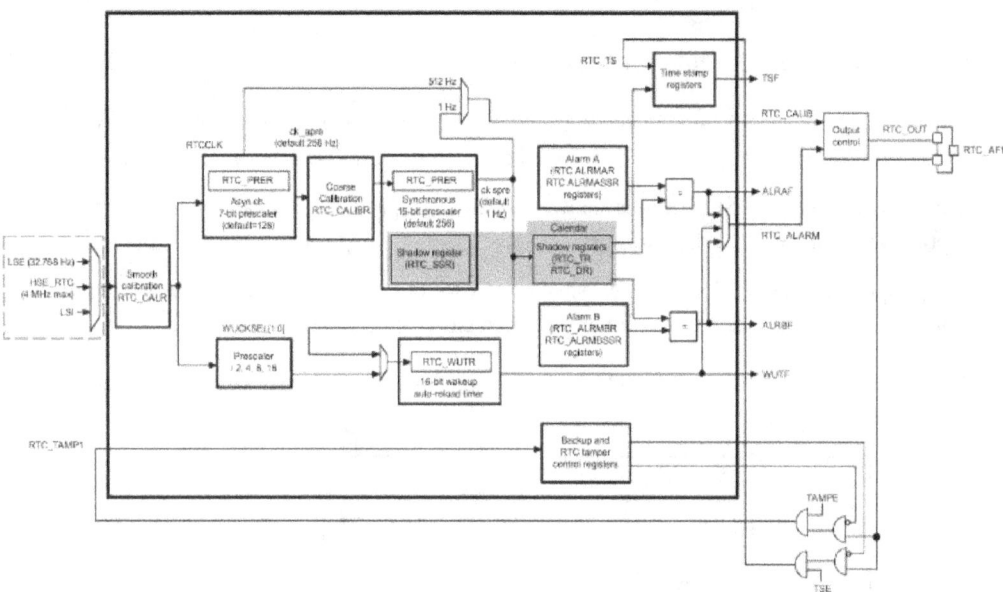

Figure 15.1: RTC block diagram with clock sources highlighted

🔍 **Quick tip**: Need to see a high-resolution version of this image? Open this book in the next-gen Packt Reader or view it in the PDF/ePub copy.

🔓 **The next-gen Packt Reader** and a **free PDF/ePub copy** of this book are included with your purchase. Unlock them by scanning the QR code below or visiting `https://www.packtpub.com/unlock/9781835460818`.

The STM32F4 RTC can use multiple clock sources:

- **Low-speed external (LSE)**: A 32.768 kHz crystal oscillator known for its stability and low power consumption. This is typically the preferred clock source for accurate timekeeping.

- **Low-speed internal (LSI)**: An internal RC oscillator that provides a less accurate but convenient option when an external crystal is not available.

- **High-speed external (HSI)**: A high-speed clock source that can be used but is less common for RTC applications due to its higher power consumption.

The selected clock source feeds into the RTC's prescalers, which are responsible for dividing the clock frequency into suitable levels for timekeeping:

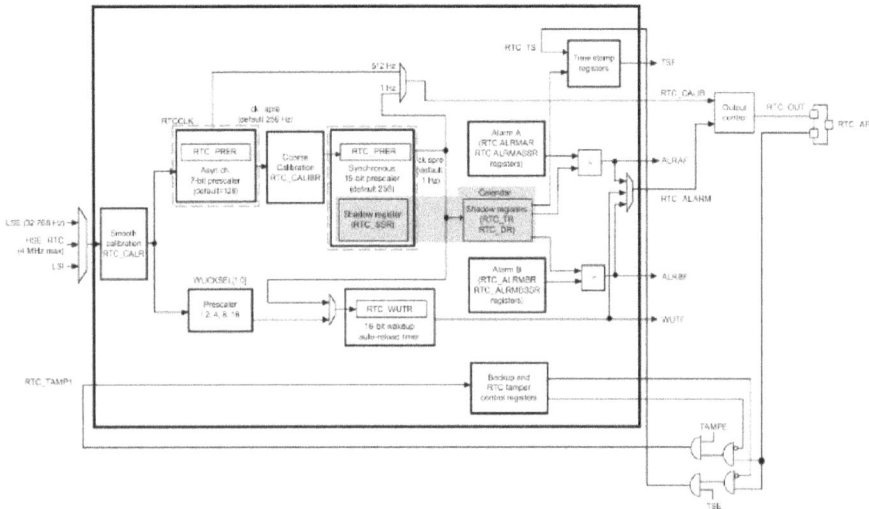

Figure 15.2: RTC block diagram – asynchronous and synchronous prescalers

Prescalers

The RTC module employs two types of prescalers:

- **Asynchronous prescaler**: This prescaler, typically set to divide by **128**, reduces the clock frequency to a lower rate that can be managed by the synchronous prescaler. It helps balance power consumption and accuracy.

- **Synchronous prescaler**: Often configured to divide by **256**, this prescaler further reduces the clock frequency to generate a precise **1 Hz clock**, which is essential for updating the time and date registers accurately.

These prescalers ensure the RTC can operate efficiently, providing the necessary timekeeping precision while conserving power. Next, we have the time and date registers.

Time and date registers

Figure 15.3 highlights the time and date registers of the RTC block:

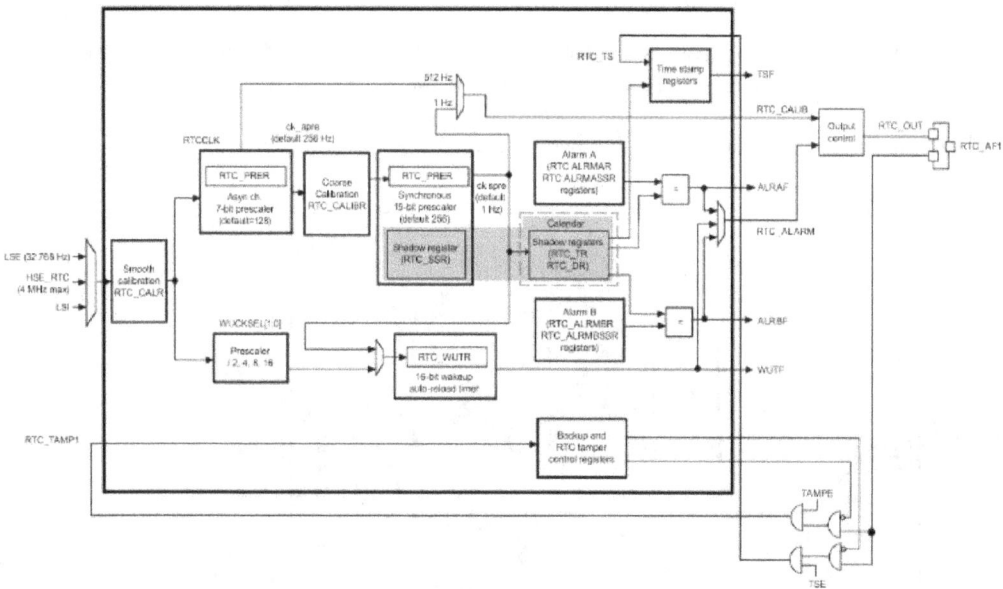

Figure 15.3: RTC block diagram – time and date registers

At the core of the RTC's functionality are the time and date registers:

- **Time register (RTC_TR)**: This register holds the current time in hours, minutes, and seconds, stored in **Binary-Coded Decimal (BCD)** format. It is updated every second by the 1 Hz clock from the prescalers.

- **Date register (RTC_DR)**: This register maintains the current date, including the year, month, and day, also in BCD format.

These registers are crucial for maintaining accurate time and date information, which can be read and adjusted as needed. The next key component is the RTC alarm.

Alarms

The RTC module features two programmable alarms, Alarm A and Alarm B. These alarms can be set to trigger at specific times, providing a powerful tool for scheduling tasks:

- **Alarm registers**: Each alarm has its own set of registers (RTC_ALRMAR and RTC_ALRMBR) to store the alarm time and date.

- **Interrupts**: When an alarm is triggered, it can generate an interrupt, waking up the microcontroller from a low-power state or initiating a specific function.

The alarm modules are indicated in the following figure:

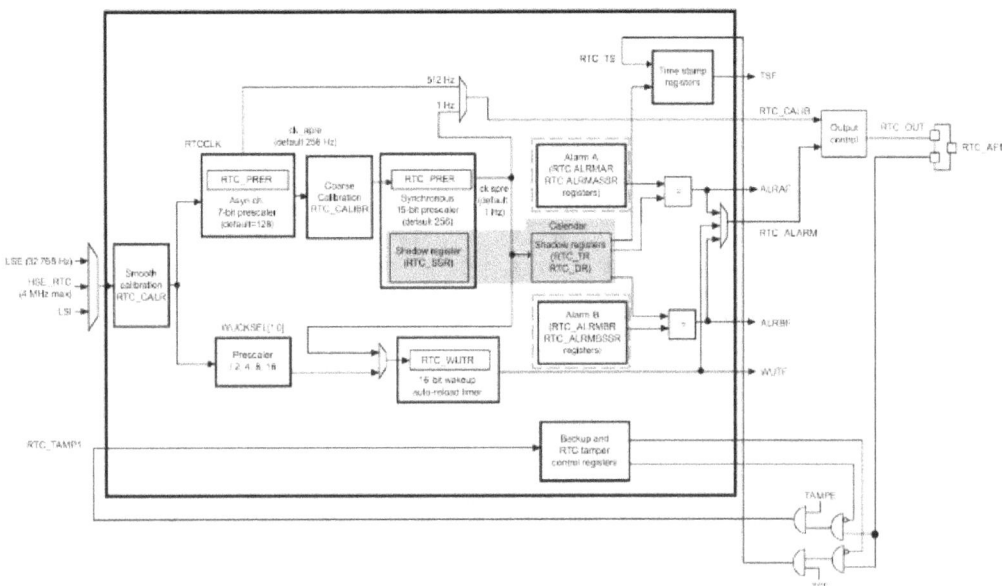

Figure 15.4: RTC block diagram – alarms

Next, we have the wakeup timer.

Wakeup timer

Another key feature of the RTC module is the wakeup timer, which is managed by the `RTC_WUTR` register:

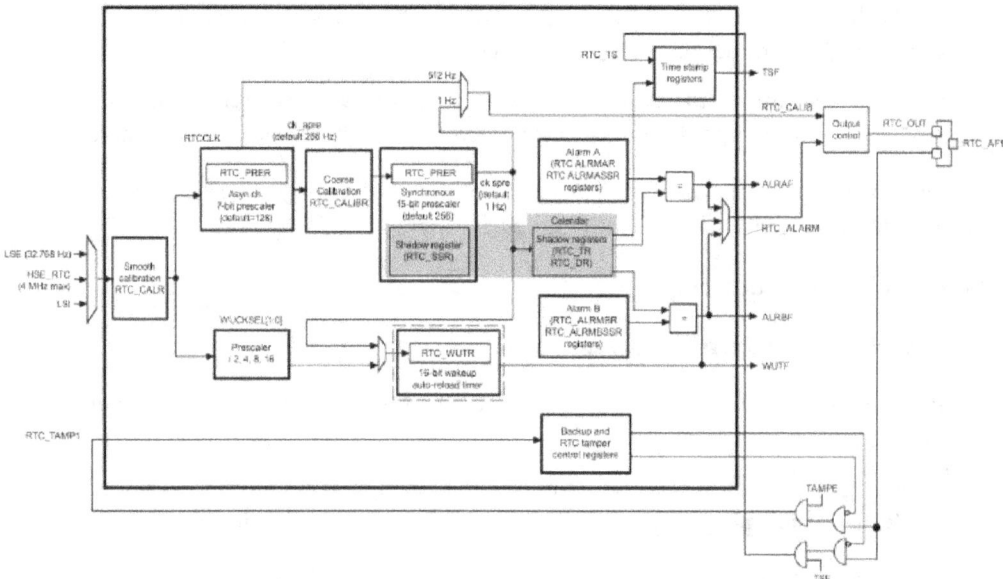

Figure 15.5: RTC block diagram – wakeup timer

This **16-bit** auto-reload timer can generate periodic wakeup events, bringing the system out of low-power modes at regular intervals. It's ideal for tasks such as sensor readings or system checks, ensuring efficient power usage.

There's also the tamper detection module. Let's take a look.

Tamper detection

Security is a vital aspect of many applications, and the RTC module includes tamper detection features. The tamper detection circuitry can log events when a tamper attempt is detected, using the timestamp registers to record the exact time and date. This adds an extra layer of security, especially in applications requiring reliable timekeeping and event logging. Next, we have the calibration register features.

Calibration and synchronization

To maintain high accuracy, the RTC module includes calibration features:

- **Digital calibration**: The RTC_CALR register allows for fine adjustments to the clock frequency, compensating for any deviations in the crystal oscillator

- **External clock synchronization**: The RTC can synchronize with an external clock source, enhancing accuracy by periodically adjusting the internal clock so that it matches the external reference

These features ensure the RTC maintains precise timekeeping, even in varying environmental conditions. We also have the backup and control registers module.

Backup and control registers

The RTC module includes several backup and control registers:

- **Backup registers**: These registers store critical data that must be retained even when the main power supply is off

- **Control registers**: These registers (such as RTC_CR) manage the configuration and operation of the RTC, including enabling the clock, setting alarms, and configuring wakeup events

The backup and control registers module is indicated in the following figure:

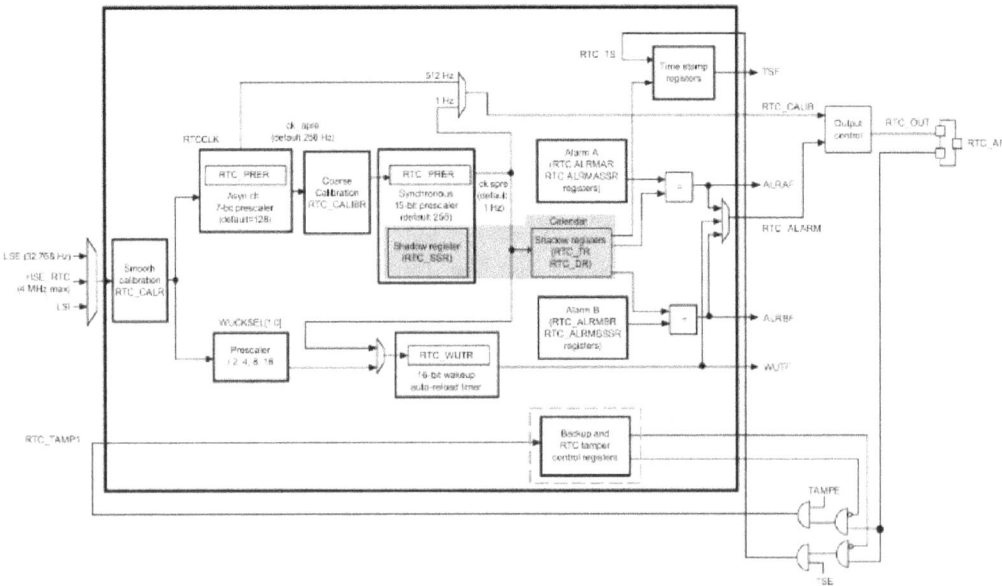

Figure 15.6: RTC block diagram – backup and control registers

Finally, there is the output control block.

Output control

The RTC module can output specific signals, such as a calibration clock or alarm outputs, through the `RTC_AF1` pin. This allows the RTC module to interact with other components or systems, providing synchronized signals or triggering external events.

In the next section, we will analyze some of the key registers for configuring the RTC peripheral.

Some key RTC registers

In this section, we will explore the characteristics and functions of some of the important registers within the RTC module. These registers are the building blocks that allow us to configure, control, and utilize the RTC's features effectively. Let's start with the **RTC Time Register** (`RTC_TR`).

RTC Time Register (RTC_TR)

The `RTC_TR` register is responsible for keeping track of the current time. It maintains the hours, minutes, and seconds in BCD format, ensuring that time is easily readable and manipulable. Here are some of the key fields in this register:

- **Hour tens (HT) and hour units (HU)**: These bits represent the tens and units of the hour, respectively. They can handle both **24-hour** and **12-hour** formats.

- **Minute tens (MNT) and minute units (MNU)**: These bits represent the tens and units of the minutes.

- **Second tens (ST) and second units (SU)**: These bits represent the tens and units of the seconds.

- **PM**: This bit indicates the **AM/PM** notation when in **12-hour** format.

Further information about this register can be found on *page 450* of the *reference manual*. Let's move on to the **RTC Date Register** (`RTC_DR`).

RTC Date Register (RTC_DR)

The `RTC_DR` register is responsible for maintaining the current date. It keeps track of the year, month, day of the month, and day of the week, all in BCD format.

The following are the key fields in this register:

- **Year tens (YT) and year units YU)**: These bits represent the tens and units of the year

- **Month tens (MT) and month units (MU)**: These bits represent the tens and units of the month

- **Date tens (DT) and date units (DU)**: These bits represent the tens and units of the day of the month

- **Week day units (WDU)**: This bit represents the day of the week (1 to 7)

You can read more about this register on *page 451* of the *reference manual*. The next crucial register is the **RTC Control Register** (RTC_CR).

RTC Control Register (RTC_CR)

The RTC_CR register is where we control the various operational modes and features of the RTC. This register allows us to enable the RTC, configure alarms, and set up the wakeup timer.

Let's consider the key bits in this register:

- **WUTE**: Enable the wakeup timer. This bit enables the RTC wakeup timer.

- **TSE**: Enable a timestamp event. This bit enables the timestamping of events.

- **ALRAE and ALRBE**: Enable Alarm A and Alarm B. These bits enable the respective alarms.

- **DCE**: Enable digital calibration. This bit enables digital calibration of the RTC clock.

- **FMT**: Hour format. This bit sets the hour format to either 24-hour or 12-hour (AM/PM).

Further details about this register can be found on *page 453* of the *reference manual*. Next, we have the **RTC Initialization and Status Register** (RTC_ISR).

RTC Initialization and Status Register (RTC_ISR)

The RTC_ISR register plays a dual role in both initializing the RTC and monitoring its status. This register is crucial during the setup process and for checking the RTC's current state. Here are the key bits in this register:

- **INIT**: Initialization mode. Setting this bit puts RTC into initialization mode.

- **RSF**: Registers synchronization flag. This bit indicates that the calendar registers are synchronized.

- **INITS**: Initialization status flag. This bit indicates whether the RTC calendar has been initialized.

- **ALRAF and ALRBF**: Alarm A and Alarm B flags. These bits indicate whether an alarm has been triggered.

Next, we have the **RTC Prescaler Register** (RTC_PRER).

RTC Prescaler Register (RTC_PRER)

The RTC_PRER register manages the prescalers that divide the RTC clock source to produce the 1 Hz clock necessary for accurate timekeeping. There are two key fields in this register:

- **PREDIV_A**: Asynchronous prescaler. This field sets the value for the asynchronous prescaler.

- **PREDIV_S**: Synchronous prescaler. This field sets the value for the synchronous prescaler.

Configuring the `RTC_PRER` register properly is vital for maintaining the accuracy of the RTC.

Next, we'll look at the RTC Alarm Registers, `RTC_ALRMAR` and `RTC_ALRMBR`.

RTC Alarm Registers (RTC_ALRMAR and RTC_ALRMBR)

These registers handle the configuration of Alarms A and B. They allow us to set specific times when the alarms should trigger. Here are the key fields in these registers:

- **ALRMASK**: Alarm mask bits. These bits allow you to mask certain parts of the alarm time, providing flexibility in how and when the alarms trigger.

- **ALRH, ALRMN, and ALRS**: Hour, minute, and second fields. These fields set the specific time for the alarm.

The `RTC_ALRMAR` and `RTC_ALRMBR` registers are vital for applications requiring reliable, time-based event triggering. Lastly, let's explore the **RTC Wakeup Timer Register** (RTC_WUTR).

RTC Wakeup Timer Register (RTC_WUTR)

The `RTC_WUTR` register configures the wakeup timer, enabling the RTC to periodically wake the system from low-power modes. The key field in this register is the **wakeup auto-reload value** (**WUT**). This field sets the interval for the wakeup timer.

In the next section, we will apply everything we've learned in this section to develop a driver for the RTC peripheral.

Developing the RTC driver

In this section, we will develop the RTC calendar driver so that we can configure and keep track of the time and date.

As always, we will create a copy of our previous project while following the steps outlined in earlier chapters. We rename this copied project to `RTC_Calendar`. Next, create a new file named `rtc.c` in the `Src` folder and another file named `rtc.h` in the `Inc` folder. The RTC configuration can be quite elaborate, so we will create several helper functions to modularize the initialization process.

The RTC implementation file

Let's begin by populating the `rtc.c` file, starting with the helper functions necessary for the initialization function. Here are the macro definitions that we will use in the RTC configuration:

```
#include "rtc.h"
#define PWREN       (1U << 28)
#define CR_DBP      (1U << 8)
#define CSR_LSION   (1U << 0)
```

```
#define CSR_LSIRDY    (1U << 1)
#define BDCR_BDRST    (1U << 16)
#define BDCR_RTCEN    (1U << 15)

#define RTC_WRITE_PROTECTION_KEY_1 ((uint8_t)0xCAU)
#define RTC_WRITE_PROTECTION_KEY_2 ((uint8_t)0x53U)
#define RTC_INIT_MASK              0xFFFFFFFFU
#define ISR_INITF                  (1U << 6)

#define WEEKDAY_FRIDAY             ((uint8_t)0x05U)
#define MONTH_DECEMBER             ((uint8_t)0x12U)
#define TIME_FORMAT_PM             (1U << 22)
#define CR_FMT                     (1U << 6)
#define ISR_RSF                    (1U << 5)

#define RTC_ASYNCH_PREDIV          ((uint32_t)0x7F)
#define RTC_SYNCH_PREDIV           ((uint32_t)0x00F9)
```

Let's break them down:

- PWREN (1U << 28): This macro enables the clock for the PWR module by setting bit 28 in the APB1 peripheral clock enable register.

- CR_DBP (1U << 8): This enables access to the backup domain by setting bit 8 in the PWR register.

- CSR_LSION (1U << 0): This macro enables the LSI oscillator by setting bit 0 in the Clock Control & Status Register.

- CSR_LSIRDY (1U << 1): This macro is used to read the state of the LSI register. The LSIRDY bit is set to 1 when the LSI is stable and ready to be used.

- BDCR_BDRST (1U << 16): This macro forces a reset of the backup domain by setting bit 16 in the Backup Domain Control Register.

- BDCR_RTCEN (1U << 15): This enables RTC by setting bit 15 in the Backup Domain Control Register.

- RTC_WRITE_PROTECTION_KEY_1 ((uint8_t)0xCAU): This key is used to disable write protection on the RTC registers.

- RTC_WRITE_PROTECTION_KEY_2 ((uint8_t)0x53U): This is the second key needed to disable write protection on the RTC registers.

- RTC_INIT_MASK (0xFFFFFFFFU): This mask is used to enter initialization mode in the RTC peripheral.

- `ISR_INITF` (1U << 6): This bit in the `ISR` register indicates that the RTC peripheral is in initialization mode.

- `WEEKDAY_FRIDAY` ((uint8_t) 0x05U): This macro is used to configure the weekday of the calendar to Friday.

- `MONTH_DECEMBER` ((uint8_t) 0x12U): This macro is used to configure the month of the calendar to December.

- `TIME_FORMAT_PM` (1U << 22): This macro sets the time format to PM in the 12-hour format.

- `CR_FMT` (1U << 6): This macro sets the hour format to 24-hour format in the RTC control register (RTC->CR).

- `ISR_RSF` (1U << 5): This bit in the `ISR` register indicates that the RTC registers are synchronized.

- `RTC_ASYNCH_PREDIV` ((uint32_t) 0x7F): This value sets the asynchronous prescaler for the RTC peripheral and is used to divide the clock frequency.

- `RTC_SYNCH_PREDIV` ((uint32_t) 0x00F9): This value sets the synchronous prescaler for the RTC peripheral and is used to further divide the clock frequency for timekeeping accuracy.

Let's examine the two functions responsible for setting the prescaler values for the RTC peripheral.

First, we have `rtc_set_asynch_prescaler`:

```
static void rtc_set_asynch_prescaler(uint32_t AsynchPrescaler)
{
  MODIFY_REG(RTC->PRER, RTC_PRER_PREDIV_A, AsynchPrescaler << RTC_
  PRER_PREDIV_A_Pos);
}
```

This function sets the asynchronous prescaler value for the RTC peripheral. Let's break it down:

- `MODIFY_REG`: This macro modifies specific bits in a register. It is defined in the `stm32f4xx.h` header file.

- `RTC->PRER`: The PRER register of the RTC.

- `RTC_PRER_PREDIV_A`: The mask for the asynchronous prescaler bits in the PRER register.

- `AsynchPrescaler << RTC_PRER_PREDIV_A_Pos`: This snippet shifts the `AsynchPrescaler` value to the correct position within the PRER register.

In short, this function configures the asynchronous prescaler by updating the appropriate bits in the PRER register. Next, we have the function for setting the synchronous prescaler – that is, `rtc_set_synch_prescaler`:

```
static void rtc_set_synch_prescaler(uint32_t SynchPrescaler)
{
  MODIFY_REG(RTC->PRER, RTC_PRER_PREDIV_S, SynchPrescaler);
}
```

This function configures the synchronous prescaler by updating the appropriate bits in the PRER register:

- RTC->PRER: The PRER register of the RTC
- RTC_PRER_PREDIV_S: This is the mask for the synchronous prescaler bits in the PRER register
- SynchPrescaler: This directly sets the synchronous prescaler value in the PRER register

Next, we must analyze the other RTC initialization helper functions to understand how they work together to configure and synchronize the RTC peripheral.

First, we have the `_rtc_enable_init_mode` function:

```
void _rtc_enable_init_mode(void)
{
    RTC->ISR = RTC_INIT_MASK;
}
```

Simply put, this function sets the RTC peripheral to initialization mode by writing the initialization mask to the ISR register:

- RTC->ISR: This is the RTC_ISR register of the RTC
- RTC_INIT_MASK: This mask is used to enter initialization mode

Next, we have `_rtc_disable_init_mode`:

```
void _rtc_disable_init_mode(void)
{
    RTC->ISR = ~RTC_INIT_MASK;
}
```

This function disables initialization mode by clearing the initialization mask in the ISR register. Here, ~RTC_INIT_MASK clears the initialization mask, exiting initialization mode.

The next helper function is `_rtc_isActiveflag_init`:

```
uint8_t _rtc_isActiveflag_init(void)
{
    return ((RTC->ISR & ISR_INITF) == ISR_INITF);
}
```

This function returns 1 if the RTC peripheral is in initialization mode by checking the `ISR_INITF` bit, which indicates that the RTC peripheral is in initialization mode.

Next, we have `_rtc_isActiveflag_rs`:

```
uint8_t _rtc_isActiveflag_rs(void)
{
    return ((RTC->ISR & ISR_RSF) == ISR_RSF);
}
```

This function returns 1 if the RTC registers are synchronized by checking the `ISR_RSF` bit, which indicates that the RTC registers are synchronized.

There's also the `rtc_init_seq` function:

```
static uint8_t rtc_init_seq(void)
{
    /* Start init mode */
    _rtc_enable_init_mode();

    /* Wait till we are in init mode */
    while (_rtc_isActiveflag_init() != 1) {}

    return 1;
}
```

This function starts the RTC initialization by enabling initialization mode and waiting until the RTC peripheral enters initialization mode:

- `_rtc_enable_init_mode`: This line puts the RTC peripheral into initialization mode
- `_rtc_isActiveflag_init`: This line waits until the RTC peripheral is in initialization mode

Next, we have the `wait_for_synchro` function:

```
static uint8_t wait_for_synchro(void)
{
    /* Clear RSF */
    RTC->ISR &= ~ISR_RSF;

    /* Wait for registers to synchronize */
    while (_rtc_isActiveflag_rs() != 1) {}

    return 1;
}
```

This function clears the synchronization flag and waits until the RTC registers are synchronized.

We also have the `exit_init_seq` function:

```
static uint8_t exit_init_seq(void)
{
    /* Stop init mode */
    _rtc_disable_init_mode();

    /* Wait for registers to synchronize */
    return (wait_for_synchro());
}
```

This function exits the RTC initialization mode and waits for the registers to synchronize to ensure everything is set up correctly. Now, let's see the functions for configuring the date and time.

First, we have `rtc_date_config`:

```
static void rtc_date_config(uint32_t WeekDay, uint32_t Day, uint32_t
Month, uint32_t Year)
{
  register uint32_t temp = 0U;

  temp = (WeekDay << RTC_DR_WDU_Pos) |\
          (((Year & 0xF0U) << (RTC_DR_YT_Pos - 4U)) |
          ((Year & 0x0FU) << RTC_DR_YU_Pos)) |\
          (((Month & 0xF0U) << (RTC_DR_MT_Pos - 4U)) |
          ((Month & 0x0FU) << RTC_DR_MU_Pos)) |\
          (((Day & 0xF0U) << (RTC_DR_DT_Pos - 4U)) |
          ((Day & 0x0FU) << RTC_DR_DU_Pos));
```

```
    MODIFY_REG(RTC->DR, (RTC_DR_WDU | RTC_DR_MT | RTC_DR_MU | RTC_DR_DT
    | RTC_DR_DU | RTC_DR_YT | RTC_DR_YU), temp);
}
```

This function sets the date in the RTC peripheral.

We begin by creating a temporary variable called `temp` to hold the date value. This variable is carefully constructed by shifting and combining the weekday, day, month, and year into the appropriate positions for the RTC's date register. The `MODIFY_REG` macro is then used to update the RTC's date register with this new value. Here, we can have the following values:

- `WeekDay`: Represents the day of the week

- `Day`: Represents the day of the month

- `Month`: Represents the month of the year

- `Year`: Represents the year

In essence, `rtc_date_config` takes the date components, assembles them into a single value, and writes it to the RTC's date register, ensuring the RTC peripheral accurately tracks the current date.

We have the `rtc_time_config` function for configuring the time:

```
static void rtc_time_config(uint32_t Format12_24, uint32_t Hours,
uint32_t Minutes, uint32_t Seconds)
{
    register uint32_t temp = 0U;

    temp = Format12_24 |\
            (((Hours & 0xF0U) << (RTC_TR_HT_Pos - 4U)) |
            ((Hours & 0x0FU) << RTC_TR_HU_Pos)) |\
            (((Minutes & 0xF0U) << (RTC_TR_MNT_Pos - 4U)) |
            ((Minutes & 0x0FU) << RTC_TR_MNU_Pos)) |\
            (((Seconds & 0xF0U) << (RTC_TR_ST_Pos - 4U)) |
            ((Seconds & 0x0FU) << RTC_TR_SU_Pos));

    MODIFY_REG(RTC->TR, (RTC_TR_PM | RTC_TR_HT | RTC_TR_HU | RTC_TR_MNT
    | RTC_TR_MNU | RTC_TR_ST | RTC_TR_SU), temp);
}
```

This function sets the time in the RTC peripheral. Much like the date configuration, `rtc_time_config` begins by initializing a temporary variable, `temp`, to hold the time value. The time components – format, hours, minutes, and seconds – are then combined into this variable. The `MODIFY_REG` macro updates the RTC's time register with the newly constructed time value. Here, we have the following additional values:

- `Format12_24`: This determines whether the time is in 12-hour or 24-hour format

- `Hours`: Represents the hour value

- `Minutes`: Represents the minute value

- `Seconds`: Represents the second value

Now that we've implemented all the helper functions required for initialization, let's go ahead and implement the `rtc_init()` function:

```
void rtc_init(void)
{
    /* Enable clock access to PWR */
    RCC->APB1ENR |= PWREN;
    /* Enable Backup access to config RTC */
    PWR->CR |= CR_DBP;

    /* Enable Low Speed Internal (LSI) */
    RCC->CSR |= CSR_LSION;

    /* Wait for LSI to be ready */
    while((RCC->CSR & CSR_LSIRDY) != CSR_LSIRDY) {}

    /* Force backup domain reset */
    RCC->BDCR |= BDCR_BDRST;

    /* Release backup domain reset */
    RCC->BDCR &= ~BDCR_BDRST;

    /* Set RTC clock source to LSI */
    RCC->BDCR &= ~(1U << 8);
    RCC->BDCR |= (1U << 9);

    /* Enable the RTC */
    RCC->BDCR |= BDCR_RTCEN;

    /* Disable RTC registers write protection */
    RTC->WPR = RTC_WRITE_PROTECTION_KEY_1;
```

```
        RTC->WPR = RTC_WRITE_PROTECTION_KEY_2;

        /* Enter the initialization mode */
        if(rtc_init_seq() != 1)
        {
            // Handle initialization failure
        }

        /* Set desired date: Friday, December 29th, 2016 */
        rtc_date_config(WEEKDAY_FRIDAY, 0x29, MONTH_DECEMBER, 0x16);

        /* Set desired time: 11:59:55 PM */
        rtc_time_config(TIME_FORMAT_PM, 0x11, 0x59, 0x55);

        /* Set hour format */
        RTC->CR |= CR_FMT;

        /* Set Asynchronous prescaler */
        rtc_set_asynch_prescaler(RTC_ASYNCH_PREDIV);

        /* Set Synchronous prescaler */
        rtc_set_synch_prescaler(RTC_SYNCH_PREDIV);

        /* Exit the initialization mode */
        exit_init_seq();

        /* Enable RTC registers write protection */
        RTC->WPR = 0xFF;
}
```

Let's break it down:

```
RCC->APB1ENR |= PWREN;
```

We start by enabling the clock for the PWR module. This is crucial as it allows us to access and configure the RTC peripheral:

```
PWR->CR |= CR_DBP;
```

Next, we must enable access to the backup domain. This step is necessary to make changes to the RTC configuration:

```
RCC->CSR |= CSR_LSION;
```

Then, we enable the LSI oscillator, which serves as the clock source for the RTC peripheral:

```
while((RCC->CSR & CSR_LSIRDY) != CSR_LSIRDY) {}
```

We must wait until the LSI oscillator is stable and ready to use:

```
RCC->BDCR |= BDCR_BDRST;
```

To ensure a clean configuration, we must force a reset of the backup domain:

```
RCC->BDCR &= ~BDCR_BDRST;
```

Then, we must release the reset, allowing the backup domain to function normally:

```
RCC->BDCR &= ~(1U << 8);
RCC->BDCR |= (1U << 9);
```

After, we must configure the RTC peripheral so that it uses the LSI oscillator as its clock source:

```
RCC->BDCR |= BDCR_RTCEN;
```

Then, we must enable the RTC peripheral by setting the appropriate bit in the backup domain control register:

```
RTC->WPR = RTC_WRITE_PROTECTION_KEY_1;
RTC->WPR = RTC_WRITE_PROTECTION_KEY_2;
```

To allow changes to the RTC registers, we must disable write protection:

```
if(rtc_init_seq() != 1)
{
    // Handle initialization failure
}
```

We can enter initialization mode using our `rtc_init_seq()` helper function. If it fails, we must handle the error appropriately:

```
rtc_date_config(WEEKDAY_FRIDAY, 0x29, MONTH_DECEMBER, 0x16);
```

At this point, we must configure the RTC peripheral to the desired date using another one of our helper functions. In this example, we will set the date to *Friday, December 29, 2016*:

```
rtc_time_config(TIME_FORMAT_PM, 0x11, 0x59, 0x55);
```

Then, we must set the RTC peripheral to the desired time. In this case, we will set the time to *11:59:55 P.M.*:

```
RTC->CR |= CR_FMT;
```

Next, we must configure the RTC peripheral so that it uses a 24-hour format:

```
rtc_set_asynch_prescaler(RTC_ASYNCH_PREDIV);
rtc_set_synch_prescaler(RTC_SYNCH_PREDIV);
```

Then, we must set the asynchronous and synchronous prescaler values:

```
exit_init_seq();
```

At this point, we can exit initialization mode using the `exit_init_seq()` helper function we created earlier:

```
RTC->WPR = 0xFF;
```

Finally, we must re-enable write protection on the RTC registers to prevent accidental changes.

This `rtc_init` function meticulously sets up the RTC peripheral by enabling the necessary clock, configuring the backup domain, setting the RTC clock source, and initializing the date and time.

Before moving on to the `main.c` file, let's implement a few helper functions that are essential for handling various RTC tasks, such as converting values and getting the current date and time.

Let's start with the `rtc_convert_dec2bcd` function:

```
uint8_t rtc_convert_dec2bcd(uint8_t value)
{
    return (uint8_t)((((value) / 10U) << 4U) | ((value) % 10U));
}
```

This function takes a decimal value and returns its equivalent in BCD format, which is useful for setting RTC values.

Let's take a closer look at decimal to BCD conversion:

- First, `((value) / 10U) << 4U` shifts the tens digit to the left by 4 bits
- Then, `((value) % 10U)` gets the units digit
- The OR operation combines these two to form the BCD value

Before going any further, let's take a closer look at the BCD format and the conversion process.

Understanding BCD format

BCD is a way of representing decimal numbers in binary form. But here's the twist: instead of converting the whole number into a single binary value, each decimal digit is represented by its own binary sequence.

How does BCD work?

In BCD, each digit of a decimal number is encoded separately as a 4-bit binary number. Let's break it down with an example to make it clearer.

Say you have the decimal number *42*. In BCD, this would be represented as follows:

- *4* in decimal is *0100* in binary

- *2* in decimal is *0010* in binary

So, 42 in BCD is *0100 0010*. Notice how each decimal digit is converted into a 4-bit binary form and then combined to represent the entire number.

Why use BCD?

You might be wondering, why not just use regular binary? Well, BCD has its perks, especially in digital systems that need to display numbers or interface with human-readable formats:

- **Ease of conversion**: Converting between BCD and decimal is straightforward. Each 4-bit group corresponds directly to a decimal digit, making it easy to read and convert.

- **Display compatibility**: Devices such as digital clocks, calculators, and other numeric displays often use BCD because it simplifies the process of converting binary values into a form that can be easily shown on a screen.

Now, let's see how this relates to RTCs.

BCD in RTC configurations

When working with RTCs, BCD is particularly handy. The RTC hardware often uses BCD to store time and date values because it simplifies how these values can be displayed and manipulated. For instance, setting the time to *12:34:56* in BCD means we have the following representations:

- *12* is *0001 0010*

- *34* is *0011 0100*

- *56* is *0101 0110*

Each of these pairs is easy to interpret and convert back into decimal for display or further processing.

BCD format is a clever way of encoding decimal numbers in a binary system. By handling each decimal digit separately, BCD simplifies many operations, especially when interfacing with human-readable displays or systems that require precise decimal representation. *Figure 15.6* illustrates how BCD values can easily be mapped onto digital displays:

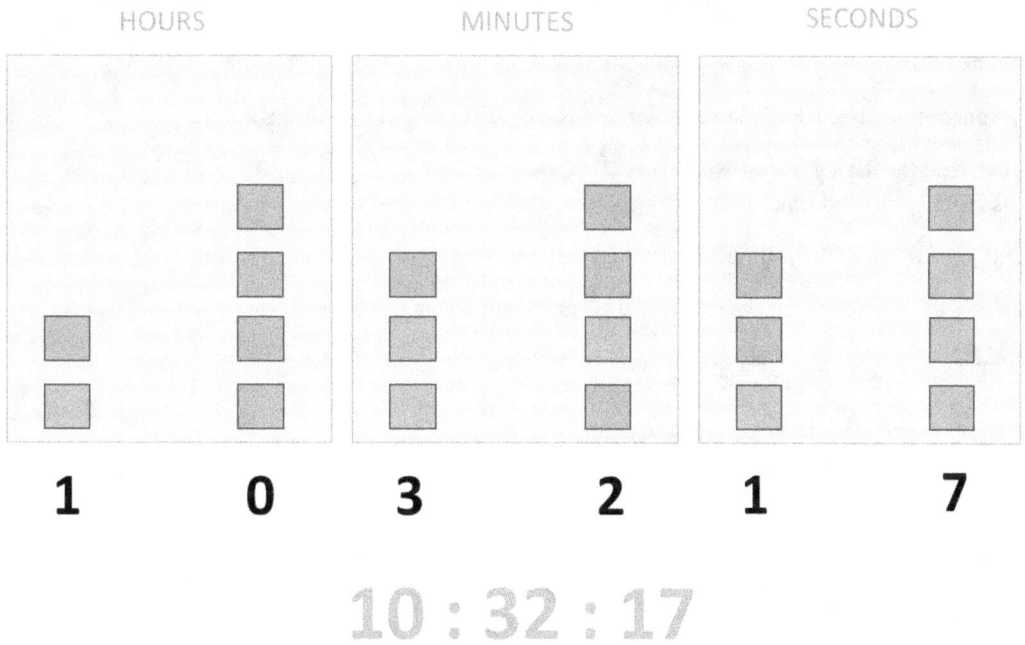

Figure 15.7: Display with BCD format

Now, let's analyze the `rtc_convert_bcd2dec` function:

```
uint8_t rtc_convert_bcd2dec(uint8_t value)
{
    return (uint8_t)(((uint8_t)((value) & (uint8_t)0xF0U) >>
    (uint8_t)0x4U) * 10U + ((value) & (uint8_t)0x0FU));
}
```

This function takes a BCD value and returns its decimal equivalent, making it easier to work with RTC data in a decimal format.

Here's the BCD to decimal conversion for this:

- First, `((value) & (uint8_t)0xF0U) >> (uint8_t)0x4U` extracts the tens digit
- Then, `((value) & (uint8_t)0x0FU)` extracts the units digit
- Multiplication and addition combine these to form the binary value

Next, we have a helper function for getting the day – that is, `rtc_date_get_day`:

```
uint32_t rtc_date_get_day(void)
{
    return (uint32_t)((READ_BIT(RTC->DR, (RTC_DR_DT | RTC_DR_DU))) >>
    RTC_DR_DU_Pos);
}
```

This function reads the RTC date register and returns the current day of the month.

We can read the day by using `READ_BIT(RTC->DR, (RTC_DR_DT | RTC_DR_DU))`, which reads the day tens and units bits.

Shifting the result to the right positions the value correctly.

We also have a function for the year – that is, `rtc_date_get_year`:

```
uint32_t rtc_date_get_year(void)
{
    return (uint32_t)((READ_BIT(RTC->DR, (RTC_DR_YT | RTC_DR_YU))) >>
    RTC_DR_YU_Pos);
}
```

This function reads the RTC date register and returns the current year.

Here, `READ_BIT(RTC->DR, (RTC_DR_YT | RTC_DR_YU))` reads the year tens and units bits.

We also have functions for retrieving the month, second, minute, and hour, all of which are implemented using the same approach as the other getter functions:

- First, we have `rtc_date_get_month`:

  ```
  uint32_t rtc_date_get_month(void)
  {
      return (uint32_t)((READ_BIT(RTC->DR, (RTC_DR_MT | RTC_DR_
      MU))) >> RTC_DR_MU_Pos);
  }
  ```

- Then, we have `rtc_time_get_second`:

  ```
  uint32_t rtc_time_get_second(void)
  {
      return (uint32_t)(READ_BIT(RTC->TR, (RTC_TR_ST | RTC_TR_SU))
      >> RTC_TR_SU_Pos);
  }
  ```

- Next, there's `rtc_time_get_minute`:

```
uint32_t rtc_time_get_minute(void)
{
    return (uint32_t)(READ_BIT(RTC->TR, (RTC_TR_MNT | RTC_TR_
    MNU)) >> RTC_TR_MNU_Pos);
}
```

- Finally, there's `rtc_time_get_hour`:

```
uint32_t rtc_time_get_hour(void)
{
    return (uint32_t)((READ_BIT(RTC->TR, (RTC_TR_HT | RTC_TR_
    HU))) >> RTC_TR_HU_Pos);
}
```

With all of these functions implemented, our `rtc.c` file is complete. Our next task is to populate the `rtc.h` file.

The header file

Here's the code:

```
#ifndef RTC_H__
#define RTC_H__
#include <stdint.h>
#include "stm32f4xx.h"

void rtc_init(void);
uint8_t rtc_convert_bcd2dec(uint8_t value);
uint32_t rtc_date_get_day(void);
uint32_t rtc_date_get_year(void);
uint32_t rtc_date_get_month(void);
uint32_t rtc_time_get_second(void);
uint32_t rtc_time_get_minute(void);
uint32_t rtc_time_get_hour(void);

#endif
```

Here, we're simply exposing the functions implemented in `rtc.c`, making them callable from other files. Let's go ahead and test our driver's `main.c` file.

The main file

Let's update the main.c file so that it looks like this:

```c
#include <stdio.h>
#include "rtc.h"
#include "uart.h"


#define BUFF_LEN        20

uint8_t time_buff[BUFF_LEN] = {0};
uint8_t date_buff[BUFF_LEN] = {0};

static void display_rtc_calendar(void);

int main(void)
{

    /*Initialize debug UART*/
    uart_init();

    /*Initialize rtc*/
    rtc_init();

    while(1)
    {
        display_rtc_calendar();

    }
}

static void display_rtc_calendar(void)
{
    /*Display format :  hh : mm : ss*/
    sprintf((char *)time_buff,"%.2d :%.2d :%.2d",rtc_convert_
    bcd2dec(rtc_time_get_hour()),
            rtc_convert_bcd2dec(rtc_time_get_minute()),
            rtc_convert_bcd2dec(rtc_time_get_second()));

    printf("Time : %.2d :%.2d :%.2d\n\r",rtc_convert_bcd2dec(rtc_time_
    get_hour()),
            rtc_convert_bcd2dec(rtc_time_get_minute()),
```

```
            rtc_convert_bcd2dec(rtc_time_get_second()));



    /*Display format :   mm : dd : yy*/
    sprintf((char *)date_buff,"%.2d - %.2d - %.2d",rtc_convert_
    bcd2dec(rtc_date_get_month()),
            rtc_convert_bcd2dec(rtc_date_get_day()),
            rtc_convert_bcd2dec(rtc_date_get_year()));

    printf("Date : %.2d - %.2d - %.2d     ",rtc_convert_bcd2dec(rtc_
    date_get_month()),
            rtc_convert_bcd2dec(rtc_date_get_day()),
            rtc_convert_bcd2dec(rtc_date_get_year()));
}
```

Let's break down the unique aspects of the code, starting with the `display_rtc_calendar` function. This function retrieves the current time and date from the RTC peripheral, formats these values, prints them out, and stores them in a buffer for further processing.

The following are our buffer definitions:

```
#define BUFF_LEN 20

uint8_t time_buff[BUFF_LEN] = {0};
uint8_t date_buff[BUFF_LEN] = {0};
```

Here, we can see the following:

- BUFF_LEN: This defines the length of the buffer for storing the time and date string
- **Buffers**:
  - time_buff: This is an array that holds the formatted time string
  - date_buff: This is an array that holds the formatted date string

Let's take a closer look at the time formatting and display block:

```
sprintf((char *)time_buff, "%.2d :%.2d :%.2d",
        rtc_convert_bcd2dec(rtc_time_get_hour()),
        rtc_convert_bcd2dec(rtc_time_get_minute()),
        rtc_convert_bcd2dec(rtc_time_get_second()));

printf("Time : %.2d :%.2d :%.2d\n\r",rtc_convert_bcd2dec(rtc_time_get_
hour()),
            rtc_convert_bcd2dec(rtc_time_get_minute()),
            rtc_convert_bcd2dec(rtc_time_get_second()));
```

Here, we can see the following:

- `sprintf`: This is used to format the time string into `time_buff`

- `rtc_convert_bcd2dec`: This converts the BCD values from RTC into decimal

- `rtc_time_get_hour`, `rtc_time_get_minute`, and `rtc_time_get_second`: These retrieve the current hour, minute, and second from the RTC peripheral, respectively

- `printf`: This function is used to print formatted output to the serial port

- `%.2d`: This format specifier means that the integer will be printed with at least 2 digits, padding with leading zeros if necessary

We're now ready to test our RTC calendar driver on the microcontroller. We can test the project by following these steps:

1. Compile and upload the project to your microcontroller.

2. Launch RealTerm or your preferred serial terminal program.

3. Configure the appropriate port and baud rate.

4. You should see the time and date values printed and updating in real time, as shown in *Figure 15.7*:

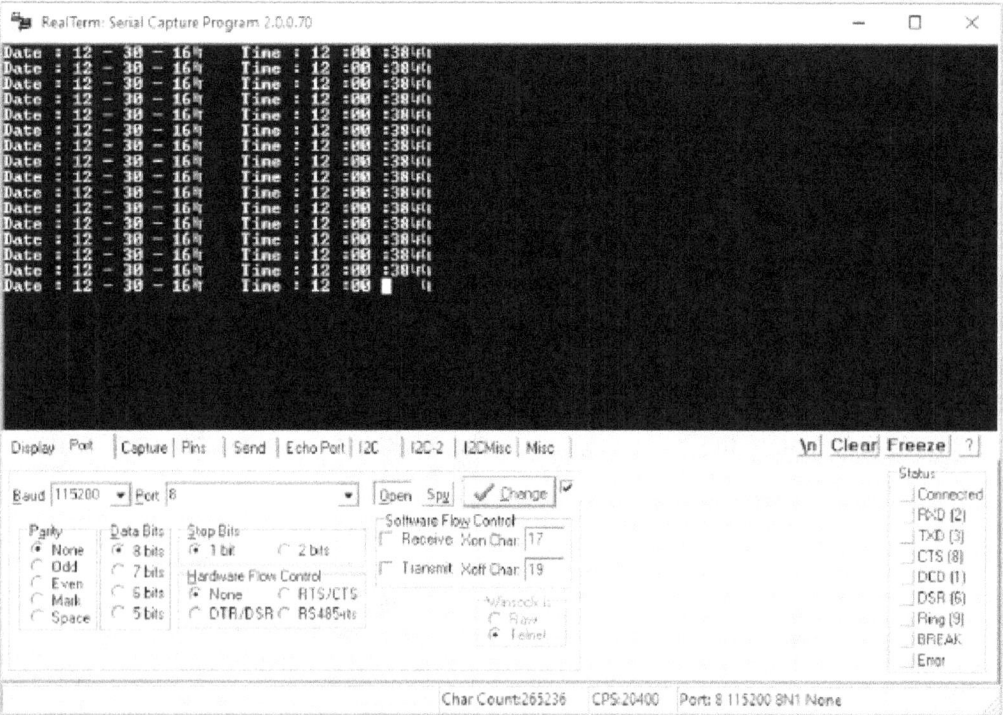

Figure 15.8: Expected results

Summary

In this chapter, we explored the RTC peripheral, a component for timekeeping in embedded systems. This peripheral is essential for applications requiring precise time and date maintenance, making it fundamental for a wide range of embedded applications.

We began by introducing RTCs and understanding their functionality. This included a deep dive into how RTCs operate, which involved focusing on the crystal oscillator, counters, time and date registers, and the importance of battery backup. We illustrated these concepts with case studies, showcasing the practical use of RTCs in data logging, alarm clocks, time-stamping transactions, and calendar functions.

Following this, we examined the STM32 RTC module, highlighting its key features and capabilities. We discussed the calendar in terms of sub-seconds accuracy, dual programmable alarms, low power consumption, backup domain, daylight saving time adjustments, automatic wakeup, tamper detection, digital calibration, and synchronization with external clocks. Each feature was detailed to show its application and importance in maintaining accurate timekeeping.

Next, we analyzed the relevant registers from the STM32 reference manual, providing a detailed understanding of the configuration and operation of the RTC. We covered the RTC_TR, RTC_DR, RTC_CR, RTC_ISR, RTC_PRER, RTC_ALRMAR, RTC_ALRMBR, and RTC_WUTR registers. Each register's role and key fields were explained to ensure you have a comprehensive grasp of how they contribute to the RTC's functionality.

Finally, we applied this knowledge to develop an RTC driver. We walked through the steps to create and configure the RTC driver, starting with the initialization sequence and covering functions to set the date and time. We also implemented helper functions for converting values between decimal and BCD formats, as well as retrieving current time and date values from the RTC peripheral.

In the next chapter, we will delve into another useful peripheral, expanding our knowledge and toolkit for embedded systems development.

Independent Watchdog (IWDG)

In this chapter, we'll learn about the **Independent Watchdog** (**IWDG**) timer, a unique component for enhancing the reliability of embedded systems. IWDG is essential for monitoring the system's operation and ensuring it can recover from unexpected faults or malfunctions.

We will begin by exploring the general concept of **watchdog timers** (**WDTs**) and their importance in embedded systems. Following this, we will examine how WDTs function and the unique features of IWDG. Next, we will focus specifically on the STM32 implementation of the IWDG, looking at its key registers and configuration. Finally, we will apply this knowledge to develop a bare-metal IWDG driver, providing practical code examples to solidify our understanding.

In this chapter, we will cover the following main topics:

- Understanding WDTs
- Types of WDTs
- The STM32 IWDG
- Developing the IWDG driver

By the end of this chapter, you will have a comprehensive understanding of IWDG timers and their critical role in embedded systems. You will also gain skills to develop and implement IWDG drivers for STM32 microcontrollers, ensuring your systems can maintain robustness and reliability.

Technical requirements

All code examples for this chapter can be found on GitHub at the following link:

```
https://github.com/PacktPublishing/Bare-Metal-Embedded-C-Programming
```

Understanding WDTs

WDTs are one of the unsung heroes of embedded systems. They quietly monitor the system's health, ensuring it can recover gracefully from unexpected hitches. Imagine them as vigilant guards, always on the lookout for system malfunctions, ready to reset the microcontroller if something goes wrong. In this section, we'll explore what WDTs are and how they function, and dive into some common use cases to illustrate their importance.

What are WDTs?

WDTs are like guardians for your microcontroller. Imagine you're using a device, and something goes wrong—a bug in the developer's code causes an infinite loop, or a hardware glitch freezes the system. Without a watchdog, your device would be stuck, potentially causing significant problems, especially in critical applications such as medical devices or automotive systems.

A WDT is a hardware or software timer that resets the system if the main program fails to reset the timer before it expires. It's a simple yet powerful mechanism to ensure that your system can recover from unexpected issues. Let's see how they work.

How WDTs work

Think of a WDT as an hourglass that you need to turn over regularly to prevent it from running out of sand. Here's a step-by-step breakdown of how it works:

1. **Initialization**: When your system starts, you initialize the WDT with a specific timeout period. This period is the maximum time your system can run without resetting the timer.

2. **Countdown**: The WDT starts counting down from the set timeout value. If it reaches zero, it assumes something went wrong and triggers a system reset.

3. **Resetting the timer**: Your main program needs to periodically reset the WDT before it reaches zero. This action is often called **feeding the watchdog** or **kicking the dog**. If the program is running correctly, it will continue to reset the timer, preventing a reset.

4. **System reset**: If your program fails to reset the WDT in time—perhaps because it got stuck in an infinite loop or encountered a critical error—the WDT will expire and reset the system, bringing it back to a known state.

Now that we understand the basics, let's look at some real-world applications where WDTs play a crucial role.

Common use cases

Following is a list of some real-world applications.

Industrial automation

In industrial automation, reliability is paramount. Machines and processes need to run continuously and without failure. WDTs ensure that if a **Programmable Logic Controller** (**PLC**) or other control systems hang or crash, they can quickly recover.

Example: Imagine a conveyor belt system in a manufacturing plant. The PLC controlling the conveyor belt has a WDT set to 1 second. If the PLC software fails to reset the watchdog within 1 second due to a software bug or external interference, the WDT will reset the PLC. This reset ensures that the conveyor belt can resume operation with minimal downtime, preventing potential production losses.

Automotive systems

Modern vehicles rely heavily on embedded systems for various functions, from engine control to infotainment. WDTs are vital in ensuring these systems operate reliably.

Example: Consider an **engine control unit** (**ECU**) in a car. The ECU monitors and controls critical engine parameters. A WDT in the ECU might be set to 500 milliseconds. If the ECU software fails to reset the watchdog due to a fault, the WDT resets the ECU. This reset can prevent engine misbehavior, ensuring the vehicle operates safely.

Medical devices

In medical devices, WDTs can be life-saving. Devices such as pacemakers, infusion pumps, and patient monitors must operate without failure.

Example: Take a patient monitor that tracks vital signs such as heart rate and blood pressure. The monitor's software includes a WDT set to 2 seconds. If the software encounters a problem and fails to reset the watchdog, the device will reset. This reset ensures the monitor can quickly recover and continue providing accurate, real-time data, which is crucial for patient care.

Consumer electronics

Even in consumer electronics, WDTs help maintain system reliability and enhance user experience. Think of smartphones, smart home devices, and gaming consoles.

Example: In a smart thermostat, the software manages temperature settings and connectivity. A WDT ensures that if the software freezes, the system resets and continues operating. This functionality prevents users from experiencing extended downtime, maintaining comfort and convenience in their homes.

These examples illustrate the crucial role that WDTs play in modern systems. When implementing WDTs, it's essential to consider several key factors to ensure their effectiveness. Let's see some of these factors.

Practical considerations

When implementing WDTs, you must consider the following:

- **Timeout period**: Choose an appropriate timeout period based on your application's needs. Too short, and you risk unnecessary resets; too long, and you might not recover quickly enough from faults.

- **Reset mechanism**: Ensure that resetting the WDT (feeding the dog) is done in a part of your code that runs regularly and indicates the system is operating correctly.

- **Recovery strategy**: Plan how your system should recover after a watchdog reset. Ensure critical data is preserved and the system returns to a safe state.

- **Testing**: Thoroughly test your WDT implementation to ensure it behaves as expected under various fault conditions.

Next, let's see the types of WDTs available.

Types of WDTs

WDTs can be categorized into several types based on their functionality and integration. Let's explore the most common types.

Internal WDTs

Internal WDTs are built into the microcontroller. They are a convenient option because they don't require additional external components. These timers are directly integrated into the microcontroller's architecture and can be configured through software.

They have the following features:

- **Integration**: No need for external circuitry

- **Configuration**: Typically configured using the microcontroller's registers

- **Power**: They can continue to operate in low-power modes, making them suitable for battery-powered applications

Example use case: In a small IoT device, an internal WDT can monitor the microcontroller's operation without adding extra hardware, ensuring the device can reset itself if it encounters an error.

External WDTs

External WDTs are separate components connected to the microcontroller. These timers provide additional flexibility and can be used when the internal WDT isn't sufficient or if redundancy is required.

Here is a list of their features:

- **Flexibility**: Can be chosen based on specific requirements (for example, longer timeout periods)
- **Redundancy**: Adding an external watchdog provides an extra layer of safety
- **Independence**: Operate independently of the microcontroller's clock and power

Example use case: In a critical automotive system, an external WDT can provide an additional safeguard, ensuring the system resets even if the internal timer fails.

Windowed WDTs

Windowed WDTs (**WWDTs**) add an extra layer of control by introducing a *window* period. The system must reset the timer within a specific window period; too early or too late resets the system. This prevents scenarios where the software gets stuck in a loop resetting the watchdog too frequently (which could mask a malfunction).

Their features include the following:

- **Precision**: Require the timer to be reset within a specific time window
- **Fault detection**: Can detect both early and late watchdog resets, offering improved fault detection
- **Security**: Enhance system security by ensuring the timer is reset at appropriate intervals

Example use case: In a medical device, a WWDT ensures the control software operates correctly within defined time intervals, adding an extra layer of reliability.

IWDGs

IWDGs are designed to be robust and reliable. They run from a separate clock source, usually a **low-speed internal** (**LSI**) clock, and operate independently of the main system clock. This independence ensures they continue to function even if the main clock fails.

Their features include the following:

- **Independence**: Operate from a separate clock source
- **Robustness**: Continue functioning even if the main system clock fails
- **Minimal configuration**: Typically simple to configure and use

Example use case: In an **industrial control system** (**ICS**), an IWDG ensures the system can recover from malfunctions, even if the main clock source is disrupted.

Selecting the appropriate WDT depends on several factors, including the criticality of the application, power constraints, and required reliability.

Choosing the right WDT

Here's a quick guide to help you choose the right WDT:

- **For low-power applications**: Consider internal WDTs due to their integration and low power consumption

- **For high-reliability systems**: Use external WDTs for redundancy

- **For applications requiring precise timing**: WWDTs provide enhanced fault detection

- **For systems needing robust operation**: IWDGs offer continued functionality even if the main clock fails

Understanding the different types of WDTs and their features allows us to choose the right one for our applications. Whether it's an internal WDT for simplicity, an external one for redundancy, a WWDT for precise control, or an IWDG for robustness, there's a WDT suited for every need.

In the upcoming section, we will delve into the IWDG embedded within the STM32F411 microcontroller, examining its features and how to leverage it for enhanced system reliability.

The STM32 IWDG

In this section, we'll analyze the STM32 IWDG module, exploring its main features and other relevant information to help you understand how to leverage this powerful feature in your embedded applications.

STM32 microcontrollers feature two types of WDTs: the IWDG and the **Window Watchdog** (**WWDG**). Both are essential for detecting and correcting software malfunctions by initiating a system reset, but they each have unique characteristics and applications.

The IWDG operates using a dedicated LSI clock, ensuring it continues to function even if the main system clock fails. This makes it highly reliable for applications that require continuous monitoring, regardless of the main clock's state. In contrast, the WWDG derives its clock from the APB1 clock and features a configurable time window. The system must refresh the WWDG within this time window; failing to do so, either too early or too late, will trigger a system reset.

The IWDG is best suited for applications needing an independent watchdog process with lower timing accuracy constraints, while the WWDG is ideal for applications requiring precise timing windows. Let's see some key features of the IWDG.

Key features of the IWDG

The IWDG in STM32 microcontrollers boasts several key features:

- **Free-running downcounter**: The IWDG operates as a free-running downcounter, continuously counting down from a preset value

- **Independent clock source**: It uses an independent **resistor-capacitor** (**RC**) oscillator, allowing it to function in low-power modes such as Standby and Stop

- **System reset on timeout**: If the WDT is activated and the downcounter reaches zero (0x000), a system reset is triggered

- **Write access protection**: To modify critical registers, a specific sequence of operations is required, ensuring protection against accidental or malicious modifications

Let's see how it works.

How the IWDG works

The IWDG module operates as an independent safeguard for the microcontroller, ensuring the system can recover from software malfunctions. *Figure 16.1* presents a block diagram of the IWDG, sourced from the reference manual:

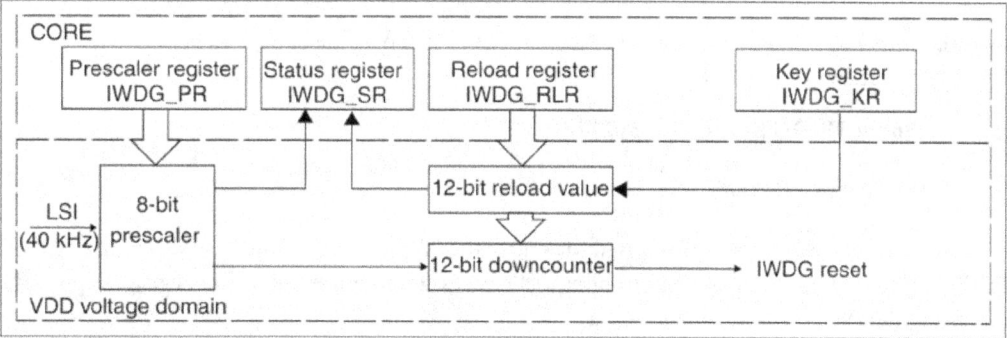

Figure 16.1: IWDG block diagram

Let's break down its functional blocks and how it operates:

- **Starting the IWDG**: To start the IWDG, we write the value `0xCCCC` to the Key Register (`IWDG_KR`). This action initiates the downcounter, which begins counting down from the reset value of `0xFFF`.

- **Preventing a reset**: To prevent the WDT from reaching zero and triggering a reset, the counter must be periodically reloaded. This is done by writing the value `0xAAAA` to the `IWDG_KR` register, which reloads the counter with the value from the Reload Register (`IWDG_RLR`) and prevents a reset.

- **Hardware watchdog feature**: If enabled, the IWDG is automatically activated at power-on. In this mode, the WDT will generate a reset unless the Key Register is written with the appropriate value before the counter reaches zero.

- **Register access protection**: To modify the prescaler (`IWDG_PR`) and reload (`IWDG_RLR`) registers, we must temporarily disable write access protection by writing the code `0x5555` to the `IWDG_KR` register. After doing this, any changes to these registers must be made immediately; otherwise, access protection will be re-enabled.

With this general overview of the IWDG block in mind, let's analyze the key registers one by one.

IWDG registers

In this section, we will explore the characteristics and functions of some of the crucial registers within the IWDG peripheral.

Let's start with the **IWDG Key Register** (`IWDG_KR`).

IWDG Key Register (IWDG_KR)

The `IWDG_KR` register is a key register used to control the IWDG's operations, including starting the watchdog, reloading the counter, and disabling write access to other registers. This register is pivotal in managing the IWDG's functionality.

Key operations in this register include the following:

- `0xCCCC`: Start the IWDG. Writing this value to `IWDG_KR` starts the IWDG timer.

- `0xAAAA`: Reload the counter. Writing this value reloads the IWDG counter, preventing it from reaching zero and triggering a system reset.

- `0x5555`: Disable write protection. This value allows modifications to the prescaler and reload registers.

Next, let's discuss the **IWDG Prescaler Register** (`IWDG_PR`).

IWDG Prescaler Register (IWDG_PR)

The IWDG_PR register is used to set the prescaler value, which determines the frequency of the IWDG clock by dividing the LSI clock. Adjusting this register helps configure the countdown speed of the WDT.

Key bits in this register is PR[2:0]: Prescaler value. These bits can be set to divide the LSI clock by 4, 8, 16, 32, 64, 128, or 256, allowing flexibility in setting the WDT interval.

Next, we move on to the **IWDG Reload Register** (IWDG_RLR).

IWDG Reload Register (IWDG_RLR)

The IWDG_RLR register defines the reload value for the IWDG counter. This value determines the timeout period before the watchdog triggers a system reset if not reloaded in time.

The key field in this register is RL[11:0]; this means reload value. It is a 12-bit value that sets the counter's reload value, which can range from 0x000 to 0xFFF.

Finally, we examine the **IWDG Status Register** (IWDG_SR).

IWDG Status Register (IWDG_SR)

The IWDG_SR register provides status information about the IWDG, indicating whether updates to the prescaler or reload registers are ongoing.

Key bits in this register include the following:

- PVU: **Prescaler value update (PVU)**. This bit indicates that the prescaler value is being updated.

- RVU: **Reload value update (RVU)**. This bit indicates that the reload value is being updated.

With a clear understanding of the IWDG's functionality and its registers, we can now move on to the next section, where we will develop the IWDG driver.

Developing the IWDG driver

In this section, we'll use what we've learned so far in this chapter to develop the IWDG driver.

Let's start by setting up the project.

Create a copy of your previous project in your IDE, following the steps outlined in earlier chapters. Rename this copied project to IWDG. Next, create a new file named iwdg.c in the Src folder and another file named iwdg.h in the Inc folder.

The IWDG implementation file

Populate your `iwdg.c` file with the following code:

```c
#include "iwdg.h"

#define IWDG_KEY_ENABLE                    0x0000CCCCU
#define IWDG_KEY_WR_ACCESS_ENABLE           0x00005555U
#define IWDG_PRESCALER_4                   0x00000000U
#define IWDG_RELOAD_VAL                 0xFFF

static uint8_t isIwdg_ready(void);

void iwdg_init(void)
{
    /*Enable the IWDG by writing 0x0000CCCC in the IWDG_KR register*/
    IWDG->KR = IWDG_KEY_ENABLE;

    /*Enable register access by writing 0x0000 5555 in the IWDG_KR
    register*/
    IWDG->KR = IWDG_KEY_WR_ACCESS_ENABLE;

    /*Set the IWDG Prescaler*/
    IWDG->PR =  IWDG_PRESCALER_4;

    /*Set the reload register (IWDG_RLR) to the largest value 0xFFF*/
    IWDG->RLR = IWDG_RELOAD_VAL;

    /*Wait for the registers to be updated (IWDG_SR = 0x0000 0000)*/
    while(isIwdg_ready() != 1){}

    /*Refresh the counter value with IWDG_KR (IWDG_KR = 0x0000 AAAA)*/
    IWDG->KR = IWDG_KEY_RELOAD;
}


static uint8_t isIwdg_ready(void)
{
 return ((READ_BIT(IWDG->SR, IWDG_SR_PVU | IWDG_SR_RVU) == 0U) ? 1UL :
 0UL);
}
```

Let's break it down. Let's look at the **macro definitions** first:

```
#include "iwdg.h"

#define IWDG_KEY_ENABLE                0x0000CCCCU
#define IWDG_KEY_WR_ACCESS_ENABLE      0x00005555U
#define IWDG_PRESCALER_4               0x00000000U
#define IWDG_RELOAD_VAL                0xFFF
```

- `IWDG_KEY_ENABLE`: This macro defines the key to enable the IWDG

- `IWDG_KEY_WR_ACCESS_ENABLE`: This macro defines the key to enable write access to IWDG registers

- `IWDG_PRESCALER_4`: This macro defines the prescaler value for the IWDG

- `IWDG_RELOAD_VAL`: This macro sets the reload value to the maximum, 0xFFF, providing the longest timeout period

Next, we have the `iwdg_init()` function.

```
IWDG->KR = IWDG_KEY_ENABLE;
```

> 💡 **Quick tip**: Enhance your coding experience with the **AI Code Explainer** and **Quick Copy** features. Open this book in the next-gen Packt Reader. Click the **Copy** button (**1**) to quickly copy code into your coding environment, or click the **Explain** button (**2**) to get the AI assistant to explain a block of code to you.
>
> | | Copy | Explain |
> |---|---|---|
> | `function calculate(a, b) {` | **1** | **2** |
> | `  return {sum: a + b};` | | |
> | `};` | | |
>
> 🔒 **The next-gen Packt Reader** is included for free with the purchase of this book. Unlock it by scanning the QR code below or visiting `https://www.packtpub.com/unlock/9781835460818`.
>
>

This line writes 0x0000CCCC to the IWDG **Key Register** (**KR**) to enable the IWDG.

```
IWDG->KR = IWDG_KEY_WR_ACCESS_ENABLE;
```

This line writes 0x00005555 to the IWDG Key Register to enable write access to the prescaler and reload registers.

```
IWDG->PR = IWDG_PRESCALER_4;
```

This line sets the prescaler register to divide the clock by 4, as defined by `IWDG_PRESCALER_4`.

```
IWDG->RLR = IWDG_RELOAD_VAL;
```

This sets the reload register to the maximum value of 0xFFF to get the longest possible timeout period.

```
while (isIwdg_ready() != 1) {}
```

This waits until the IWDG **status register** (**SR**) indicates that the prescaler and reload registers have been updated and are ready.

```
IWDG->KR = IWDG_KEY_RELOAD;
```

This writes 0x0000AAAA to the IWDG Key Register to reload the counter and prevent the IWDG from resetting the system.

The `isIwdg_ready()` function checks the Status Register to see if the PVU and RVU bits are cleared. It returns 1 if both bits are cleared, indicating that the IWDG is ready, or 0 otherwise.

We are now ready to test inside `main.c`.

The main file

Update your `main.c` file as shown next:

```
#include <stdio.h>
#include "adc.h"
#include "uart.h"
#include "gpio.h"
#include "iwdg.h"
#include "gpio_exti.h"

uint8_t g_btn_press;
static void check_reset_source(void);

int main(void)
{
```

```
        /*Initialize debug UART*/
        uart_init();

        /*Initialize LED*/
        led_init();

        /*Initialize EXTI*/
        pc13_exti_init();

        /*Find reset source*/
        check_reset_source();

        /*Initialize IWDG*/
        iwdg_init();
        while(1)
        {

            if( g_btn_press != 1)
              {
                  /*Refresh IWDG down-counter to default value*/
                  IWDG->KR = IWDG_KEY_RELOAD;
                  led_toggle();
                  for(int i = 0; i < 90000; i++){}


              }
          }
    }
```

The main function kicks off by setting up the **universal asynchronous receiver-transmitter (UART)** for serial communication, ensuring that we can send debugging information to the serial port. Next, it initializes the LED and configures the external interrupt (EXTI) on pin PC13, which is connected to the blue push button. The function then calls check_reset_source to determine if the last reset was caused by the IWDG. After that, it initializes the IWDG itself, ensuring that the system can recover from software malfunctions. The main loop continually checks the state of g_btn_press; if the button hasn't been pressed, it refreshes the IWDG to prevent a system reset and toggles the LED, providing a visual indicator of system activity.

Let's look at the next part of the code:

```
static void check_reset_source(void)
{
if ((RCC->CSR & RCC_CSR_IWDGRSTF) == (RCC_CSR_IWDGRSTF))
      {
            /*Clear IWDG Reset flag*/
```

```
RCC->CSR = RCC_CSR_RMVF;

/*Turn LED On*/
led_on();
printf("RESET was caused by IWDG.....\n\r");

while( g_btn_press != 1)
{
}
g_btn_press =  0;

}

}
```

The `check_reset_source` function determines if the most recent system reset was triggered by the IWDG. It begins by checking the **Reset and Clock Control (RCC)** status register (CSR) for the IWDG reset flag (RCC_CSR_IWDGRSTF). If this flag is set, it confirms that the watchdog initiated the reset. The function then clears this flag by writing to the RCC_CSR_RMVF bit, ensuring the flag is reset for future detection. As a visual indication of the IWDG-triggered reset, the LED is turned on, and a message is printed to the UART. The function enters a loop, waiting for the user to press the button (detected by the g_btn_press variable) before clearing the button press state and exiting.

Then, there is the callback:

```
static void exti_callback(void)
{
    g_btn_press = 1;

}
```

This is followed by the interrupt handler:

```
void EXTI15_10_IRQHandler(void) {
    if((EXTI->PR & LINE13)!=0)
    {
        /*Clear PR flag*/
        EXTI->PR |=LINE13;
        //Do something...
        exti_callback();
    }

}
```

The `exti_callback` function is a simple yet vital part of our interrupt-handling mechanism. Its sole purpose is to set the `g_btn_press` flag to 1, indicating that a button press has been detected. This flag is later used in the main loop to control the program flow. The `EXTI15_10_IRQHandler` function is the interrupt handler for external interrupts on lines 10 to 15. When an interrupt is triggered on *line 13*, this handler first checks the pending register (`PR`) to confirm the interrupt source. Once verified, it clears the pending flag by writing back to the `PR` register. After clearing the interrupt, the handler calls `exti_callback` to update the `g_btn_press` flag.

Testing the project

To test the project on the microcontroller, follow these steps:

1. **Build and run the project**: Compile the code and upload it to your microcontroller. Once running, you should observe the green LED blinking, indicating that the system is functioning correctly.

2. **Monitor the serial output**: Open RealTerm or any other serial terminal application. Configure it with the appropriate port and baud rate to view the debug messages. This will allow you to confirm when the system restarts due to the IWDG.

3. **Trigger a watchdog reset**: Press the blue push button to stop the IWDG timer from being refreshed. After the IWDG timeout period elapses, the system will reset, and you should see a corresponding message in the serial terminal.

Summary

In this chapter, we explored the IWDG timer, an important component for enhancing the reliability of embedded systems. We began by discussing the general concept of WDTs, emphasizing their role in ensuring systems can recover from unexpected faults or malfunctions. We explored how WDTs function and highlighted the unique features of the IWDG.

Next, we focused on the STM32 implementation of the IWDG, examining its key registers and configuration options. We detailed the purpose and usage of essential registers such as the Key Register (`IWDG_KR`), Prescaler Register (`IWDG_PR`), Reload Register (`IWDG_RLR`), and Status Register (`IWDG_SR`). This provided a comprehensive understanding of how to configure and control the IWDG for robust system operation.

We also provided practical examples to solidify our understanding, including the development of a bare-metal IWDG driver. This involved initializing the IWDG, configuring its prescaler and reload values, and implementing the necessary functions to ensure the system can recover from software malfunctions.

In the next chapter, we will learn about the **Direct Memory Access** (**DMA**) module, an advanced feature for transferring data.

Unlock this book's exclusive benefits now

This book comes with additional benefits designed to elevate your learning experience.

Note: Have your purchase invoice ready before you begin.

https://www.packtpub.com/
unlock/9781835460818

17

Direct Memory Access (DMA)

In this chapter, we will explore **Direct Memory Access (DMA)**, a powerful feature in microcontrollers that allows peripherals to transfer data to and from memory without involving the CPU. This functionality is critical for improving data throughput and freeing up the CPU to handle other tasks, making it fundamental to high-performance embedded system development.

We will begin by understanding the basic principles of DMA and its significance in embedded systems. We will then delve into the specifics of the DMA controller in STM32F4 microcontrollers, examining its structure and features and how it manages data transfers. Following this, we will apply this theoretical knowledge to develop practical DMA drivers for various use cases, including memory-to-memory transfers, **Analog-to-Digital Converter (ADC)** data transfers, and **Universal Asynchronous Receiver-Transmitter (UART)** communications.

In this chapter, we will cover the following main topics:

- An overview of DMA
- The STM32F4 DMA
- Developing the DMA ADC driver
- Developing the DMA UART driver
- Developing the DMA memory-to-memory driver

By the end of this chapter, you will have a comprehensive understanding of how DMA works and how to implement it in your projects. You will be able to develop efficient DMA drivers to handle data transfers in various scenarios, significantly enhancing the performance and responsiveness of your embedded systems.

Technical requirements

All the code examples for this chapter can be found on GitHub at `https://github.com/PacktPublishing/Bare-Metal-Embedded-C-Programming`.

Understanding Direct Memory Access (DMA)

DMA is a feature that can significantly elevate the performance of your embedded systems. If you've been dealing with data transfers in your microcontroller projects, you know how taxing it can be on the CPU to handle all that data movement. This is where DMA steps in as a game-changer, offloading the data transfer tasks from the CPU and allowing it to focus on more critical functions. Let's see how it works.

How DMA works

So, what exactly is DMA, and how does it work? In simple terms, DMA is a method that allows peripherals within a microcontroller to transfer data directly to and from memory, without requiring continuous CPU intervention. Imagine it as a dedicated assistant that takes over the tedious task of moving boxes (data) so that you (the CPU) can focus on more important work, such as solving complex problems or managing other peripherals.

A typical DMA controller in a microcontroller has multiple channels, each capable of handling a specific data transfer operation. Each channel can be configured independently to manage transfers between various peripherals and memory.

Here's a step-by-step look at how DMA generally operates:

1. **Initialization**: The DMA controller and channels are configured. This setup includes specifying the source and destination addresses, the direction of data transfer, and the number of data units to transfer.

2. **Trigger**: The data transfer is initiated by a trigger, which can be an event such as a peripheral signaling that it's ready to send or receive data, or a software command.

3. **Data transfer**: Once triggered, the DMA controller takes over, reading data from the source address and writing it to the destination address. This process continues until the specified number of data units is transferred.

4. **Completion**: Upon completing the transfer, the DMA controller can generate an interrupt to notify the CPU that the transfer is done, allowing the system to perform any necessary post-transfer processing.

Next, let's take a look at some key features of DMA controllers.

Key features

DMA controllers are packed with features that make them versatile and powerful. Let's break down some of the key specifications you'll often encounter:

- **Channels and streams**: DMA controllers typically have multiple channels and streams, each capable of handling a different transfer. For instance, the STM32F4 microcontroller has up to 16 streams in its DMA controllers.

- **Priorities**: Channels can be assigned different priority levels, ensuring that more critical transfers get precedence over less critical ones.

- **Transfer Types**: DMA can handle various types of transfers, including memory-to-memory, peripheral-to-memory, and memory-to-peripheral.

- **FIFO**: Some DMA controllers come with a **First-In-First-Out** (**FIFO**) buffer, which helps manage data flow and improve efficiency, especially in burst transfers.

- **Circular mode**: This mode allows the DMA to continuously transfer data in a loop, which is particularly useful for peripherals that need constant data streaming, such as audio or video feeds.

- **Interrupts**: DMA controllers can generate interrupts on transfer completion, half-transfer completion, and transfer errors, allowing the CPU to react appropriately to different stages of the transfer.

To understand the real power of DMA, let's look at some common use cases where DMA shines.

Common use cases

Here are some common use cases of DMA:

- **Audio streaming**: DMA is heavily used in audio applications where continuous data streaming is essential. For instance, in a digital audio player, the audio samples must be continuously sent to a **Digital-to-Analog Converter** (**DAC**). Using DMA, the audio data can be streamed from memory to the DAC without CPU intervention, ensuring smooth playback and freeing up the CPU to manage the user interface and other tasks.

- **Sensor data acquisition**: In applications such as environmental monitoring or industrial automation, sensors often need to sample data at precise intervals. For example, an ADC can be configured to continuously sample temperature data, with DMA transferring the sampled data directly to memory. This setup ensures that the CPU isn't bogged down with handling each individual sample, thus maintaining efficient and timely data collection.

- **Communication interfaces**: DMA is a lifesaver when dealing with high-speed communication protocols such as UART, SPI, or I2C. Consider a scenario where an embedded system needs to log data received over UART to an SD card. Without DMA, the CPU would need to handle each byte of data, process it, and then write it to the SD card, which can be highly inefficient. With DMA, the data received over UART can be directly written to memory, and another DMA channel can transfer it to the SD card, all with minimal CPU intervention.

- **Graphics processing**: DMA is also crucial in applications involving graphics, such as updating a display buffer. In a system where the display needs to be refreshed continuously, the DMA can handle the transfer of image data from memory to the display controller. This ensures smooth and flicker-free graphics rendering, while the CPU can focus on generating the next frame or managing user inputs.

With this in mind, let's compare some DMA solutions to non-DMA solutions.

Case study 1 – audio streaming

Scenario: You are developing an audio playback system that streams digital audio data from a microcontroller to a DAC.

Without DMA: The CPU is responsible for fetching each audio sample from memory and sending it to the DAC. Given the high sampling rate required for audio applications (e.g., 44.1 kHz for CD-quality audio), the CPU must handle tens of thousands of interrupts per second just to maintain the audio stream. This constant load significantly limits the CPU's ability to perform other tasks, potentially leading to audio glitches and reduced system responsiveness.

With DMA: The DMA controller is configured to transfer audio data directly from memory to the DAC. The CPU sets up the DMA transfer and then handles higher-level tasks, only occasionally checking the status of the transfer. This setup ensures smooth and uninterrupted audio playback while freeing up the CPU to manage other aspects of the system, such as user interface and control logic.

Case study 2 – high-speed data acquisition

Scenario: You are developing a data acquisition system that continuously samples data from multiple sensors via ADCs and stores the data for later analysis.

Without DMA: The CPU must handle each ADC conversion, read the data, and store it in memory. If the sampling rate is high, the CPU can become overwhelmed, leading to missed samples and unreliable data collection. This approach can also complicate real-time data processing and analysis, as the CPU is bogged down with managing the data flow.

With DMA: The ADC is configured to generate DMA requests. Each time a conversion is complete, the DMA controller transfers the data from the ADC to memory without CPU intervention. The CPU can then process the collected data in batches, ensuring that no samples are missed and enabling real-time data analysis and decision-making.

Case study 3 – an LCD display refresh

Scenario: You are developing a graphical application that continuously updates an LCD display with new data.

Without DMA: The CPU must update the display by sending each pixel or line of data directly to the LCD controller. This process can be very CPU-intensive, especially for high-resolution displays, leading to sluggish performance and reduced responsiveness in the user interface.

With DMA: The DMA controller is configured to transfer display data from memory to the LCD controller. The CPU sets up the DMA transfer and then focuses on generating new graphical data or handling user inputs. The DMA controller ensures that the display is updated smoothly and efficiently.

In each of the case studies, we've seen how using DMA can transform a system's capabilities, freeing up the CPU to handle more critical tasks and ensuring that data transfers are handled efficiently and reliably. Understanding and implementing DMA in your projects can lead to more robust, responsive, and high-performance embedded systems.

In the following section, we will delve deeper into the specifics of the STM32F4 DMA controller, exploring its architecture, key registers, and practical implementation techniques.

The DMA modules of the STM32F4 microcontroller

Each STM32F4 microcontroller is equipped with **two DMA** controllers, each supporting up to **8 streams**. Each stream can manage multiple requests, providing up to **16 streams** in total to handle various data transfer tasks. A **stream** is a unidirectional pathway that facilitates data transfer between a source and a destination. The architecture includes an **arbiter** to prioritize these DMA requests, ensuring that high-priority transfers are handled promptly.

Let's see some key features of the STM32F4 DMA controller.

The key features of the STM32F4 DMA controller

The following are the key features of the STM32F4 DMA controller:

- **Independent FIFO**: Each stream includes a **four-word FIFO buffer**, which can operate in either direct mode or FIFO mode. In direct mode, data transfers occur immediately upon request, while FIFO mode allows for **threshold-level buffering**, enhancing efficiency for burst data transfers.

- **Flexible configuration**: Each stream can be configured to handle the following:

  - Peripheral-to-memory

  - Memory-to-peripheral

  - Memory-to-memory transfers

Additionally, streams can be set up for regular or double-buffer transfers, the latter enabling seamless data handling by swapping memory buffers automatically:

- **Prioritization and arbitration**: DMA stream requests are prioritized via software with four levels – **very high**, **high**, **medium**, and **low**. If two streams have the same priority, hardware prioritization based on the stream number ensures orderly data transfer.

- **Incremental and burst transfers**: The DMA controller supports both incremental and non-incremental addressing for source and destination. It can manage burst transfers of **4, 8, or 16 beats**, optimizing bandwidth usage. The term **beat** refers to the individual units of data that are transferred in a single DMA transaction.

- **Interrupts and error handling**: Each stream supports multiple event flags such as transfer complete, half-transfer, transfer error, FIFO error, and direct mode error. These flags can trigger interrupts, providing robust error handling and status monitoring.

To fully utilize the STM32F DMA controller, it's important to understand DMA transactions and channel selection. Let's break these concepts down.

- **DMA transactions**: A typical DMA transaction involves the following:

  - Loading data from the source address (either a peripheral or memory)

  - Storing data to the destination address

  - Decrementing the `DMA_SxNDTR` register to track the number of remaining data items

- **Channel selection**: Each stream can be associated with up to eight different channels, selectable via the `CHSEL` bits in the `DMA_SxCR` register. This flexibility allows various peripherals to initiate DMA requests efficiently.

Previously, we discussed that the STM32F4 DMA controller supports three distinct transfer modes. Now, let's explore the characteristics of each mode.

Transfer modes

There are three transfer modes:

- **Peripheral-to-memory mode**: This mode is activated by setting the EN bit in the DMA_SxCR register. The stream transfers data from the peripheral to memory, using the FIFO buffer if enabled.

- **Memory-to-peripheral mode**: This is similar to peripheral-to-memory, but the transfer direction is reversed. Data is loaded from memory and sent to the peripheral.

- **Memory-to-memory mode**: This mode is unique, as it does not require peripheral requests. The DMA stream transfers data between two memory locations, using the FIFO buffer if enabled. This is particularly useful for large data transfers within memory.

The DMA controller can automatically increment source and destination pointers, facilitating efficient data transfers across different memory regions. This is configurable via the PINC and MINC bits in the DMA_SxCR register.

The STM32F4 DMA controller also provides data mode options.

DMA data modes

This data mode options include the following:

- **Circular mode**: Circular mode allows the DMA to handle continuous data flows by automatically reloading the initial values in the DMA_SxNDTR register. This is especially useful for applications such as ADC sampling, where data needs to be continuously recorded.

- **Double buffer mode**: Double buffer mode enhances efficiency by allowing the DMA controller to swap between two memory buffers automatically. This ensures continuous data processing. While the CPU works on one buffer, the DMA can load the next set of data into the other buffer.

In the next section, we will examine the STM32F4 DMA block diagram from the reference manual. This will help us better understand the key characteristics and functionalities of the DMA controller.

The STM32F4 DMA block diagram

The DMA has **two ports** for data transfer – one peripheral port and one memory port, as shown in *Figure 17.1*.

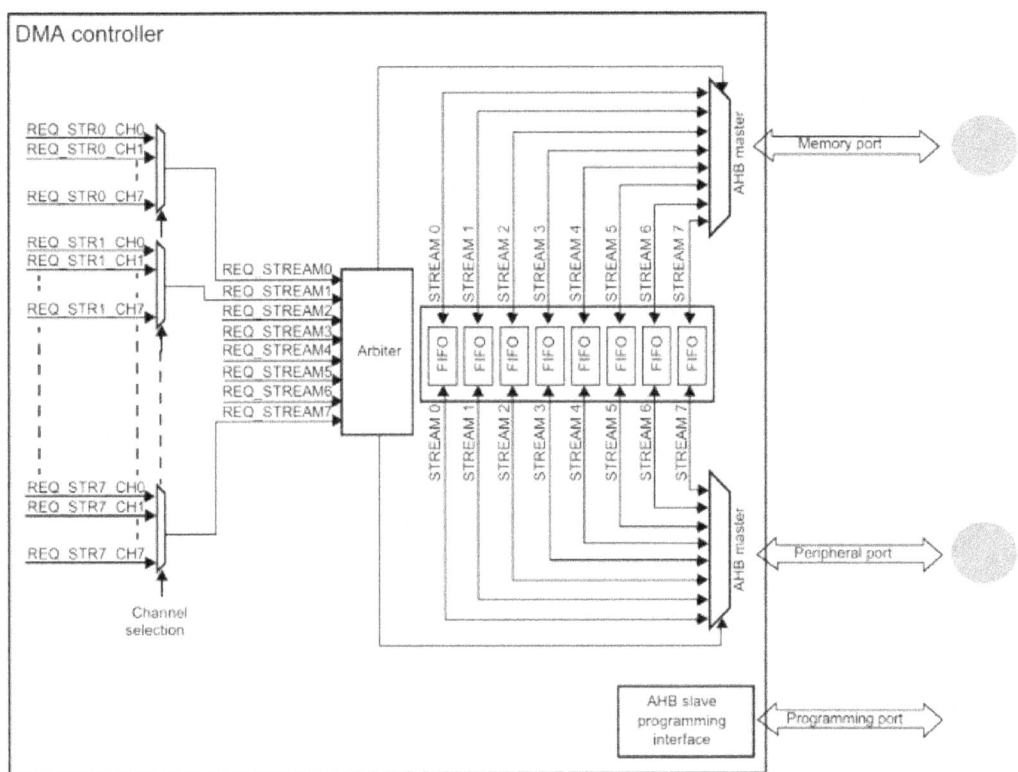

Figure 17.1: The DMA module, indicating the data transfer ports

Each of the two DMA modules features **eight distinct streams**, with each stream dedicated to handling memory access requests from various peripherals.

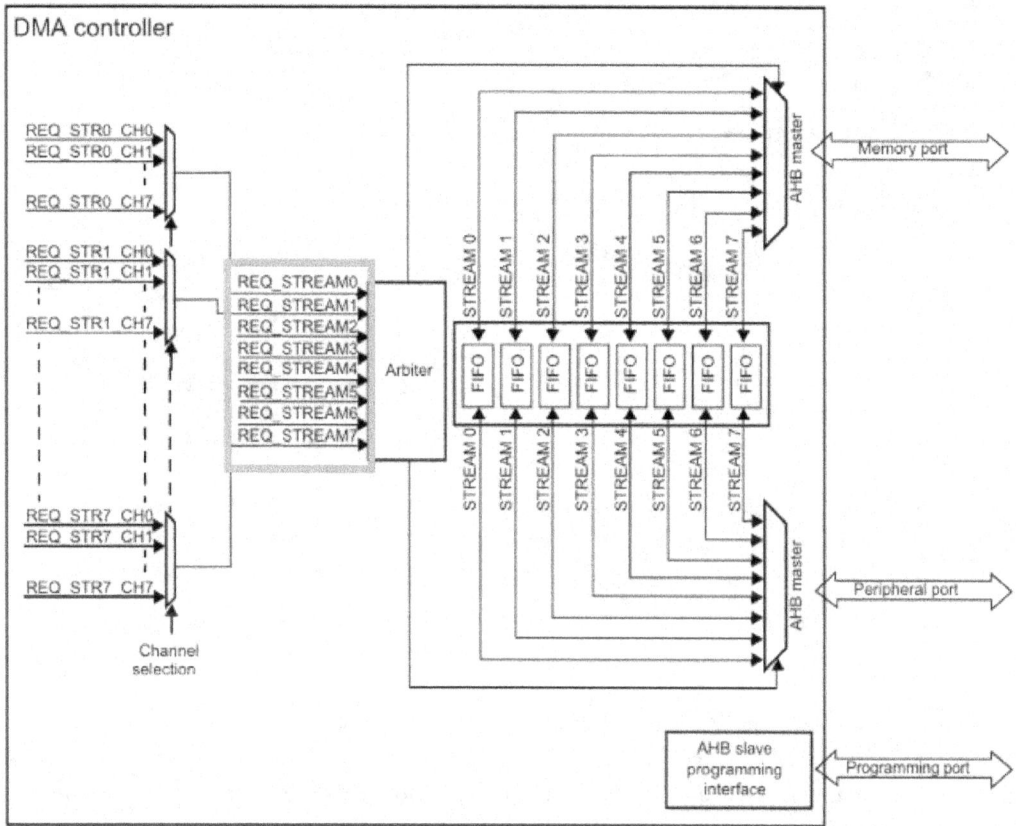

Figure 17.2: The DMA module, indicating the streams

Each stream can accommodate up to **eight selectable channels**, which are software-configurable to enable multiple peripherals to initiate DMA requests. However, within any given stream, only one channel can be active at a time.

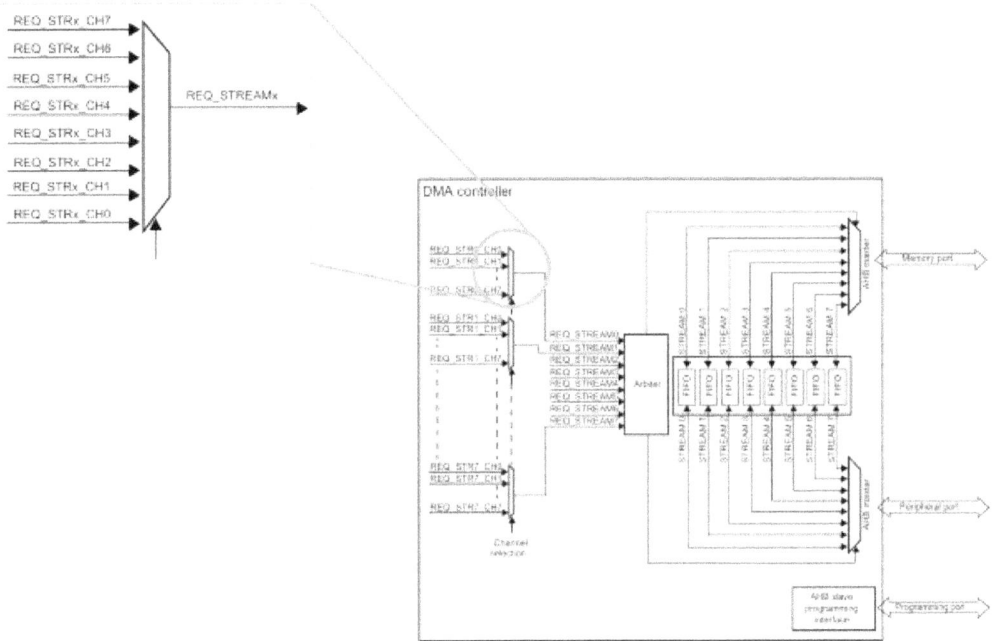

Figure 17.3: The DMA module, with channels zoomed in

To find the mappings of DMA channels and streams to the various peripherals of the microcontroller, refer to *page 170* of the reference manual (**RM0383**).

Before we start developing our DMA drivers, the final piece of the puzzle involves familiarizing ourselves with the key DMA registers.

The key STM32 DMA registers

In this section, we will explore the characteristics and functions of some of the crucial registers within the DMA peripheral, starting with the DMA Stream Configuration Register.

The DMA Stream Configuration Register (DMA_SxCR)

The DMA Stream Configuration Register (DMA_SxCR) is one of the primary registers used to configure a DMA stream's operational settings. This register allows us to set up various parameters, such as the **data direction**, the **size** of the data items, and the **priority** level of the stream. The key bits in this register include the following:

- **EN**: Stream enable. Setting this bit activates the stream.
- **CHSEL[2:0]**: Channel selection. These bits select the DMA channel for the stream.
- **DIR[1:0]**: Data transfer direction. These bits specify the direction of the data transfer (peripheral-to-memory, memory-to-peripheral, or memory-to-memory).
- **CIRC**: Circular mode. Setting this bit enables circular mode, which allows continuous data transfers.
- **PINC**: Peripheral increment mode. When set, this bit increments the peripheral address after each data transfer.
- **MINC**: Memory increment mode. When set, this bit increments the memory address after each data transfer.
- **PSIZE[1:0]**: Peripheral data size. These bits specify the size of the data items read from or written to the peripheral (8-bit, 16-bit, or 32-bit).
- **MSIZE[1:0]**: Memory data size. These bits specify the size of the data items read from or written to memory.
- **PL[1:0]**: Priority level. These bits set the priority level of the stream (low, medium, high, or very high).

You can find detailed information about this register on *page 190* of the STM32F411 reference manual (RM0383). Next, we have the DMA Stream Number of Data Register (DMA_SxNDTR).

DMA Stream Number of Data Register (DMA_SxNDTR)

The DMA Stream Number of Data Register (DMA_SxNDTR) specifies the number of data items to be transferred by the DMA stream. This register is crucial for controlling the length of the data transfer.

The only field in this register is **NDT[15:0]**. This stands for the number of data items. This field specifies the total number of data items to be transferred. The value in this register is decremented after each transfer until it reaches zero.

Further information about this register can be found on *page 193* of the reference manual. Let's move on to the DMA Stream Peripheral Address Register (DMA_SxPAR).

DMA Stream Peripheral Address Register (DMA_SxPAR)

The DMA Stream Peripheral Address Register (DMA_SxPAR) holds the address of the peripheral data register that the data will be read to or written from.

The only field in this register is **PA[31:0]**. This stands for **peripheral address**. This field contains the address of the peripheral data register involved in the data transfer.

More details about this register can be found on *page 194* of the reference manual. Finally, we have the DMA Stream Memory Address Registers (DMA_SxM0AR and DMA_SxM1AR).

DMA Stream Memory Address Registers (DMA_SxM0AR and DMA_SxM1AR)

These registers store the addresses of the memory locations used for data transfers. The DMA_SxM0AR register is used for single buffer mode, while both DMA_SxM0AR and DMA_SxM1AR are used in double buffer mode.

The only field in these registers is **MA[31:0]**. This stands for **memory address**. This field contains the address of the memory location involved in the data transfer. For detailed information, refer to *pages 194* of the reference manual.

Developing the ADC DMA driver

In this section, we will develop three distinct DMA drivers – one for transferring ADC data, another for UART data, and a third for transferring data between memory locations. Let's begin with the ADC DMA driver.

The ADC DMA driver

Create a copy of your previous project in your IDE, following the steps outlined in earlier chapters. Rename this copied project ADC_DMA. Next, create a new file named adc_dma.c in the Src folder and another file named adc_dma.h in the Inc folder.

Populate your adc_dma.c file with the following code:

```
#include "adc_dma.h"

#define GPIOAEN             (1U<<0)
#define ADC1EN              (1U<<8)
#define CR1_SCAN          (1U<<8)
#define CR2_DMA             (1U<<8)
#define CR2_DDS             (1U<<9)
#define CR2_CONT          (1U<<1)
#define CR2_ADCON         (1U<<0)
```

```c
#define CR2_SWSTART         (1U<<30)

#define DMA2EN               (1U<<22)
#define DMA_SCR_EN          (1U<<0)
#define DMA_SCR_MINC        (1U<<10)
#define DMA_SCR_PINC        (1U<<9)
#define DMA_SCR_CIRC        (1U<<8)
#define DMA_SCR_TCIE        (1U<<4)
#define DMA_SCR_TEIE        (1U<<2)
#define DMA_SFCR_DMDIS       (1U<<2)


uint16_t adc_raw_data[NUM_OF_CHANNELS];

void adc_dma_init(void)
{
    /************GPIO Configuration**********/
    /*Enable clock access to ADC GPIO Pin's Port*/
    RCC->AHB1ENR |= GPIOAEN;

    /*Set PA0 and PA1 mode to analog mode*/
    GPIOA->MODER |= (1U<<0);
    GPIOA->MODER |= (1U<<1);

    GPIOA->MODER |= (1U<<2);
    GPIOA->MODER |= (1U<<3);

    /************ADC Configuration**********/
    /*Enable clock access to ADC*/
    RCC->APB2ENR |= ADC1EN;

    /*Set sequence length*/
    ADC1->SQR1 |= (1U<<20);
    ADC1->SQR1 &= ~(1U<<21);
    ADC1->SQR1 &= ~(1U<<22);
    ADC1->SQR1 &= ~(1U<<23);

    /*Set sequence*/
    ADC1->SQR3 = (0U<<0) | (1U<<5);

    /*Enable scan mode*/
    ADC1->CR1 = CR1_SCAN;
```

```
/*Select to use DMA*/
ADC1->CR2 |=CR2_CONT |CR2_DMA|CR2_DDS;


/***********DMA Configuration**********/
/*Enable clock access to DMA*/
RCC->AHB1ENR |=DMA2EN;


/*Disable DMA stream*/
DMA2_Stream0->CR &=~DMA_SCR_EN;

/*Wait till DMA is disabled*/
while((DMA2_Stream0->CR & DMA_SCR_EN)){}

/*Enable Circular mode*/
DMA2_Stream0->CR |=DMA_SCR_CIRC;

/*Set MSIZE i.e Memory data size to half-word*/

DMA2_Stream0->CR |= (1U<<13);
DMA2_Stream0->CR &= ~(1U<<14);

/*Set PSIZE i.e Peripheral data size to half-word*/
DMA2_Stream0->CR |= (1U<<11);
DMA2_Stream0->CR &= ~(1U<<12);

/*Enable memory addr increment*/
DMA2_Stream0->CR |=DMA_SCR_MINC;

/*Set periph address*/
DMA2_Stream0->PAR = (uint32_t)(&(ADC1->DR));
/*Set mem address*/
DMA2_Stream0->M0AR = (uint32_t)(&adc_raw_data);

/*Set number of transfer*/
DMA2_Stream0->NDTR = (uint16_t)NUM_OF_CHANNELS;

/*Enable DMA stream*/
DMA2_Stream0->CR |= DMA_SCR_EN;

/***********ADC Configuration*********/
/*Enable ADC*/
```

```
    ADC1->CR2 |=CR2_ADCON;

    /*Start ADC*/
    ADC1->CR2 |=CR2_SWSTART;


}
```

Let's go through each part of the code step by step to understand its purpose and functionality.

1. **Including the header file and defining constants**: We start by including the adc_dma.h header file, which calls the stm32f4xx.h file and contains a macro for the number of channels of our DMA driver. We then define several constants using #define statements. These constants represent bit masks for various registers and control flags, making the code more readable and maintainable.

2. **Defining the ADC data array**: We declare an array, adc_raw_data, that stores the raw ADC data. The size of this array is determined by a predefined constant, NUM_OF_CHANNELS.

3. **The ADC DMA initialization function**: In the adc_dma_init function, we begin by enabling the clock for GPIOA, which is necessary for configuring the GPIO pins used by the ADC. We then set the mode of the PA0 and PA1 pins to analog, as they are connected to the ADC channels.

4. Next, we enable the clock for ADC1 and configure the ADC sequence length and channel sequence. We set the ADC to scan mode, allowing it to convert multiple channels sequentially. Additionally, we enable DMA and continuous conversion mode for the ADC.

5. **DMA configuration**: We then enable the clock for DMA2 and ensure that the DMA stream is disabled before making any configurations. We configure the DMA stream for circular mode, which allows continuous data transfer. We set the memory and peripheral data sizes to half-word (16 bits). We enable memory address increment to correctly move through the adc_raw_data array and set the peripheral address to the ADC data register. We specify the number of data items to transfer and, finally, enable the DMA stream.

6. **Enable and start ADC**: In the final steps, we simply enable the ADC and start the conversion process by setting the appropriate control bits.

Our next task is to populate the adc_dma.h file. Here is the code:

```
#ifndef ADC_DMA_H__
#define ADC_DMA_H__
#include <stdint.h>
#include "stm32f4xx.h"
void adc_dma_init(void);
```

```
#define NUM_OF_CHANNELS        2

#endif
```

Let's move on to the `main.c` file. Update your `main.c` file, as shown here:

```c
#include <stdio.h>
#include "uart.h"
#include "adc_dma.h"

extern uint16_t adc_raw_data[NUM_OF_CHANNELS];

int main(void)
{
    /*Initialize debug UART*/
    uart_init();
    /*Initialize ADC DMA*/
    adc_dma_init();
    while(1)
    {
        printf("Value from sensor one : %d \n\r ",adc_raw_data[0]);
        printf("Value from sensor two : %d \n\r ",adc_raw_data[1]);
        for( int i = 0; i < 90000; i++){}
    }
}
```

This code initializes the UART for debugging, and it sets up the ADC with DMA to continuously read sensor data from ADC channels connected to GPIO pins. In the main function, we start by initializing the UART for communication, and then we call the `adc_dma_init` function to configure the ADC and DMA for data transfers. In the infinite loop, we repeatedly print the values from two sensors stored in the `adc_raw_data` array to the console.

To test the project, follow the steps we outlined in *Chapter 11*. Let's proceed by developing the UART DMA driver.

Developing the UART DMA driver

Create a copy of your previous project in your IDE. Rename this copied project UART_DMA. Next, create a new file named `uart_dma.c` in the Src folder and another file named `uart_dma.h` in the Inc folder. Update your `uart_dma.c` file, as shown here:

```c
#include "uart_dma.h"

#define UART2EN            (1U<<17)
```

```c
#define GPIOAEN              (1U<<0)

#define CR1_TE               (1U<<3)
#define CR1_RE               (1U<<2)
#define CR1_UE               (1U<<13)
#define SR_TXE               (1U<<7)

#define CR3_DMAT             (1U<<7)
#define CR3_DMAR             (1U<<6)
#define SR_TC                (1U<<6)
#define CR1_TCIE             (1U<<6)

#define UART_BAUDRATE        115200
#define CLK                  16000000

#define DMA1EN               (1U<<21)
#define DMA_SCR_EN           (1U<<0)
#define DMA_SCR_MINC         (1U<<10)
#define DMA_SCR_PINC         (1U<<9)
#define DMA_SCR_CIRC         (1U<<8)
#define DMA_SCR_TCIE         (1U<<4)
#define DMA_SCR_TEIE         (1U<<2)
#define DMA_SFCR_DMDIS       (1U<<2)

#define HIFCR_CDMEIF5        (1U<<8)
#define HIFCR_CTEIF5         (1U<<9)
#define HIFCR_CTCIF5         (1U<<11)

#define HIFCR_CDMEIF6        (1U<<18)
#define HIFCR_CTEIF6         (1U<<19)
#define HIFCR_CTCIF6         (1U<<21)

#define HIFSR_TCIF5          (1U<<11)
#define HIFSR_TCIF6          (1U<<21)

static uint16_t compute_uart_bd(uint32_t periph_clk, uint32_t
baudrate);
static void uart_set_baudrate(uint32_t periph_clk, uint32_t baudrate);

char uart_data_buffer[UART_DATA_BUFF_SIZE];

uint8_t g_rx_cmplt;
uint8_t g_tx_cmplt;
```

```
uint8_t g_uart_cmplt;

void uart2_rx_tx_init(void)
{
    /*************Configure UART GPIO pin*******************/
    /*1.Enable clock access to GPIOA*/
    RCC->AHB1ENR |= GPIOAEN;

    /*2.Set PA2 mode to alternate function mode*/
    GPIOA->MODER &= ~(1U<<4);
    GPIOA->MODER |=    (1U<<5);

    /*3.Set PA3 mode to alternate function mode*/
    GPIOA->MODER &= ~(1U<<6);
    GPIOA->MODER |=    (1U<<7);

    /*4.Set PA2 alternate function function type to AF7(UART2_TX)*/
    GPIOA->AFR[0] |= (1U<<8);
    GPIOA->AFR[0] |= (1U<<9);
    GPIOA->AFR[0] |= (1U<<10);
    GPIOA->AFR[0] &= ~(1U<<11);

    /*5.Set PA3 alternate function function type to AF7(UART2_TX)*/
    GPIOA->AFR[0] |= (1U<<12);
    GPIOA->AFR[0] |= (1U<<13);
    GPIOA->AFR[0] |= (1U<<14);
    GPIOA->AFR[0] &= ~(1U<<15);
    /*************Configure UART Module*******************/

    /*6. Enable clock access to UART2*/
    RCC->APB1ENR |= UART2EN;

    /*7. Set baudrate*/
    uart_set_baudrate(CLK,UART_BAUDRATE);

    /*8. Select to use DMA for TX and RX*/
    USART2->CR3 = CR3_DMAT |CR3_DMAR;

    /*9. Set transfer direction*/
    USART2->CR1 = CR1_TE |CR1_RE;

    /*10.Clear TC flag*/
    USART2->SR &=~SR_TC;
```

```
    /*11.Enable TCIE*/
    USART2->CR1 |=CR1_TCIE;

    /*12. Enable uart module*/
    USART2->CR1 |= CR1_UE;

    /*13.Enable USART2 interrupt in the NVIC*/
    NVIC_EnableIRQ(USART2_IRQn);

}
```

Next, we have the initialization function:

```
void dma1_init(void)
{
    /*Enable clock access to DMA*/
    RCC->AHB1ENR |=DMA1EN;

    /*Enable DMA Stream6 Interrupt in NVIC*/
    NVIC_EnableIRQ(DMA1_Stream6_IRQn);
}
```

And then, the function for configuring the rx stream:

```
void dma1_stream5_uart_rx_config(void)
{
    /*Disable DMA stream*/
    DMA1_Stream5->CR &=~DMA_SCR_EN;

    /*Wait till DMA Stream is disabled*/
    while((DMA1_Stream5->CR & DMA_SCR_EN)){}

    /*Clear interrupt flags for stream 5*/
    DMA1->HIFCR = HIFCR_CDMEIF5 |HIFCR_CTEIF5|HIFCR_CTCIF5;
    /*Set periph address*/
    DMA1_Stream5->PAR = (uint32_t)(&(USART2->DR));

    /*Set mem address*/
    DMA1_Stream5->M0AR = (uint32_t)(&uart_data_buffer);

    /*Set number of transfer*/
    DMA1_Stream5->NDTR = (uint16_t)UART_DATA_BUFF_SIZE;
```

```
/*Select Channel 4*/
DMA1_Stream5->CR &= ~(1u<<25);
DMA1_Stream5->CR &= ~(1u<<26);
DMA1_Stream5->CR |= (1u<<27);

/*Enable memory addr increment*/
DMA1_Stream5->CR |=DMA_SCR_MINC;

/*Enable transfer complete interrupt*/
DMA1_Stream5->CR |= DMA_SCR_TCIE;

/*Enable Circular mode*/
DMA1_Stream5->CR |=DMA_SCR_CIRC;

/*Set transfer direction : Periph to Mem*/
DMA1_Stream5->CR &=~(1U<<6);
DMA1_Stream5->CR &=~(1U<<7);

/*Enable DMA stream*/
DMA1_Stream5->CR |= DMA_SCR_EN;

/*Enable DMA Stream5 Interrupt in NVIC*/
NVIC_EnableIRQ(DMA1_Stream5_IRQn);

}
```

And then, the one for the tx stream

```
void dma1_stream6_uart_tx_config(uint32_t msg_to_snd, uint32_t msg_
len)
{
    /*Disable DMA stream*/
    DMA1_Stream6->CR &=~DMA_SCR_EN;

    /*Wait till  DMA Stream is disabled*/
    while((DMA1_Stream6->CR & DMA_SCR_EN)){}

    /*Clear interrupt flags for stream 6*/
    DMA1->HIFCR = HIFCR_CDMEIF6 |HIFCR_CTEIF6|HIFCR_CTCIF6;

    /*Set periph address*/
    DMA1_Stream6->PAR = (uint32_t)(&(USART2->DR));
```

```c
    /*Set mem address*/
    DMA1_Stream6->M0AR = msg_to_snd;

    /*Set number of transfer*/
    DMA1_Stream6->NDTR = msg_len;

    /*Select Channel 4*/
    DMA1_Stream6->CR &= ~(1u<<25);
    DMA1_Stream6->CR &= ~(1u<<26);
    DMA1_Stream6->CR |= (1u<<27);

    /*Enable memory addr increment*/
    DMA1_Stream6->CR |=DMA_SCR_MINC;

    /*Set transfer direction :Mem to Periph*/
    DMA1_Stream6->CR |=(1U<<6);
    DMA1_Stream6->CR &=~(1U<<7);

    /*Set transfer complete interrupt*/
    DMA1_Stream6->CR |= DMA_SCR_TCIE;

    /*Enable DMA stream*/
    DMA1_Stream6->CR |= DMA_SCR_EN;

}
```

Next, the function for computing the baudrate value for the UART:

```c
static uint16_t compute_uart_bd(uint32_t periph_clk, uint32_t
baudrate)
{
    return ((periph_clk +( baudrate/2U ))/baudrate);
}
```

And then, the function for writing the baudrate value to the baudrate register:

```c
static void uart_set_baudrate(uint32_t periph_clk, uint32_t baudrate)
{
    USART2->BRR  = compute_uart_bd(periph_clk,baudrate);
}
```

Next, we have the interrupt handler for Stream6:

```c
void DMA1_Stream6_IRQHandler(void)
{
    if((DMA1->HISR) & HIFSR_TCIF6)
    {
        //do_ssomething
        g_tx_cmplt = 1;
        /*Clear the flag*/
        DMA1->HIFCR |= HIFCR_CTCIF6;
    }
}
```

And then, the interrupt handler for Stream5:

```c
void DMA1_Stream5_IRQHandler(void)
{
    if((DMA1->HISR) & HIFSR_TCIF5)
    {

        g_rx_cmplt = 1;


        /*Clear the flag*/
        DMA1->HIFCR |= HIFCR_CTCIF5;
    }
}
void USART2_IRQHandler(void)
{
    g_uart_cmplt  = 1;

    /*Clear TC interrupt flag*/
    USART2->SR &=~SR_TC;
}
```

Let's go through each part of the code step by step.

UART initialization

In the `uart2_rx_tx_init` function, we start by configuring the GPIO pins for UART2 communication. We enable the clock for GPIOA, ensuring that the PA2 and PA3 pins can be used. Setting PA2 and PA3 to alternate function mode allows them to serve as UART2_TX and UART2_RX, respectively. We further specify the alternate function type to AF7, which is the type for UART2 operations.

With the GPIO configuration complete, we proceed to enable the clock for UART2. We set the baud rate using the `uart_set_baudrate` function. By enabling DMA for both transmission and reception, we offload data handling from the CPU, allowing for more efficient data transfers. We set the transfer direction to both transmit and receive, clear any pending transmission complete flags, and enable the transmission complete interrupt. Finally, we enable the UART module and configure the NVIC to handle UART2 interrupts, ensuring that the system can respond to UART events promptly.

DMA initialization

Here, we start by enabling the clock for DMA1, ensuring that the DMA controller is powered and ready for configuration. Additionally, we enable the **DMA Stream6** interrupt in the NVIC, preparing the system to handle DMA-related interrupts efficiently.

DMA configuration for UART reception

In `dma1_stream5_uart_rx_config`, we configure **DMA1 Stream5** to receive data via UART2. First, we disable the DMA stream to safely configure its parameters. After ensuring that the stream is disabled, we clear any existing interrupt flags. We set the peripheral address to the UART2 data register (USART2-DR) and the memory address to our `uart_data_buffer`, where incoming data will be stored.

We specify the number of data items to transfer and select the appropriate DMA channel. Enabling memory address increment mode ensures that the data buffer is filled sequentially. Circular mode is enabled to allow continuous data reception, and the transfer direction is set **from peripheral to memory**. We enable the DMA stream and configure the NVIC to handle Stream5 interrupts, ensuring that the system is prepared for DMA events.

DMA configuration for UART transmission

The `dma1_stream6_uart_tx_config` function configures DMA1 Stream6 to transmit data via UART2. Similar to the reception configuration, we start by disabling the DMA stream and clearing any existing interrupt flags. We set the peripheral address to the UART2 data register and the memory address to the data buffer that will be transmitted.

We specify the number of data items to transfer and select the appropriate DMA channel. Memory address increment mode is enabled to ensure that data is transmitted sequentially from the buffer. We set the transfer direction **from memory to peripheral** and enable the transfer complete interrupt. Finally, we enable the DMA stream to start the data transmission process.

Helper functions

The `compute_uart_bd` function calculates the UART baud rate setting based on the peripheral clock and desired baud rate. The `uart_set_baudrate` function uses this computed value to set the baud rate in the UART's BRR register, ensuring that UART communication occurs at the correct speed.

Interrupt handlers

Let's break down the interrupt handlers:

- `DMA1_Stream6_IRQHandler`: This handler responds to DMA Stream6 interrupts. When a transfer completes, it sets the `g_tx_cmplt` flag and clears the interrupt flag, ensuring that the system is aware that the transmission is complete.

- `DMA1_Stream5_IRQHandler`: This handler responds to DMA Stream5 interrupts. When a transfer completes, it sets the `g_rx_cmplt` flag and clears the interrupt flag, indicating that new data has been received.

- `USART2_IRQHandler`: This handler manages UART2 interrupts. It sets the `g_uart_cmplt` flag when a UART event occurs and clears the transmission complete flag, maintaining the proper flow of UART communication.

Next, we populate the `uart_dma.h` file. Here is the code:

```
#ifndef UART_DMA_H__
#define UART_DMA_H__
#include <stdint.h>
#include "stm32f4xx.h"
#define UART_DATA_BUFF_SIZE        5
void uart2_rx_tx_init(void);
void dma1_init(void);
void dma1_stream5_uart_rx_config(void);
void dma1_stream6_uart_tx_config(uint32_t msg_to_snd, uint32_t msg_
len);

#endif
```

And then, we populate the `main.c` file:

```
#include <stdio.h>
#include <string.h>

#include "uart.h"
#include "uart_dma.h"

extern uint8_t g_rx_cmplt;
```

```
extern uint8_t g_uart_cmplt;
extern uint8_t g_tx_cmplt;

extern char uart_data_buffer[UART_DATA_BUFF_SIZE];
char msg_buff[150] ={'\0'};


int main(void)
{

    uart2_rx_tx_init();
    dma1_init();
    dma1_stream5_uart_rx_config();
    sprintf(msg_buff,"Initialization...cmplt\n\r");
    dma1_stream6_uart_tx_config((uint32_t)msg_buff,strlen(msg_buff));


    while(!g_tx_cmplt){}


    while(1)
    {

        if(g_rx_cmplt)
        {
            sprintf(msg_buff, "Message received : %s \r\n",uart_data_
            buffer);
            g_rx_cmplt = 0;
            g_tx_cmplt = 0;
            g_uart_cmplt = 0;
            dma1_stream6_uart_tx_config((uint32_t)msg_buff,strlen(msg_
            buff));
            while(!g_tx_cmplt){}


        }
    }
}
```

In the main function, we start by initializing the UART and DMA, and then we configure DMA1 Stream5 for UART reception and DMA1 Stream6 for UART transmission. We prepare a message that indicates initialization completion and initiate its transmission via DMA. The main loop continuously checks whether a UART message has been received. When a message is received, it formats the received data into a response message, resets the completion flags, and transmits the response using DMA.

Testing the project

To test the project, compile the code and upload it to your microcontroller. Open RealTerm or any other serial terminal application, and then configure it with the appropriate port and baud rate to view the debug messages. Press the black push button on the development board to reset the microcontroller. Ensure the output area of RealTerm is active by clicking on it. Then, type any five keys on your keyboard. You should see these keys appear in the output area of RealTerm. The microcontroller receives the typed keys through the `dma1_stream5_uart_rx_config` function, stores them in the `msg_buff`, and transmits them to your host computer's serial port via the `dma1_stream6_uart_tx_config` function. The last received data remains in `msg_buff` for further processing if needed. We type five characters because the `UART_DATA_BUFF_SIZE` is set to 5 in the `uart_dma.h` file.

In the next section, we will develop our final DMA driver – the DMA memory-to-memory driver.

Developing the DMA memory-to-memory driver

Create a copy of your previous project in your IDE and rename it `DMA_MemToMem`. Next, create a new file named `dma.c` in the `Src` folder and another file named `dma.h` in the `Inc` folder. Update your `dma.c` file, as shown here:

```
#include "dma.h"

#define DMA2EN              (1U<<22)
#define DMA_SCR_EN          (1U<<0)
#define DMA_SCR_MINC        (1U<<10)
#define DMA_SCR_PINC        (1U<<9)
#define DMA_SCR_TCIE        (1U<<4)
#define DMA_SCR_TEIE        (1U<<2)
#define DMA_SFCR_DMDIS      (1U<<2)

void dma2_mem2mem_config(void)
{
    /*Enable clock access to the dma module*/
    RCC->AHB1ENR |= DMA2EN;

    /*Disable dma stream*/
    DMA2_Stream0->CR = 0;

    /*Wait until stream is disabled*/
    while((DMA2_Stream0->CR & DMA_SCR_EN)){}

    /*Configure dma parameters*/

    /*Set MSIZE i.e Memory data size to half-word*/
```

```
    DMA2_Stream0->CR  |=  (1U<<13);
    DMA2_Stream0->CR  &=  ~(1U<<14);

    /*Set PSIZE i.e Peripheral data size to half-word*/
    DMA2_Stream0->CR  |=  (1U<<11);
    DMA2_Stream0->CR  &=  ~(1U<<12);

    /*Enable memory addr increment*/
    DMA2_Stream0->CR  |=DMA_SCR_MINC;

    /*Enable peripheral addr increment*/
    DMA2_Stream0->CR  |=DMA_SCR_PINC;

    /*Select mem-to-mem transfer*/
    DMA2_Stream0->CR  &=  ~(1U<<6);
    DMA2_Stream0->CR  |=  (1U<<7);

    /*Enable transfer complete interrupt*/
    DMA2_Stream0->CR  |=  DMA_SCR_TCIE;

    /*Enable transfer error interrupt*/
    DMA2_Stream0->CR  |=  DMA_SCR_TEIE;

    /*Disable direct mode*/
    DMA2_Stream0->FCR  |=DMA_SFCR_DMDIS;

    /*Set DMA FIFO threshold*/
    DMA2_Stream0->FCR  |=(1U<<0);
    DMA2_Stream0->FCR  |=(1U<<1);

    /*Enable DMA interrupt in NVIC*/
    NVIC_EnableIRQ(DMA2_Stream0_IRQn);
}
```

This function sets up the DMA2 controller for memory-to-memory data transfers. It begins by enabling the clock for the DMA2 module and ensures that the DMA stream is disabled before making any configuration changes. The function configures the data size for both memory and peripheral to half-word (16-bit) and enables automatic incrementing of the memory and peripheral addresses. It sets the transfer direction to memory-to-memory and enables interrupts for transfer completion and transfer errors, ensuring robust error handling and efficient operation. Direct mode is disabled to use FIFO mode, and the FIFO threshold is set to full. Finally, the function enables the DMA stream and configures the NVIC to handle DMA interrupts, ensuring that the system can respond to DMA events appropriately. We also have the dma_transfer_start function:

```c
void dma_transfer_start(uint32_t src_buff, uint32_t dest_buff,
uint32_t len)
{
    /*Set peripheral address*/
    DMA2_Stream0->PAR = src_buff;

    /*Set memory address*/
    DMA2_Stream0->M0AR = dest_buff;

    /*Set transfer length*/
    DMA2_Stream0->NDTR = len;

    /*Enable dma stream*/
    DMA2_Stream0->CR |= DMA_SCR_EN;

}
```

This function initiates the DMA transfer by configuring the source and destination addresses and the length of the data transfer. It begins by setting the peripheral address of the value passed in `src_buff` and the memory address of the value passed in `dest_buff`. The transfer length is then specified by setting the NDTR register to `len`, indicating the number of data items to transfer. Finally, the function enables the DMA stream by setting the EN bit in the CR register, thereby starting the data transfer from the source to the destination. This is the dma.h file:

```c
#ifndef DMA_H__
#define DMA_H__
#include <stdint.h>
#include "stm32f4xx.h"

#define LISR_TCIF0        (1U<<5)
#define LIFCR_CTCIF0       (1U<<5)
#define LISR_TEIF0        (1U<<3)
#define LIFCR_CTEIF0       (1U<<3)

void dma2_mem2mem_config(void);
void dma_transfer_start(uint32_t src_buff, uint32_t dest_buff,
uint32_t len);

#endif
```

And this is the main.c file:

```c
#include <stdio.h>
#include <string.h>
```

```
#include "uart.h"
#include "dma.h"
#include "uart.h"


#define BUFFER_SIZE          5


uint16_t sensor_data_arr[BUFFER_SIZE] = {892,731,1234,90,23};
uint16_t temp_data_arr[BUFFER_SIZE];

volatile uint8_t g_transfer_cmplt;

int main(void)
{
    g_transfer_cmplt = 0;

    uart_init();
    dma2_mem2mem_config();

    dma_transfer_start((uint32_t)sensor_data_arr,(uint32_t) temp_data_
    arr, BUFFER_SIZE);

    /*Wait until transfer complete*/
    while(!g_transfer_cmplt){}

    for( int i = 0; i < BUFFER_SIZE; i++)
    {
        printf("Temp buffer[%d]: %d\r\n",i,temp_data_arr[i]);
    }

    g_transfer_cmplt = 0;


    while(1)
    {

    }
}
```

The main function sets up and initiates a memory-to-memory DMA transfer, transferring data from the globally declared and initialized sensor_data_arr to the uninitialized temp_data_arr. It starts by initializing the transfer complete flag, g_transfer_cmplt, to 0, ensuring that we can monitor the transfer status. The function then initializes UART for debugging purposes and configures the DMA using dma2_mem2mem_config. The DMA transfer is started by calling dma_transfer_start, specifying the source (sensor_data_arr), destination (temp_data_arr), and the length of the transfer (BUFFER_SIZE). The function then enters a loop, waiting until the transfer is complete, indicated by g_transfer_cmplt being set to 1. Once the transfer is complete, it prints the contents of the temp_data_arr to the console, confirming that the data has been successfully transferred.

Our main.c file also contains the DMA stream's IRQHandler:

```
void DMA2_Stream0_IRQHandler(void)
{
    /*Check if transfer complete interrupt occurred*/
    if((DMA2->LISR) & LISR_TCIF0)
    {
        g_transfer_cmplt = 1;

        /*Clear flag*/
        DMA2->LIFCR |=LIFCR_CTCIF0;
    }
    /*Check if transfer error occurred*/
    if((DMA2->LISR) & LISR_TEIF0)
    {
        /*Do something...*/
        /*Clear flag*/
        DMA2->LIFCR |= LIFCR_CTEIF0;

    }

}
```

The handler manages both transfer completion and error events. The function first checks whether the transfer complete interrupt flag (TCIF0) is set in the low interrupt status register (LISR). If this flag is set, it indicates that the DMA transfer has successfully finished. The function then sets the g_transfer_cmplt flag to 1 to signal to the main function that the transfer is complete, and it clears the interrupt flag by writing to the low interrupt flag clear register (LIFCR). Additionally, the function checks for a transfer error interrupt (TEIF0). If a transfer error is detected, it performs any necessary error handling and clears the error flag in LIFCR. This interrupt handler ensures smooth operation by promptly handling the completion of data transfers and addressing any errors that might occur during the process.

Now, it's time to test the project. To test the project, compile the code and upload it to your microcontroller. Open RealTerm or another serial terminal application, and then configure it with the appropriate port and baud rate to view the debug messages. Press the black push button on the development board to reset the microcontroller. You should see the sensor values printed, indicating that the values have been successfully copied from `sensor_data_arr` to `temp_data_arr`, as our code only prints the contents of `temp_data_arr`.

Summary

In this chapter, we learned about DMA, an important feature in microcontrollers for enhancing data throughput and offloading the CPU from routine data transfer tasks. We began by discussing the basic principles of DMA, emphasizing its role in high-performance embedded systems. We explored how DMA works and its significance in improving system efficiency, by allowing peripherals to transfer data directly to and from memory without continuous CPU intervention.

Then, we focused on the STM32F4 implementation of the DMA controller, examining its key features and configuration options. We detailed the structure of the DMA controller, including its channels, streams, and key registers, such as the Stream Configuration Register (`DMA_SxCR`), Stream Number of Data Register (`DMA_SxNDTR`), Stream Peripheral Address Register (`DMA_SxPAR`), and Stream Memory Address Registers (`DMA_SxM0AR` and `DMA_SxM1AR`). This provided a comprehensive understanding of how to configure and control DMA for various data transfer operations.

We provided practical examples to solidify our understanding, including the development of DMA drivers for different use cases, such as ADC data transfers, UART communications, and memory-to-memory transfers. These examples involved initializing DMA, setting up the necessary parameters, and implementing functions to handle data transfers efficiently.

In the next chapter, we will learn about power management energy efficiency techniques in embedded systems.

18

Power Management and Energy Efficiency in Embedded Systems

In this chapter, we will delve into power management and energy efficiency in embedded systems, a critical aspect in today's technology-driven world. Efficient power management is vital for prolonging battery life and ensuring optimal performance in embedded devices. This chapter aims to equip you with the necessary knowledge and skills to implement effective power management techniques in your designs.

We will begin by exploring various power management techniques, laying the foundation to understand how to reduce power consumption in embedded systems. Following this, we will examine the different sleep modes and low-power states available in STM32F4 microcontrollers, providing detailed insights into their configurations and applications. Then, we will discuss the wake-up sources and triggers in the STM32F4, which are essential to ensure that the microcontroller can respond promptly to external events. Finally, we will put theory into practice by developing a driver to enter standby mode and wake up the microcontroller, demonstrating how to apply these concepts in real-world scenarios.

In this chapter, we will cover the following main topics:

- An overview of power management techniques
- Low-power modes in STM32F4
- Wake-up sources and triggers in STM32F4
- Developing a driver to enter standby mode and wake up

By the end of this chapter, you will have a thorough understanding of power management in embedded systems and be able to implement energy-efficient designs using STM32F4 microcontrollers. This knowledge will enable you to create embedded systems that optimize power consumption and extend battery life, essential for modern applications.

Technical requirements

All the code examples for this chapter can be found on GitHub at `https://github.com/PacktPublishing/Bare-Metal-Embedded-C-Programming`.

An overview of power management techniques

In this section, we will explore the world of power management techniques, a crucial aspect of embedded systems design. As our devices become more advanced and our expectations for battery life increase, understanding how to manage power effectively is more important than ever. Let's dive into the various power management techniques and how they are implemented, taking a look at some case studies to see these techniques in action.

Power management in embedded systems involves a combination of hardware and software strategies designed to reduce energy consumption. This is particularly important for battery-powered devices, where efficient power usage can significantly extend battery life. The main techniques we'll cover include **Dynamic Voltage and Frequency Scaling** (**DVFS**), clock gating, power gating, and utilizing low-power modes.

Let's start with DVFS.

Dynamic Voltage and Frequency Scaling (DVFS)

DVFS is a method where the voltage and frequency of a microcontroller are adjusted based on the workload. By lowering the voltage and frequency during periods of low activity, power consumption can be greatly reduced. Conversely, during periods of high demand, the voltage and frequency are increased to ensure performance.

How is DVFS implemented?

In STM32 microcontrollers, DVFS can be managed through specific power control registers. These registers allow a system to dynamically adjust the operating points based on the required performance levels. For example, the STM32F4 series has several power modes that can be configured to adjust the system clock and core voltage.

An example use case – mobile phones

Mobile phones are a prime example of DVFS in action. When a phone is idle, it reduces the CPU frequency and voltage to save the battery. As soon as you start using an app or playing a game, the CPU ramps up its frequency and voltage to provide the necessary performance. This balance between performance and power savings is what makes modern smartphones so efficient.

Another common technique is clock gating.

Clock gating

Clock gating is a technique where the clock signal to certain parts of a microcontroller is turned off when they are not in use. This prevents unnecessary switching of transistors, which in turn saves power.

How is clock gating implemented?

Clock gating is typically controlled through clock control registers. In the STM32 series, each peripheral's clock can be enabled or disabled individually using these registers. For instance, if a particular peripheral such as the ADC is not needed, its clock can be disabled to save power.

An example use case – smart home devices

Smart home devices such as smart thermostats or lights use clock gating to manage power efficiently. These devices spend a significant amount of time in a low-power state, waking up only to perform specific tasks. By gating the clock to unused peripherals, these devices can conserve energy and extend their battery life.

Another technique is power gating.

Power gating

Power gating takes power savings a step further by completely shutting off the power to certain parts of a microcontroller. This technique ensures zero power consumption for the powered-down sections.

How is power gating implemented?

Power gating is more complex than clock gating and often involves dedicated power management units within a microcontroller. These units control the power supply to various domains of the microcontroller. In STM32 microcontrollers, power gating can be configured using the power control registers to turn off specific peripherals, or even entire sections of the microcontroller.

An example use case – wearable devices

Wearable devices, such as fitness trackers, benefit greatly from power gating. These devices need to operate for extended periods on a single charge. By powering down sensors and other components when they are not in use, wearables can achieve longer battery life without compromising functionality.

Next, let's discuss low-power modes.

Low-power modes

Low-power modes are predefined states within microcontrollers that significantly reduce power consumption by disabling or reducing the functionality of various components. These modes range from simple CPU sleep modes to more complex deep sleep or standby modes.

How are low-power modes implemented?

Low-power modes are implemented through the power control registers. The STM32F4 microcontrollers, for example, offer several low-power modes, including **sleep**, **stop**, and **standby**. Each mode provides a different balance between power savings and wake-up time.

An example use case – remote sensors

Remote sensors used in agriculture or environmental monitoring often use low-power modes. These sensors might spend the majority of their time in a low-power state, waking up periodically to take measurements and transmit data. By leveraging low-power modes, these sensors can operate for months or even years on a single battery charge.

Now, let's take a closer look at a couple of case studies that illustrate how a combination of these power management techniques is used in real-world applications.

Case study 1 – an energy-efficient smartwatch

Smartwatches are a great example of a device that relies heavily on power management techniques. These devices need to balance performance with battery life, as users expect them to run for days on a single charge. Let's break down the roles the different techniques play in an energy-efficient smartwatch design:

- **DVFS**: The smartwatch uses DVFS to adjust the CPU frequency based on the current workload. When the user interacts with the watch, the CPU frequency increases to provide a smooth experience. When the watch is idle, the frequency is lowered to save power.

- **Clock gating**: Peripherals such as the GPS or heart rate monitor are only powered when needed. When these features are not in use, their clocks are gated to conserve energy.

- **Power gating**: Components like the display driver are powered down completely when the display is off.

- **Low-power modes**: The watch enters deep sleep mode during periods of inactivity, waking up only to check for notifications or user interactions.

By combining these techniques, smartwatches can achieve impressive battery life without compromising on functionality. Another excellent example is solar-powered environmental monitoring.

Case study 2 – a solar-powered environmental monitor

A solar-powered environmental monitor deployed in remote locations must operate efficiently to ensure continuous data collection and transmission. The roles are as follows:

- **DVFS**: The monitor adjusts its operating frequency based on the intensity of sunlight and battery charge. During peak sunlight hours, it operates at a higher frequency to process more data.

- **Clock gating**: Sensors such as temperature, humidity, and air quality are only active during data collection intervals. The clocks to these sensors are gated when not in use.

- **Power gating**: Non-essential components are completely powered down during nighttime or cloudy periods to conserve energy.

- **Low-power modes**: The monitor enters deep sleep mode between data collection intervals, waking up periodically to take measurements and transmit data.

With these power management techniques, the monitor can operate autonomously for extended periods, relying solely on solar power.

Power management is a vital aspect of embedded system design, especially as devices become more portable and battery-dependent. By understanding and implementing techniques such as DVFS, clock gating, power gating, and low-power modes, we can design embedded systems that are both powerful and energy-efficient. Whether it's a smartwatch, a remote sensor, or any other battery-powered device, effective power management ensures longer battery life and better overall performance. As we continue to push the boundaries of what embedded systems can do, mastering these power management techniques will be more important than ever.

In the next section, we will explore the low-power modes in our STM32F4 microcontroller.

Low-power modes in STM32F4

In this section, we will learn about the low-power modes available in STM32F4 microcontrollers. We'll cover the various low-power modes, how to configure them, and the practical aspects of using them in your projects.

Let's start by understanding these low-power modes. Low-power modes in the STM32F4 microcontrollers are designed to reduce power consumption by disabling or limiting the functionality of certain components. The STM32F4 offers several low-power states, each providing a different balance between power savings and wake-up latency. These modes include sleep, stop, and standby modes.

We can put our system into low-power mode by executing the **Wait For Interrupt** (**WFI**) or **Wait For Event** (**WFE**) instructions, or setting the SLEEPONEXIT bit in the **Cortex®-M4 with FPU system control register** on return from an ISR.

Let's dive into the details of each low-power mode, starting with sleep mode.

Sleep mode

Sleep mode is the most basic low-power mode, where the CPU clock is stopped but peripherals continue to operate. This mode offers a **quick wake-up time**, making it ideal for applications that require frequent transitions between active and low-power states.

To enter sleep mode, we need to clear the SLEEPDEEP bit in the **System Control Register (SCR)** and then execute the WFI or WFE instruction, as shown in the following snippet:

```
void enter_sleep_mode(void) {
    // Clear the SLEEPDEEP bit to enter Sleep mode
    SCB->SCR &= ~SCB_SCR_SLEEPDEEP_Msk;

    // Request Wait For Interrupt
    __WFI();
}
```

The microcontroller **exits Sleep mode upon any interrupt or event**. Since the peripherals remain active, any configured interrupt from a peripheral can wake the CPU.

An example use case is **sensor monitoring**.

For applications such as continuous sensor monitoring, sleep mode provides an efficient way to reduce power consumption without sacrificing responsiveness. The microcontroller can wake up quickly to process sensor data and then return to sleep mode.

The next mode is stop mode.

Stop mode

Stop mode offers a **deeper power-saving state than sleep mode** by stopping the main internal regulator and halting the system clock. Only the low-speed clock (LSI or LSE) remains active. This mode provides a **moderate wake-up time** and significant power savings.

To enter stop mode, set the SLEEPDEEP bit in the **Power Control Register** (PWR_CR), and then execute the WFI or WFE instruction, as shown in the following snippet. Additional configuration can also be applied to further reduce power consumption in stop mode:

```
void enter_stop_mode(void) {
    // Set SLEEPDEEP bit to enable deep sleep mode
    SCB->SCR |= SCB_SCR_SLEEPDEEP_Msk;
    // Request Wait For Interrupt
    __WFI();
}
```

The MCU exits stop mode upon any **external interrupt** or **wake-up event from configured EXTI lines**, **RTC alarms**, or other configured wake-up sources. The wake-up time from stop mode is longer than from sleep mode, but it still allows for a relatively quick return to full operation.

An example use case is **periodic data logging**.

In applications such as data logging, a microcontroller can remain in stop mode and wake up periodically, based on RTC alarms, to log data, and then return to stop mode. This significantly reduces power consumption while ensuring regular data logging.

The final mode is standby mode.

Standby mode

Standby mode provides the **highest power savings** by turning off most internal circuitry, including the main regulator. Only a small portion of the microcontroller remains powered to monitor wake-up sources. This mode has the **longest wake-up time** but offers the lowest power consumption.

To enter standby mode, set the PDDS and SLEEPDEEP bits in the Power Control (PWR_CR) register, and then configure the wake-up sources. This snippet demonstrates how to enter standby mode:

```
void enter_standby_mode(void) {
    // Clear Wakeup flag
    PWR->CR |= PWR_CR_CWUF;

    // Set the PDDS bit to enter Standby mode
    PWR->CR |= PWR_CR_PDDS;

    // Set the SLEEPDEEP bit to enable deep sleep mode
    SCB->SCR |= SCB_SCR_SLEEPDEEP_Msk;

    // Request Wait For Interrupt
    __WFI();
}
```

The microcontroller exits standby mode upon a **wake-up event from an external wake-up pin (WKUP)**, an RTC alarm, or a reset event. When the microcontroller wakes up from standby mode, it **undergoes a full reset sequence**, and the execution starts from the reset vector.

An example use case is **remote IoT devices**.

Standby mode is perfect for remote IoT devices that need to operate for extended periods on battery power. These devices can remain in standby mode most of the time and wake up only for critical events or scheduled tasks, thus maximizing battery life.

Now that we understand how to enter the various low-power modes, we will look at how to wake up from them.

Wake-up sources and triggers from low-power modes in STM32F4

While low-power modes help conserve energy, ensuring that a microcontroller can wake up promptly when needed is equally important. The STM32F4 microcontroller series provides a variety of wake-up sources and triggers to handle this effectively. In this section, we'll explore these wake-up sources, how they function, and their practical applications.

Understanding wake-up sources

Wake-up sources are mechanisms that bring a microcontroller out of a low-power state. The STM32F4 offers several types of wake-up sources, each suited for different scenarios. These include external interrupts, RTC alarms, watchdog timers, and various internal events. By understanding these triggers, we can design systems that balance power efficiency with responsiveness.

The wake-up sources can be grouped as follows:

- External interrupts
- **Real-Time Clock (RTC)** alarms
- Internal events

Let's delve into each of these wake-up sources to understand how they work and their typical use cases.

External interrupts

External interrupts are one of the primary wake-up sources for STM32F4 microcontrollers. These interrupts can be triggered by events on specific GPIO pins. When a microcontroller is in a low-power mode, an external signal, such as a button press or sensor output, can wake it up.

Here's how it works:

- **GPIO configuration**: Configure the GPIO pins to act as interrupt sources. This involves setting pin mode and enabling the interrupt on the desired edge (rising, falling, or both).
- **EXTI configuration**: Each GPIO pin can be mapped to an EXTI line, which can be configured to generate an interrupt.
- **NVIC configuration**: Enable the EXTI line interrupt in the **Nested Vectored Interrupt Controller (NVIC)** to ensure that the microcontroller responds to the external event.

Example use cases are a **smart doorbell system and smart lighting**.

Imagine a smart doorbell system. The microcontroller remains in a low-power mode to conserve battery life. When someone presses the doorbell button (connected to a GPIO pin), an external interrupt is triggered, waking the microcontroller to process the event and send a notification to the homeowner. Another excellent example is smart home lighting systems.

A smart home lighting system needs to conserve energy while being responsive to user inputs. The microcontroller stays in a low-power mode until an external interrupt (e.g., a motion sensor detecting movement) wakes it up. Upon waking, the microcontroller processes the event, turns on the lights, and then goes back to sleep after a predefined period of inactivity.

The next wake-up source we will examine is the RTC Alarm.

RTC alarms

The RTC is a versatile peripheral that can generate wake-up events at specific intervals or predefined times. It is particularly useful for applications requiring periodic wake-ups, such as data logging or scheduled tasks.

Here's how it works:

- **RTC configuration**: Configure the RTC to generate alarms or periodic wake-up events. This involves setting the RTC clock source, enabling the wake-up timer, and setting the alarm time.
- **Interrupt handling**: Enable the RTC alarm or wake-up interrupt in the NVIC to ensure that the microcontroller wakes up when the alarm or timer event occurs.

An example use case is an **environmental monitoring system**.

Consider a remote environmental monitoring system that logs temperature and humidity data. The microcontroller can be put into low-power mode, waking up at regular intervals (e.g., every hour) using RTC alarms to read sensors and log data, and then return to the low-power state.

The final wake-up sources we will examine are **internal events**.

Internal events

Apart from external triggers, internal events can also wake up a microcontroller from low-power modes. These events include the following:

- **Peripheral events**: Events generated by internal peripherals, such as ADC conversions or communication interface activity
- **System events**: Internal system events such as power voltage detection or clock stability issues

Here's how it works:

- **Peripheral configuration**: Configure the peripheral to generate interrupts upon specific events. For instance, an ADC can generate an interrupt when a conversion is complete.

- **Event handling**: Enable the relevant interrupts in the NVIC to handle these internal events and wake the microcontroller.

An example use case is a **fitness tracker**

A wearable fitness tracker that monitors heart rate can use the ADC to read sensor data. The microcontroller stays in a low-power mode and wakes up when the ADC completes a conversion, allowing it to process and store the heart rate data.

Before we conclude this section, let's summarize some key practical considerations to keep in mind when configuring wake-up sources.

Practical considerations

When configuring wake-up sources, you must consider the following practical aspects:

- **Response time**: Ensure the chosen wake-up source can provide the required response time for your application. External interrupts typically offer the fastest wake-up times.

- **Power consumption**: Balance power consumption with wake-up requirements. RTC alarms and watchdog timers can be configured for periodic wake-ups with minimal power overhead.

- **Reliability**: Choose reliable wake-up sources for critical applications. Watchdog timers are essential for safety-critical systems to ensure that a microcontroller can recover from faults.

- **Peripheral configuration**: Ensure that peripherals needed for wake-up are properly configured and their clocks remain enabled, even in low-power states.

Understanding and properly configuring these wake-up sources ensures that your embedded systems are both energy-efficient and reliable.

In the next section, we will learn how to develop a driver to enter standby mode and subsequently wake up the system.

Developing a driver to enter standby mode and wake up

Create a copy of your previous project in your IDE, following the steps outlined in earlier chapters. Rename this copied project `StandByModeWithWakeupPin`. Next, create a new file named `standby_mode.c` in the `Src` folder, and then another file named `standby_mode.h` in the `Inc` folder.

Populate your `standby_mode.c` file with the following code:

```c
#include "standby_mode.h"
#define PWR_MODE_STANDBY          (PWR_CR_PDDS)
#define WK_PIN                    (1U<<0)


static void set_power_mode(uint32_t pwr_mode);

void wakeup_pin_init(void)
{
    //Enable clock for GPIOA
    RCC->AHB1ENR |= RCC_AHB1ENR_GPIOAEN;

    //Set PA0 as input pin
    GPIOA->MODER &= ~(1U<<0);
    GPIOA->MODER &= ~(1U<<1);

    //No pull
    GPIOA->PUPDR &= ~(1U<<0);
    GPIOA->PUPDR &= ~(1U<<1);
}
```

This function is responsible for configuring PA0 to be used as a wake-up pin for exiting low-power mode. It sets PA0 as an input pin by clearing the corresponding bits in the GPIOA mode register, and then it configures the pin with no pull-up or pull-down resistors:

```c
void standby_wakeup_pin_setup(void)
{

    /*Wait for wakeup pin to be released*/
    while(get_wakeup_pin_state() == 0){}

    /*Disable wakeup pin*/
    PWR->CSR &=~(1U<<8);

    /*Clear all wakeup flags*/
    PWR->CR |=(1U<<2);

    /*Enable wakeup pin*/
    PWR->CSR |=(1U<<8);

    /*Enter StandBy mode*/
    set_power_mode(PWR_MODE_STANDBY);
```

```
    /*Set SLEEPDEEP bit in the CortexM System Control Register*/
    SCB->SCR |=(1U<<2);

    /*Wait for interrupt*/
    __WFI();
}
```

This function prepares the microcontroller to enter standby mode and ensures that it can wake up via the configured wake-up pin. It begins by waiting for the wake-up pin to be released, ensuring that the pin is in a stable state before proceeding. The function then disables the wake-up pin to clear any residual settings, followed by clearing all wake-up flags to reset the wake-up status. After re-enabling the wake-up pin, the function sets the power mode to Standby by configuring the appropriate power control register. Finally, the function executes the WFI instruction, placing the microcontroller into standby mode until an interrupt occurs, triggering the wake-up process:

```
uint32_t get_wakeup_pin_state(void)
{
        return ((GPIOA->IDR & WK_PIN) == WK_PIN);
}
```

This function checks the current state of the wake-up pin (PA0). It reads the input data register (IDR) for GPIOA and performs a bitwise AND operation with the wake-up pin's bit mask (WK_PIN). This operation isolates the state of PA0. The function then compares this result to the bit mask itself to determine whether the pin is high. If PA0 is high, the function returns true; otherwise, it returns false:

```
static void set_power_mode(uint32_t pwr_mode)
{
  MODIFY_REG(PWR->CR, (PWR_CR_PDDS | PWR_CR_LPDS | PWR_CR_FPDS | PWR_
  CR_LPLVDS | PWR_CR_MRLVDS), pwr_mode);

}
```

This function configures the power mode of the STM32F4 microcontroller by modifying specific bits in PWR_CR. This function takes a parameter, pwr_mode, which specifies the desired power mode settings. It uses the MODIFY_REG macro to update the PWR_CR register, specifically targeting the bits related to different power modes such as **PDDS (Power Down Deepsleep)**, **LPDS (Low-Power Deepsleep)**, **FPDS (Flash Power Down in Stop Mode)**, **LPLVDS (Low-Power Regulator in Low Voltage in Deepsleep)**, and **MRLVDS (Main Regulator in Low Voltage in Deepsleep)**.

Next, we will populate the standby_mode.h file:

```
#ifndef STANDBY_MODE_H__
#define STANDBY_MODE_H__

#include <stdint.h>
```

```
#include "stm32f4xx.h"

uint32_t get_wakeup_pin_state(void);
void wakeup_pin_init(void);
void standby_wakeup_pin_setup(void);
#endif
```

We are now ready to test inside main.c. Update your main.c file, as shown here:

```
#include <stdio.h>
#include <string.h>
#include "standby_mode.h"
#include "gpio_exti.h"


#include "uart.h"
uint8_t g_btn_press;
static void check_reset_source(void);


int main(void)
{

    uart_init();

    wakeup_pin_init();

    /*Find reset source*/
    check_reset_source();

    /*Initialize EXTI*/
    pc13_exti_init();


    while(1)
    {

    }
}
```

The main function starts by initializing the UART for serial communication with uart_init, ensuring that we can send debugging information to the serial port. Next, it configures the wake-up pin by calling wakeup_pin_init, preparing the microcontroller to respond to external wake-up signals. The check_reset_source function is then called to determine the cause of the microcontroller's reset, whether from standby mode or another source, and to handle any necessary flag clearing. Following this, the **external interrupt (EXTI)** for `PC13` is initialized with `pc13_exti_init`. The function then enters an infinite **loop**, `while(1)`, maintaining the program's operational state and waiting for interrupts or events to occur:

```c
static void check_reset_source(void)
{

        /*Enable clock access to PWR*/
        RCC->APB1ENR |= RCC_APB1ENR_PWREN;

    if ((PWR->CSR & PWR_CSR_SBF) == (PWR_CSR_SBF))
    {
        /*Clear Standby flag*/
        PWR->CR |= PWR_CR_CSBF;

        printf("System resume from Standby.....\n\r");

        /*Wait for wakeup pin to be released*/
        while(get_wakeup_pin_state() == 0){}

    }

    /*Check and Clear Wakeup flag*/
    if((PWR->CSR & PWR_CSR_WUF) == PWR_CSR_WUF )
    {
        PWR->CR |= PWR_CR_CWUF;
    }
}
```

This function determines the cause of the microcontroller's reset and handles the necessary flags accordingly. It begins by enabling the clock for the power control (`PWR`) peripheral to ensure access to the power control and status registers. It then checks whether the **standby flag (SBF)** is set in the `PWR_CSR` register, which indicates that the system has resumed from standby mode. If the flag is set, it clears the SBF and prints a message, indicating that the system has resumed from standby. The function also waits for the wake-up pin to be released, ensuring that the pin is in a stable state. Additionally, it checks whether the Wakeup flag (`WUF`) is set and, if so, clears the flag to reset the wake-up status.

This is the interrupt callback function:

```
static void exti_callback(void)
{
    standby_wakeup_pin_setup();


}
```

And finally, we have the interrupt handler:

```
void EXTI15_10_IRQHandler(void) {
    if((EXTI->PR & LINE13)!=0)
    {
        /*Clear PR flag*/
        EXTI->PR |=LINE13;

        //Do something...
        exti_callback();
    }

}
```

The exti_callback function, coupled with EXTI15_10_IRQHandler, ensures that the microcontroller properly handles external interrupts from the wake-up pin. The exti_callback function is a straightforward handler that calls standby_wakeup_pin_setup. The EXTI15_10_IRQHandler function is an interrupt service routine specifically for EXTI lines 15 to 10. It checks whether the interrupt was triggered by line 13 (associated with the wake-up pin), and if so, it clears the interrupt pending flag to acknowledge the interrupt. After clearing the flag, it calls exti_callback to handle the wake-up event.

Now, let's test the project!

To test the project, start by pressing the blue push button to enter standby mode. Remember that PA0 is configured as the wake-up pin and is active low. In normal mode, connect a jumper wire from PA0 to the ground. To trigger a wake-up event, pull out the jumper wire and connect it to 3.3V, causing a change in logic that will wake the microcontroller from standby mode.

To test on the microcontroller, simply build the project and run it.

Open RealTerm and configure the appropriate port and baud rate to view the printed message that confirms the system has resumed from standby mode.

Summary

In this chapter, we delved into the critical aspects of power management and energy efficiency in embedded systems. Efficient power management is essential for prolonging battery life and ensuring optimal performance in embedded devices. We began by exploring various power management techniques, laying the foundation for understanding how to reduce power consumption in embedded systems.

We then examined the different low-power modes available in STM32F4 microcontrollers, providing detailed insights into their configurations and applications. Then, we discussed the wake-up sources and triggers in STM32F4, which are essential to ensure that a microcontroller can promptly come out of low-power modes.

Finally, we put theory into practice by developing a driver to enter standby mode and wake up the microcontroller.

With this journey into bare-metal embedded C programming now complete, it's important to acknowledge the profound expertise you've gained. By mastering the nuances of microcontroller architecture and the discipline of register-level programming, you've equipped yourself with the tools to create efficient and reliable embedded systems from the ground up. This book was designed to offer more than just technical instruction; it also aimed to instill a deeper understanding of the hardware and a methodical approach to firmware development. As you move forward, remember that true mastery in this field lies in the continuous application and refinement of these principles.

Unlock this book's exclusive benefits now

This book comes with additional benefits designed to elevate your learning experience.

Note: Have your purchase invoice ready before you begin.

```
https://www.packtpub.com/
     unlock/9781835460818
```

19

Unlock Your Book's Exclusive Benefits

Your copy of *Bare-Metal Embedded C Programming* comes with the following exclusive benefits:

- Next-gen Packt Reader
- AI assistant (beta)
- DRM-free PDF/ePub downloads

Use the following guide to unlock them if you haven't already. The process takes just a few minutes and needs to be done only once.

How to unlock these benefits in three easy steps

Step 1

Have your purchase invoice for this book ready, as you'll need it in *Step 3*. If you received a physical invoice, scan it on your phone and have it ready as either a PDF, JPG, or PNG.

For more help on finding your invoice, visit `https://www.packtpub.com/unlock-benefits/help`.

> **Note**
> Bought this book directly from Packt? You don't need an invoice. After completing *Step 2*, you can jump straight to your exclusive content.

Step 2

Scan the following QR code or visit `https://www.packtpub.com/unlock/9781835460818`:

Step 3

Sign in to your Packt account or create a new one for free. Once you're logged in, upload your invoice. It can be in PDF, PNG, or JPG format and must be no larger than 10 MB. Follow the rest of the instructions on the screen to complete the process.

Need help?

If you get stuck and need help, visit `https://www.packtpub.com/unlock-benefits/help` for a detailed FAQ on how to find your invoices and more. The following QR code will take you to the help page directly:

> **Note**
>
> If you are still facing issues, reach out to `customercare@packt.com`.

Index

packtpub.com

Subscribe to our online digital library for full access to over 7,000 books and videos, as well as industry leading tools to help you plan your personal development and advance your career. For more information, please visit our website.

Why subscribe?

- Spend less time learning and more time coding with practical eBooks and Videos from over 4,000 industry professionals
- Improve your learning with Skill Plans built especially for you
- Get a free eBook or video every month
- Fully searchable for easy access to vital information
- Copy and paste, print, and bookmark content

Did you know that Packt offers eBook versions of every book published, with PDF and ePub files available? You can upgrade to the eBook version at packtpub.com and as a print book customer, you are entitled to a discount on the eBook copy. Get in touch with us at customercare@packtpub.com for more details.

At www.packtpub.com, you can also read a collection of free technical articles, sign up for a range of free newsletters, and receive exclusive discounts and offers on Packt books and eBooks.

Other Books You May Enjoy

If you enjoyed this book, you may be interested in these other books by Packt:

Learn C Programming

Jeff Szuhay

ISBN: 978-1-80107-845-0

- Implement fundamental programming concepts through C programs
- Understand the importance of creating complex data types and the functions to manipulate them
- Develop good coding practices and learn to write clean code
- Validate your programs before developing them further
- Use the C Standard Library functions and understand why it is advantageous
- Build and run a multi-file program with Make
- Get an overview of how C has changed since its introduction and where it is going

Embedded Systems Architecture

Daniele Lacamera

ISBN: 978-1-80323-954-5

- Participate in the design and definition phase of an embedded product
- Get to grips with writing code for ARM Cortex-M microcontrollers
- Build an embedded development lab and optimize the workflow
- Secure embedded systems with TLS
- Demystify the architecture behind the communication interfaces
- Understand the design and development patterns for connected and distributed devices in the IoT
- Master multitasking parallel execution patterns and real-time operating systems
- Become familiar with Trusted Execution Environment (TEE)

Packt is searching for authors like you

If you're interested in becoming an author for Packt, please visit `authors.packtpub.com` and apply today. We have worked with thousands of developers and tech professionals, just like you, to help them share their insight with the global tech community. You can make a general application, apply for a specific hot topic that we are recruiting an author for, or submit your own idea.

Share Your Thoughts

Now you've finished *Bare-Metal Embedded C Programming*, we'd love to hear your thoughts! Scan the QR code below to go straight to the Amazon review page for this book and share your feedback or leave a review on the site that you purchased it from.

`https://packt.link/r/183546081X`

Your review is important to us and the tech community and will help us make sure we're delivering excellent quality content.